Biochemistry: Concepts and Applications

Biochemistry: Concepts and Applications

Luke Allen

STATES
ACADEMIC PRESS
www.statesacademicpress.com

Biochemistry: Concepts and Applications
Luke Allen
ISBN: 978-1-63989-074-3 (Hardback)

STATES
ACADEMIC PRESS

Published by States Academic Press,
109 South 5th Street,
Brooklyn, NY 11249, USA

Cataloging-in-Publication Data

Biochemistry : concepts and applications / Luke Allen.
p. cm.
Includes bibliographical references and index.
ISBN 978-1-63989-074-3
1. Biochemistry. 2. Biology. 3. Chemistry. I. Allen, Luke.
QH345 .B56 2022
572--dc23

Table of Contents

Preface

Biochemistry is a sub-discipline of both biology and chemistry, which focuses on the various chemical processes related to living organisms. It attempts to understand the structure of tissues and organs by studying the chemical basis of biological mechanisms that occur within and in between the living cells. Some common biomolecules studied under this field are carbohydrates, lipids, proteins and nucleic acids. Metabolic activities like gluconeogenesis, glycolysis and aerobic respiration are also an important part of this discipline. The subject of biochemistry can be further classified into structural biology and enzymology. Structural biology deals with the molecular structure of various biological macromolecules such as membranes, DNA, RNA and proteins. These structures are divided into primary structure, secondary structure, tertiary structure and quaternary structure. This book unfolds the innovative aspects of biochemistry which will be crucial for the holistic understanding of the subject matter. Some of the diverse topics covered herein address the varied branches that fall under this category. Those in search of information to further their knowledge will be greatly assisted by this book.

A foreword of all Chapters of the book is provided below:

Chapter 1 - The branch of biology which deals with the study of chemical substances and processes occurring in living organisms is known as biochemistry. A few important areas of study related to this field are animal biochemistry, plant biochemistry, molecular biology, cell biology, metabolism, immunology and genetics. This is an introductory chapter which will briefly introduce all these significant aspects of biochemistry; **Chapter 2** - Carbohydrates are the biomolecules which are made up of carbon, hydrogen and oxygen atoms. They are the primary sources of energy in all the living organisms. This chapter has been carefully written to provide an easy understanding of the varied facets of carbohydrate biochemistry; **Chapter 3** - Proteins are the macromolecules which are made up of one or more long chains of amino acid residues. Proteins are crucial for providing structure to the cells. This chapter closely examines the key concepts of protein biochemistry to provide an extensive understanding of the subject; **Chapter 4** - Chemically, a lipid is a macromolecule which is insoluble in water and is soluble in non-polar solvents such as chloroform, ether and alcohol. This chapter has been carefully written to provide an easy understanding of the varied facets of lipids such as their types, functions and metabolism; **Chapter 5** - Nucleic acids are the chemical compounds which are made up of complex nucleotides and carry primary information related to various activities of the cell. Some of the major focus areas related to nucleic acids are nucleic acid design, nucleic acid structure, nucleic acid structure determination and nucleic acid structure prediction, etc. This chapter discusses in detail the design, structure and metabolism of nucleic acids; **Chapter 6** - Enzymes are the biological catalysts which accelerate the rate of reactions without getting used in the reactions and are proteinaceous in nature. Some enzymes catalyze the proteolysis of proteins and are known as proteolytic enzymes. The topics elaborated in this chapter will help in gaining a better perspective about enzymes and proteolysis.

I would like to thank the entire editorial team who made sincere efforts for this book and my family who supported me in my efforts of working on this book. I take this opportunity to thank all those who have been a guiding force throughout my life.

Luke Allen

1

Biochemistry: An Introduction

The branch of biology which deals with the study of chemical substances and processes occurring in living organisms is known as biochemistry. A few important areas of study related to this field are animal biochemistry, plant biochemistry, molecular biology, cell biology, metabolism, immunology and genetics. This is an introductory chapter which will briefly introduce all these significant aspects of biochemistry.

Biochemistry is the application of chemistry to the study of biological processes at the cellular and molecular level. It emerged as a distinct discipline around the beginning of the 20th century when scientists combined chemistry, physiology, and biology to investigate the chemistry of living systems.

The Study of Life in its Chemical Processes

Biochemistry is both life science and a chemical science - it explores the chemistry of living organisms and the molecular basis for the changes occurring in living cells. It uses the methods of chemistry.

"Biochemistry has become the foundation for understanding all biological processes. It has provided explanations for the causes of many diseases in humans, animals and plants."

Physics, molecular biology, and immunology to study the structure and behaviour of the complex molecules found in biological material and the ways these molecules interact to form cells, tissues, and whole organisms.

Biochemists are interested, in mechanisms of brain function, cellular multiplication and differentiation, communication within and between cells and organs, and the chemical bases of inheritance and disease. The biochemist seeks to determine how specific molecules such as proteins, nucleic acids, lipids, vitamins, and hormones function in such processes. Particular emphasis is placed on the regulation of chemical reactions in living cells.

An Essential Science

Biochemistry has become the foundation for understanding all biological processes. It has provided explanations for the causes of many diseases in humans, animals, and plants. It can frequently suggest ways by which such diseases may be treated or cured.

A Practical Science

Because biochemistry seeks to unravel the complex chemical reactions that occur in a wide variety of life forms, it provides the basis for practical advances in medicine, veterinary medicine, agriculture, and biotechnology. It underlies and includes such exciting new fields as molecular genetics and bioengineering.

The knowledge and methods developed by biochemists are applied to in all fields of medicine, in agriculture and in many chemical and health-related industries. Biochemistry is also unique in providing teaching and research in protein structure/function and genetic engineering, the two basic components of the rapidly expanding field of biotechnology.

A Varied Science

As the broadest of the basic sciences, biochemistry includes many subspecialties such as neurochemistry, bioorganic chemistry, clinical biochemistry, physical biochemistry, molecular genetics, biochemical pharmacology, and immunochemistry. Recent advances in these areas have created links among technology, chemical engineering, and biochemistry.

Applications of Biochemistry

Applications in Medical Biochemistry

Biochemistry study is really related to the medicine industry. The concept of the biochemistry study has been widely use for the improvement in the medicine. So, here're some great applications of biochemistry in the medicine field:

- Physiology: Biochemistry is really helpful for scientist to fully understand about the physiological in human body. By using him concept of biochemistry, scientist can understand about the biochemical changes occurred in the people. The search for disease curing based on the biochemical changes has been conducted to get the best medicine for certain diseases.

- Pathology: Biochemistry is also used in the pathology. By using this study, physician can get the hint about the biochemical changes in patients' bodies based on their symptoms description. The physician can then confirm the patients' symptoms by examining the chemical reaction or process in their body. For an example, the physician may examine the uric acid level in the patient's body to understand the patient's symptoms.
- Nutrition Deficiency: As we've known that healthy life is really important to have. Nowadays, people consume multi vitamins and minerals to keep them healthy. In order to formulate the best composition of these multi vitamins or minerals, scientists use the biochemistry discipline. The function of vitamin or mineral in our body can be understood by the study of biochemistry.
- Hormonal Deficiency: Hormonal imbalance, both in men, women or children, can lead to the many diseases. In the correlation with the biochemistry study, physicians use the concept of biochemistry to understand deeply about the hormones, their functions and reactions in the human body. By using this, physician can create the proper medicine to treat that kind of hormonal imbalance.

Biochemistry Applications in Medical Treatment

Biochemistry also has the important role in the nursing discipline. Biochemistry can be used to monitor the progress of patients regularly in the hospital. This study can be used to define the best treatment to the patients based on their conditions. Biochemistry makes it easier for the nurses to do the diagnosis of clinical condition of the patients. Here're some other possible applications of biochemistry in the nursing:

- Kidney Function Test: The kidney function is really related to the biochemistry study. If you have the kidney problems, as an example, you may undergo the urine test. This urine test can help us to understand the change in PH, urine color or other containment in the urine. Biochemistry is the fundamental study that covers this kind of treatment.
- Blood Test: The blood test uses the concept of the biochemistry study. The analytical test for blood glucose level gives the understanding about the patient condition. As an example, for the diabetes patient, this kind of test can give the indication about the stages of the diabetes. There is also another test for diabetes patient called the ketones bodies in urine test. The ketone bodies or ketone urea is the indication of the last stage of diabetes. The containment of this substance in our urine can be examined by the study of biochemistry.
- Liver Function Test: Another body organ which can be understood by the biochemistry study is liver. Biochemistry helps us to understand how the liver operate, the type of disorder which likely to attack liver and the effect of certain medicines to the liver condition.
- Serum Cholesterol Test: Biochemistry can also be used to check your cholesterol level in blood. The cholesterol or other lipoproteins contained in your blood can indicate your health condition.
- Pregnancy Test: Pregnancy test uses the presence of HCG or human chorionic gonadotropin hormone as the indication of pregnancy. This type of hormone is made by the placenta

as soon as the implantation of embryo into the uterine walls. This hormone can be detected by the urine or blood sample. Thanks to biochemistry to help us discover this hormone.

- Ames Test: Ames test is the test applied to the salmonella bacteria. This test measures the growth of these bacteria in the body. This Ames test helps us to screen the chemicals to understand whether these bacteria mutate the structure of DNA. This test can also give us the indication of cancer potential in our body based on that kind chemical reaction.
- Breast Cancer Screening: Breast cancer screening is conducted by examining the mutations in two genes consisting of Breast Cancer-1 gene (BRCA1) and the Breast Cancer-2 gene (BRCA2). The examination of these genes can be conducted by the help of biochemistry concept.
- PKU Test: PKU or Phenylketonuria is a metabolic disease in which the individual is missing an enzyme named phenylalanine hydroxylase. The absence of this enzyme can build a phenylalanine, which can lead to mental retardation.

Biochemistry Applications in Agriculture

Biochemistry also has some valuable impact in the agriculture fields. This study is useful in the farming, fishery, poultry, sericulture, beekeeping and many more. So, here're some popular applications of biochemistry in the agriculture:

- Prevent Diseases: The concept of the biochemistry can be used to create the best prevention method to the diseases. It also can help to understand the treatment of those certain diseases.
- Enhance Growth: The biochemistry can help the farmer to enhance their plant growth, yield and quality of food by maximizing their fertilizers. The biochemistry helps us understand the reaction occurred between fertilizers or pesticide and the crops in the cell level. By knowing this, we can pick the right fertilizer to improve our plant growth and its quality.
- Enhance Yield: There're some different hormones used to different purposes. For example, there's the hormone helping growth, or boosting flowering, or supporting the fruit formation. In fisheries, there're hormones used to encourage the fish growth. These can only be achieved by the help of biochemistry study.
- Soils Condition: Understanding the condition of soils is really important to make sure that we get the great farming outcomes. Biochemistry can help farmer to understand the composition of soils and their deficiencies. These data can be used to define the proper actions to solve the problems in the farming soils.
- Adulteration: Adulteration is considered as the serious issues in agriculture field. Biochemistry can help us understand the composition of food and its alterations. Biochemical test can be used to prevent the contamination in the food product.
- Biochemical Tests: Biochemical test can also be conducted for the pesticide residues or other toxic waste in plant, food grain and soil. This can be useful in the import and export process so that the quality of the food can be kept well.

- In Animal Husbandry: Biochemical test can be done to check the quality. It can also help to diagnose any disease in animals.
- In Fisheries: Biochemical test is used in the fisheries to monitor the water quality regularly. Water composition and its chemical reactions may change and this change can lead to the death of fishes or prawns. In this condition, the regular monitoring by biochemical test is mandatory action to be done. There're some qualities which are examined in the water biochemical test like the salt or calcium containment, pH, waste accumulation, and so on.

Biochemistry Applications in Nutrition

Biochemistry also plays the significant role in the nutrition. Biochemistry helps us understand the chemical containment in food. This can be used for keeping us health, define the optimum intake of micro & macro nutrient, vitamin, mineral, essential fatty acid, water and all important nutrients. So, here is some good example of biochemistry application in nutrition:

- Food Containment: Biochemistry helps us to determine the food chemistry. We can understand the containment of food we consume like the carbohydrates, proteins, fats and others. We can also understand the possible physiological alteration due to the deficiency of certain nutrient.
- The Role of Nutrients: Biochemistry makes us understand the function of nutrient to human health. We can know how the vitamin, mineral and essential fatty acids impact our body. Based on its chemicals containment, physicians can also recommend the great substance for our body like the essential amino-acids, cod liver oil, salmon fish oil and many more.
- Nutrient Values Test: By using the biochemical test, we can define the percentage or values of nutrient in certain foods we consume.
- Food Limitation: By the concept of biochemistry, the physician can prescribe to limit usage of certain food like excess sugar for diabetics, excess oil for heart & lung problem prone patients etc. We know that these carbohydrate and fat diets can inhibit the recovery rate from said disorder.

Biochemistry Applications in Pharmacy

Biochemistry can also be beneficial in pharmacy industry. In this industry, many drugs are often stored. These are the following applications of biochemistry for pharmacy:

- Drug Constitution: Biochemistry helps us to understand the condition of certain drugs like its possibility to degrade due to different temperature. This study can help us increasing the efficiency of drugs and minimizing the side effect.
- The Half-Life Test: This biochemical test is the test conducted to the biochemical drugs to determine the longevity of drug. We can understand the stability of drug If it's kept in the certain temperature conditions. By this test we can get the date of expiry of the drugs.
- Drug Storage: Biochemical test can be done to determine the condition of drugs storage. We

know that drugs or other substances like enzymes and hormones are stored for dispensing. It can possibly get deteriorated due to temperature, contamination or bad storage condition.

- Drug Metabolism: Biochemistry helps us understand about the metabolism process in drug molecules since there're many biochemical reactions happened in the presence of enzymes. By knowing this, we can understand the side effect of certain drugs to the certain patients.

Biochemistry Applications in Plants

Biochemistry is also important for the plants life. We can understand the chemical reactions occurred in the plant and how we can maximize it in order to increase our productivity. So, here're some popular biochemistry applications for plants:

- Photosynthesis: Photosynthesis is one of the chemical reaction happened in the plant. By understanding this chemical process, we can get the information on how carbohydrates are produced by the help of sunlight, CO_2 and water. There're also complex enzymes that help plant in this process. Understanding the basic concept of this chemical reaction can give us idea on how to improve the quality of the photosynthesis process.

- Respiration: Respiration is the chemical reaction which is related to the photosynthesis process. As a product of photosynthesis, plant will release the oxygen while take the carbon dioxide from the air.

- Different Sugars Study: Because of biochemistry, we can also understand that there're some types of carbohydrates which are formed in the plants. There are trioses (3 carbon sugars such as glyceraldehyde), tetroses (4), pentoses (5), hexoses (6= glucose), heptuloses (7) and so on.

- Plants Secondary Metabolites: Biochemistry gives us the information about the forming process of plants' products like tannins, resins, alkaloids, gums, enzymes and phytohormones.

- Other Functions: Biochemistry also gives us some useful information like how plants fruits get ripened, how plant seed germinates, and the respiration process and so on. Indeed, there are many applications of biochemistry in our daily life. Also, biochemistry branches have been used to compound many industries whether in medical, nursing, medicine, agricultural, and more fields.

Animal Biochemistry

Animal biochemistry is a branch of science that studies about structure and function cellular components such as protein, carbohydrate, lipid, nucleic acids, and other biomolecules in animals. These days biochemistry is more focused specifically in chemical reactions that involve enzymes and the characteristics of protein. Biochemistry of cell metabolism is also studied commonly these days as well. Other disciplines in biochemistry are the study of genetic codes (DNA and RNA), protein synthesis, cell membrane transportation, and signal transduction.

Animal Biochemistry Synthesis

Here are the sub-branches of animal biochemistry synthesis:

Lipid Synthesis

Lipid can be synthesized from carbohydrate and protein since in metabolism these three compounds meet in Krebs cycle. Most of these encounters occur through the main gate of Krebs cycle which is co-enzyme A Acetyl. As the result, these three compounds can fill each other as the material that forms all of those substances. Lipid can be formed by protein and carbohydrate while carbohydrate also can be formed by lipid and protein and so on.

Lipid can be synthesized from carbohydrate. Glucose can be broken into pyruvic and this process will create glycerol. Glucose also can be transformed into phosphate sugar which will be turned into acetyl co-A which will be turned again into lipid acid. Glycerol and lipid acid will form lipid. Lipid also can be synthesized from protein as well. Protein will be broken into protease amino acid.

Protein Synthesis

Protein synthesis that occurs in the cell involves DNA, RNA, and ribosome. The merger of amino acid molecules in large amount will form polypeptide molecule. Basically, protein is a polypeptide. Each cell in organisms can synthesize certain types of protein based on their needs. Protein synthesis in cell can occur since on the nucleus there are substances that play major role as "protein synthesis regulator". These substances are DNA and RNA.

Animal Biosynthesis

Here are the brief explanations of sub-branches of animal Biosynthesis:

Biosynthesis of Glutamate and Aspartate

Glutamate and aspartate are synthesized from α-keto acid with simple transamination reaction. The catalysator of this reaction is glutamate dehydrogenase enzyme and after that by aspartate aminotransferase. Aspartate is also derived from asparagine with help from asparaginase. The important role of glutamate is as main intercellular amino donor for transamination reaction. Aspartate plays a role as ornithine precursor for urea cycle.

Biosynthesis of Alanine

Alanine is transported to circulation by various tissues but usually by muscles. Alanine is formed by pyruvic. Liver accumulate plasma alanine which is the reverse version of transamination that occurs on muscles and proportionally increase the production of urea.

Alanine is transported from muscles to the liver together with glucose transportation from liver to muscles. This process is called glucose-alanine cycle. Main key of this cycle is that in a molecule of alanine the periphery tissue export pyruvic and ammonia to liver where the carbon structure is recycled and most of nitrogen is eliminated. There are two main pathways to produce muscle alanine which are:

- Directly through protein degradation.
- Through pyruvic transamination with help from alanine transaminase enzyme (which is also known as glutamate-pyruvic transaminase serum).

Biosynthesis of Tyrosine

Tyrosine is produced in cells with phenylalanine hydroxylation. Half of phenylalanine is needed to produce tyrosine. If the diet is rich of tyrosine it will reduce the needs of phenylalanine up to 50%. Phenylalanine hydroxylase is the combination of oxygenase function. An oxygen atom is combined to water and the other id combined to hydroxyl group from tyrosine. Reductant that is produced is tetrahydrofolate cofactor tetrahydrobiopterin that is maintained in reduced status by NADH-dependent enzyme dihydropteridine reductase.

Biosynthesis of Ornithine and Proline

Glutamate is the precursor of ornithine and proline. With glutamate, semi aldehyde became intermediate branch point to one of two products or others. Ornithine is not one of 20 amino acids that are used for protein synthesis. Ornithine plays major role as carbamoyl phosphate acceptor in urea cycle. Ornithine also has additional role as precursor for polyamine synthesis. The production of ornithine from glutamate is important when arginine diet as other source for ornithine is limited.

The use of glutamate semi aldehyde depends on cellular condition. Ornithine production from semi aldehyde is through the reaction of glutamate-transamination dependent. When the concentration of arginine is increased, ornithine is acquired from urea cycle and added from glutamate semi aldehyde that block the aminotransferase reaction. As the result, semi aldehyde accumulation occurs. Semi aldehyde is recycled spontaneously into Δ^1pyrroline-5-carboxylate that later will be reducted into proline by NADPH-dependent reductase.

Biosynthesis of Serine

The main pathway for serine begins from intermediate glycolytic 3-phosphogliserate. NADH-linked dehydrogenase changes 3-phosphogliserate into a keto acid which is 3-phosphopyruvic according to subsequent transamination. The activity of aminotransferase with glutamate as donor produces 3-phosphoserin that is modified into serine by phosphoserine phosphatase.

Plant Biochemistry

Plant biochemistry involves the study of the biochemistry of autotrophic organisms such as photosynthesis and other biochemical processes specific to plants. Plant Biochemistry presents each topic from the cellular level to the ecological and environmental levels, placing it in the context of the whole plant. Biochemical pathways are represented as route maps, showing how one reaction follows another. These maps emphasize the dynamism and flexibility of the plant facing the environmental challenges. Plant Biochemistry illustrates the impact of plants on human activity and success, in terms of their importance as a food supply and as raw materials for industrial and pharmaceutical products, and considers that humans can benefit from exploiting plant biochemical pathways. So, Plant Biochemistry is the science which studies chemical and physic–chemical processes that take place in the living organisms having the role to establish the material substrate of the life's phenomena. Usually the development of the organisms is possible due to the biochemical processes that take place in them in different environmental conditions. The influence exerted by the environment is reflected in the morphological and chemical structure of the organisms.

The fundamental problems of the animal biochemistry cannot be broken from the vegetal biochemistry and from the unicellular organisms. The universal principle of nature's integrity is the base of the chemical composition of the organisms and of the processes that are continuously developing and whose meaning is life. The organic matter is forming, transforming and degrading in any living body (plant, animal or microorganism); it doesn't differ from an organism to other by the component substances that are always the same or related also differ through secondary and unessential features determined by the proportion and the conditions of the formation and transformation of the organic matter. This is the reason why in the organisms' life there are some formation and degradation developing processes of the live matter and they vary between certain limits determining the variability and heredity of the organisms. Animal, plant and environment are forming an indissoluble unit and the biochemistry shows the diversity of the live matter.

Plant biochemistry is related to the animal and human activity where there are similar, however in all plants each cell has achieved a much more complicated metabolic process comparing to the animal organisms which are specialized on different functions. Animal, plant and environment form an inseparable unit, and only this way, in a comparing and general way, biochemistry shows a material world and the different shapes that it can take. So, when studying biochemistry there has to be treated at the same time both the fundamental chemistry of animals and that one of chemical plants and their need. Biochemistry has a particular role in the acquisition of agricultural raw materials for finished products with good yields, quality, and economic conditions in high yield. Biochemistry is involved in obtaining such materials in the conduct of technological processes, to conserve packaging and finished products, waste recovery in manufacturing, etc. During their industrial transformation raw materials and agricultural products suffer biochemical processes necessary or harmful, caused by enzymes, processes that must be traced, suppressed or directed.

Any living organism is an energetic system. The amount of energy received from the environment is equal to the consumed energy, plus the accumulations. If the accumulations are positive the organism will gain weight, and if the accumulations are negative it will lose weight. In a healthy organism it is desired that the accumulation should be maintained constant in time. The totality of the energetic exchanges that take place permanently between the living body and the environment, meaning that the organism is spending (losing) energy replaced by the energy contained in food is called energetic expenses.

The energy from the nutrients is the potential chemical energy and it is contained in the C–H, C–O, and the C–N links from proteins, lipids and carbohydrates, these links being able to split in the conditions of the high temperatures, the pH and the chemical environment created by cellular activities. After the usual oxidative splitting of the nutrients, the organism stocks the potential chemical energy and transforms the energy in other energy forms necessary to it. If the energetic needs of the organism are reduced and the energetic effort is increased, the glucose, the fatty acids and the amino acids are being entirely catabolized and transformed in stocking forms such as glycogen, triglucides and proteins.

The basis of all the biological processes is the exchange of matter, information and energy, this exchange process being active between the living body and the environment. The nutrients, the environmental factor indispensable for life, are crude or processed substances designated for the nutrition and satisfaction of the energy necessary for the maintaining the vital functions, for providing the growing and development of the living body and for maintaining the aptitudes for a good physical and mental activity.

The nutrients are complex products which are capable to suffer the necessary transformations in the living organism in order to maintain life and forming the tissues. Along with the oxygen, the living body also incorporates water and minerals, proteins, lipids, carbohydrates and vitamins from environment. Except oxygen, which is essential for the respiration process and the water, which is the vital environment, the other compounds are imported. The consumption of substances coming from external sources is a fundamental biological necessity of any organism. Other substances are named nutritive principles or alimentary principles and for their definition are used the terms nutrients or trophies. All metabolic processes take place in cells where the mitochondria is the main center of the neurogenesis, and the ribosomes are the place for the proteins' synthesis. The functions of the nutritive principles found in food are:

- Catalytic–Favoring the normal development of the biological processes.
- Energetic–Providing the necessary energy for the development of the vital processes.
- Plastic–Providing the synthesis of the substances owned by the organism; etc.

Biochemistry studies the process of the incorporation of the substances which participate in the metabolic processes, the transformation of the nutrients in soluble substances capable to be absorbed by the blood and produced in the digestive apparatus. Food's nutritive factors have some complex combinations that cannot be used and necessitate their transformation in simple elements which can be easily assimilated by the organism.

The Nutritive Principles and Fundamental Substances

Modern biochemistry agrees with the existence of nine categories of nutritive principles: water, alimentary fibers, lipids, carbohydrates, macro minerals, oligominerals, proteins, bioactive substances and vitamins. The chemical substances which are part of these categories have precise functions useful to the organism that ingrates them. Thus, some of them provide energy and heat, others provide construction material for the organism helping with the growth or the repair of the degraded tissues, and others participate directly or indirectly in the regulation of the biochemical and physiological processes from the organism. Yet none of these nutritive principles activate independently. For a good metabolic activity and health of the organism the alimentation should contain all those nine nutritive principles in adequate amounts. Due to their structure and properties the organic substances are the basic material of the living bodies and activate in all living cells. The organic compounds mainly consist of: carbon, hydrogen, oxygen, nitrogen, phosphorus and sulphur. Proteins, lipids and carbohydrates are part of this category and serve as plastic and energetic material with a distinct structural and functional role in the living bodies. The living matter is formed of two main groups of substances: inorganic and organic. The inorganic substances (water, minerals) are up taken from environment and are not synthesized by plants. The organic substances are the main compounds of the living bodies. According to their role in organism they can be classified as:

- Plastic substances: (Carbohydrates, proteins, and lipids) with a structural and energetic role,
- Substances with informational role: Nucleic acids,
- Substances with catalytic role: Enzymes,
- Substances with a role in balance: Enzymes, vitamins, phytohormons, pigments, antibodies etc.,
- Substances with an energetic role: Macroergic compounds.

Intermediary and final substances of the metabolism: Glycosides, lignin, eteric oils, alkaloids, antibiotic substances, alcohols, amides, amines, esters etc. The living matter is formed especially of the compounds of 12 chemical elements called plastic macro elements or constitutive elements. These are: C, H, O, N, P, S, Cl, K, Na, Ca, Mg, and Si. These plastic elements, together with other elements a.i. oligoelements or microelements: Mo, Cu, Zn, Fe, B, Mn, Co, As, F, Br, Al etc. make less than 1% of the dry matter. He main source of supply with inorganic compounds for plants is the soil. Mineral substances, component part of the food, are important in forming tissues and participate in the biological and physiological processes of the body. They are classified into macro elements (K, Ca, P, Na, Mg, Fe, Cl, S), which exist in the body in large amounts and microelements (Zn, Cu, Cr, Mn, Co, Mo, I, F, Ni etc.) found in small quantities. The microelements have a very important functional role in the reduction reactions, in the balance of the osmotic pressure and of the acid–base balance, and in maintaining cellular integrity even if they are found in small amount. Having also an essential role in the activation of some enzymes they become catalytic elements. Studies show the presence in the vegetal organisms of more than 60 chemical elements. From these only seven are fundamental or macro elements and together with the C, H, O and N they form about 99% of the body weight. These macro elements are: Na, K, Ca, P, Mg, Cl, and S. The other mineral elements found in very small amount are called microelements. The most important microelements are: Fe, Cu, Co, Zn, I, Mn, Cd, Se, Nil, Cr, Pb, Hg, and As. Macro and microelements are indispensable for the physiological equilibrium of the vegetal, animal and human organisms and the lack of one or more minerals results in pathological manifestations more or less severe.

Carbon is part of those organic compounds existing in the organisms and represents 48–50% of the dry substance. Hydrogen represents 6–7% of the dry matter and just as carbon, is found in all organic substances and in the water from body. It also participates in the oxidation–reduction reactions. Oxygen is found in most of the organic compounds from the living bodies in a proportion of 41–43% and along with the hydrogen, takes part in the oxidatio–reduction processes. Nitrogen is one of the basic elements of protides, and represents 16% of plant protides. In living tissues the proportion of nitrogen reaches 1.4–1.6%. Phosphorus is part of some important biologic compounds such as: enzymes, complex lipids, ATP etc. and interferes in many biochemical processes (photosynthesis etc.). In living bodies is found in a proportion near to 0.1–1.1%. Sulphur is part of some proteins, vitamins and carbohydrates etc. and can be found as 0.02–1.8% of the dry matter. It has important role in the growth and development of the living bodies.

The basic components in the alimentation of living organisms are the essential amino acids, polyunsaturated fatty acids, all the vitamins and minerals. Some components with high physiological activity, such as phospholipids, glycoproteidele, phosphoproteinele and many other compounds, are considered essential substances. The basic principles of a balanced nutrition are the proper correlations based on substances and biologically active nutrients, like proteins, lipids, carbohydrates, vitamins, mineral substances, depending on age, gender, activity at work and lifestyles. Nutritional factors are subjected to a series of changes before their addition to the body tissues a.i. the burning and the disposal of unused remainders, formed as a result of metabolic processes taking place in the body.

The metabolism represents the chemical and energetic changes of the products resulted from digestion and which are necessary in the anabolic and catabolic processes in the organism. The anabolism is the synthesis and repair process of simple substances' tissues. The catabolism (decomposition, degradation) is a decomposition process of the composed substances used as an energy source.

Molecular Biochemistry and Biology

Molecular biochemistry is a sub-discipline of biochemistry, which is a branch of chemistry. Molecular biochemistry involves the investigation of chemical processes that take place within living organisms at the molecular level. Molecular biochemistry focuses on macromolecules, such as viruses, membranes or enzymes, or more specifically, their function and structure. Biochemists utilize tools such as electron microscopes, lasers, spectrometers and computer software.

Molecular biology is the study of living things at the level of the molecules which control them and make them up. While traditional biology concentrated on studying whole living organisms and how they interact within populations (a "top down" approach), molecular biology strives to understand living things by examining the components that make them up (a "bottom up" approach). Both approaches to biology are equally valid, although improvements to technology have permitted scientists to concentrate more on the molecules of life in recent years.

Molecular biology is a specialized branch of biochemistry, the study of the chemistry of molecules which are specifically connected to living processes. Of particular importance to molecular biology are the nucleic acids (DNA and RNA) and the proteins which are constructed using the genetic instructions encoded in those molecules. Other biomolecules, such as carbohydrates and lipids may also be studied for the interactions they have with nucleic acids and proteins. Molecular biology is often separated from the field of cell biology, which concentrates on cellular structures (organelles and the like), molecular pathways within cells and cell life cycles.

The molecules which form the basis of life provide scientists with a more predictable and mechanistic tool for scientists to study. Working with whole organisms (or even just whole cells) can be unpredictable, with the outcome of experiments relying on the interaction of thousands of molecular pathways and external factors. Molecular biology provides scientists with a toolkit with which they may "tinker" with the way life works. They may use them to determine the function of single genes or proteins, and find out what would happen if that gene or protein was absent or faulty. Molecular biology is used to examine when and why certain genes are switched "on" or "off". An understanding of each of the factors has granted scientists a deeper understanding of how living things work, and used this knowledge to develop treatments for when living things don't work so well.

Common Molecular Biology Techniques

The following list covers some of the more commonly used molecular biology techniques:

- Electrophoresis: A process which separates molecules such as DNA or proteins out according to their size, electrophoresis is a mainstay of molecular biology laboratories. While knowing the size of a molecule might not seem like all that much information, it can be used to identify molecules or fragments of molecules and as a check to make sure that we have the correct molecule present.
- Polymerase Chain Reaction (PCR): A process used to amplify very small amounts of DNA to amounts which can be used in further experiments. It is used as a basic tool in molecular biology to ensure that we have sufficient DNA to carry out further techniques such as genetic modification; however it has wider practical uses such as in forensics (identification using DNA profiling) and disease diagnosis. PCR can also be used to introduce small point mutations into a gene in a process called site-directed mutagenesis.
- Restriction Digest: The process of cutting DNA up into smaller fragments using enzymes which only act at a particular genetic sequence.
- Ligation: The process of joining two pieces of DNA together. Ligation is useful when introducing a new piece of DNA into another genome.
- Blotting: A technique used to specifically identify biomolecules following electrophoresis. The molecule of interest is indicated using either a labeled probe (a complementary strand of nucleic acid) or a labeled antibody rose against a specific protein.
- Cloning: The technique of introducing a new gene into a cell or organism. This can be used to see what effect the expression of that gene has on the organism, to turn the organism into a factory which will produce large quantities of the gene or the protein it codes for, or (within the inclusion of a label) to indicate where the products of that gene are expressed in the organism. Insertion of genetic material into a bacterium is called transformation, while insertion into a eukaryotic cell is called transfection. If a virus is used to introduce this material, the process is called transduction.

Cell Biology

Cells form the basis of all living things. They are the smallest single unit of life, from the simplest bacteria to blue whales and giant redwood trees. Differences in the structure of cells and they way that they carry out their internal mechanisms form the basis of the first major divisions of life, into the three kingdoms of Archaea ("ancient" bacteria), Eubacteria ("modern" bacteria) and Eukaryota (everything else, including us). An understanding of cells is therefore vital in any understanding of life itself.

Cell biology is the study of cells and how they function, from the subcellular processes which keep them functioning, to the way that cells interact with other cells. Whilst molecular biology concentrates largely on the molecules of life (largely the nucleic acids and proteins), cell biology concerns itself with how these molecules are used by the cell to survive, reproduce and carry out normal cell functions.

In biomedical research, cell biology is used to find out more about how cells normally work, and how disturbances in this normal function can result in disease. An understanding of these processes can lead to therapies which work by targeting the abnormal function.

Common Cell Biology Techniques

The following list covers some of the more commonly used cell biology techniques:

- Cell/Tissue Culture: In the same way that bacteria and other simple organisms can be grown in the laboratory outside their normal environment, cells and tissues from more complicated organisms can be cultured as well. The techniques are slightly different, and the culture media are more complex to reflect the complex internal environment inside the host from which the cells are derived, however cell and tissue culture is a powerful tool which provides an almost limitless supply of test material for researchers to use without resorting to using whole organisms. In addition, the controlled condition in cell and tissue culture allows researchers to carry out experiments with a lower number of variables which may affect the outcome of the test. Cell culture may use cells removed directly from an organism (primary culture), or it may use lines of cultured cancer cells. The benefit of the latter approach is that cancer cells continue to divide, while primary cultures cease dividing after a number of cycles.
- Microscopy: The basic tool of cell biology is microscopy. A recent advance in imaging technology has allowed an unprecedented amount of information to be gleaned from microscopic analysis. Types of microscopic techniques which are used include:
    - Bright Field: Traditional microscopy, where cells are illuminated by visible light. Bright field microscopy gives a general picture of cell function, although that information is not very detailed or specific. As animal cells lack cell walls, bright field microscopy may use special techniques such as phase contrast to show cellular structures in more detail. Bright field microscopy allows imaging of live or fixed (dead) cells and tissues).
    - Electron Microscopy: Uses a focused beam of electrons instead of light. Electron microscopy permits a much higher magnification of specimens than light microscopy and is useful in obtaining detailed information about sub-cellular structures. Electron microscopy requires extensive processing and so can only be performed on fixed specimens. Transmission electron microscopy provides a cross section of a specimen, while scanning electron microscopy gives a three-dimensional image of the surface of a specimen.
    - Fluorescence Microscopy: Uses fluorescent materials to indicate structures in a specimen. Fluorescence occurs when light of one wavelength "excites" a material and causes

it to emit light of a different wavelength. Most fluorescent materials give off visible light after excitation by ultraviolet light. Structures may be naturally fluorescent (auto fluorescence) or they may be labeled with a compound which is fluorescent (e.g. DAPI is a dye which binds onto DNA. The DNA and nuclei of cells stained with DAPI emit a blue light under ultraviolet light).

- Immunofluorescence: Antibodies are proteins made by the immune system which bind onto specific parts of proteins. Antibodies can be raised against any protein in the cell. If these antibodies are attached to a fluorescent tag, the tag will only show up where that antibody attached (i.e. where the target protein is found in the cell). Immunofluorescence allows very specific targeting of cellular structures.
- RNA Interference: RNA interference uses short sequences of RNA which are complementary to the mRNA which carries to instructions to translate proteins from the DNA to the ribosomes. The interfering RNA binds to the target sequence, preventing it from being translated. As a result, careful selection of interfering RNA can be used to silence a particular gene. This allows researchers to study what role a protein plays in a cell, by observing what happens when that protein is absent.
- Time-Lapse Microscopy: Many cellular processes (e.g. mitosis) occur over a period of time which is not practical for direct observation. Imaging cells over a period of time (e.g. a photograph is taken every 20 minutes for 24 hours) allows us to combine these images in a "movie" which compresses a long time period into a shorter one.

Metabolism

Metabolism is the total amount of the biochemical reactions involved in maintaining the living condition of the cells in an organism. All living organisms require energy for different essential processes and for producing new organic substances.

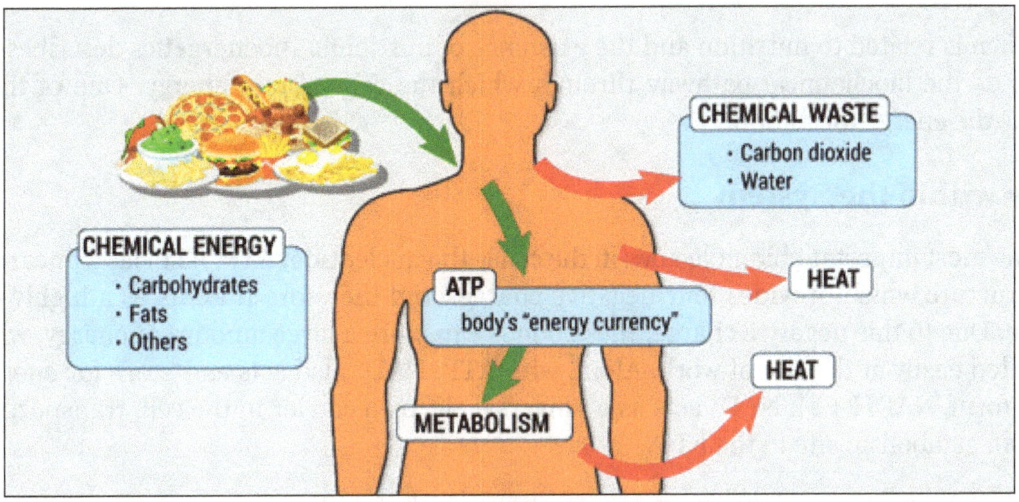

The metabolic processes help in growth and reproduction and help in maintaining the structures of living organisms. The organisms respond to the surrounding environment due to metabolic activities. All the chemical reactions occurring in the living organisms from digestion to transportation of substances from cell to cell require energy.

Metabolic Process

There are two types of metabolic process:

- Catabolism
- Anabolism

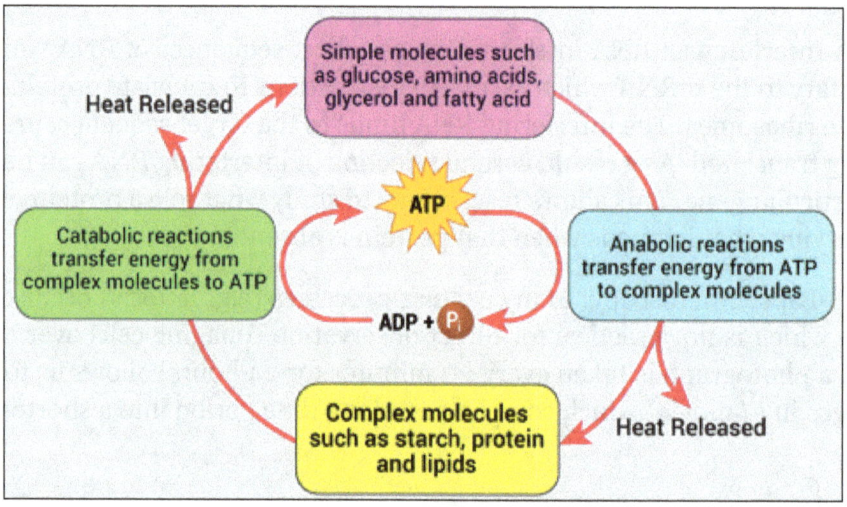

1. Catabolism: This process is mainly involved in breaking down larger organic molecules into smaller molecules. This metabolic process releases energy.
2. Anabolism: This process is mainly involved in building up or synthesizing compounds from simpler substances required by the cells. This metabolic process requires and stores energy.

Metabolism is related to nutrition and the existence of nutrients. Bioenergetics describes the metabolism as the biochemical pathway through which the cells obtain energy. One of the major aspects is the energy formation.

Energy within the System

One of the most important biomolecules in the cell is the nucleotide; ATP. ATP has a linear triphosphate structure which provides four negative charges and therefore it exists as a highly charged molecule. Due to this negative charge, these bonds can store a large amount of energy, which can be liberated easily at the site of work. Along with ATP, NAD^+ also acts as a store for energy in its reduced form, NADH + H. NAD^+ acts as a universal electron carrier in the cell, transporting electrons from catabolism site to the ETC.

Émilie du Châtelet proposed the law of conservation, which stated that energy can be neither created nor destroyed; rather, it can only be transformed from one form into another. This is linked to the first law of thermodynamics and it helps us to explain how energy flows through biological systems. The laws of thermodynamics also help us to predict if a reaction is possible and how much energy is required or released in the process. This brings us to an important concept: the role of chemical equilibria, where a reaction can be reversible. At equilibrium, there is no net reaction as both the forward and reverse reactions are moving equally fast. The system aims to bring reactions to equilibrium. However, true equilibrium is not compatible with life, as there is no longer a flow of energy.

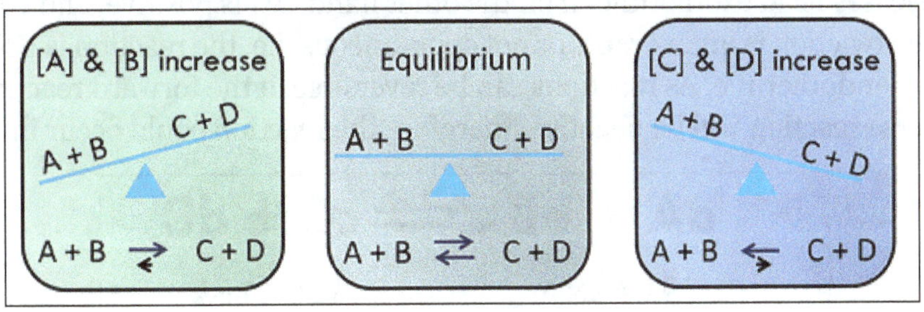

Reaching equilibrium: If you increase the concentration of A and B, this pushes the reaction to make more C and D. If you do the opposite and add more C and D, then the reverse reaction occurs. The aim is to bring the reaction back to equilibrium.

The direction of the reaction can be governed by many aspects, including the concentration of substrate or products, the energy released or required for the reaction, and in the case of metabolic pathways, the activity of the enzyme. Figure shows that simple equilibria can be influenced by the change in substrates and products and that the direction of the reaction is governed by which has a higher concentration. However, the fulcrum of the metaphorical seesaw is not always in the centre and this is described by the equilibrium constant or standard free energy change.

Gibbs free energy is used to describe whether our reaction will run in one direction or the other and is termed as the energy available to do work. This is different from total energy, which is the energy obtained from combustion. Standard free energy change is a special form of the Gibbs free energy and provides a constant energy change during the standard state – pH 7.0 at 25°C and 1 atm, when the concentration of substrates is 1.0 M. This is termed as $\Delta G^{o\prime}$ and has a characteristic constant value for a given reaction and tells us where the balancing point is. However, we still want to know if the reaction is going in one direction or the other. Change in free energy (ΔG) is a much more realistic measure of the energy available to the system under normal cellular environments. This is because in mammals internal body temperatures are not at 25°C and metabolic substrates are not at 1 M. Whilst $\Delta G^{o\prime}$ has a characteristic constant value for a given reaction, ΔG is not a constant as it is dependent on $\Delta G^{o\prime}$, the concentration of substrates, and the temperature. ΔG can be described by the following equation below:

$$\Delta G = \Delta H - T\Delta S$$

- Enthalpy (H) = total energy.
- Entropy (S) = disorder of the system.
- T = temperature (measured in Kelvin).

The second law of thermodynamics states that for a process (reaction) to take place, there must be an increase in entropy in the universe (or system). This means that for a reaction to take place spontaneously there needs to be an increase in entropy and therefore a negative change in free energy (ΔG). We can now consider what this means in terms of reaction kinetics and how ΔG can dictate the direction of the reaction, using the analogy of bicycling. If the ΔG is negative, a reaction proceeds spontaneously and with a loss of free energy. Using our bike analogy, this would be like the energy released as you cycle from the top of a hill down to the bottom. This kind of reaction is termed EXERGONIC, or if it releases heat, exothermic. If on the other hand, ΔG is positive, and you are looking up at the hill, then the reaction is unfavorable or not spontaneous, i.e. the reaction is ENDERGONIC, or in the case of heat, endothermic. As reactions can be reversible, if the forward reaction has a positive ΔG, then the reverse reaction will be negative, therefore the reaction could occur in reverse.

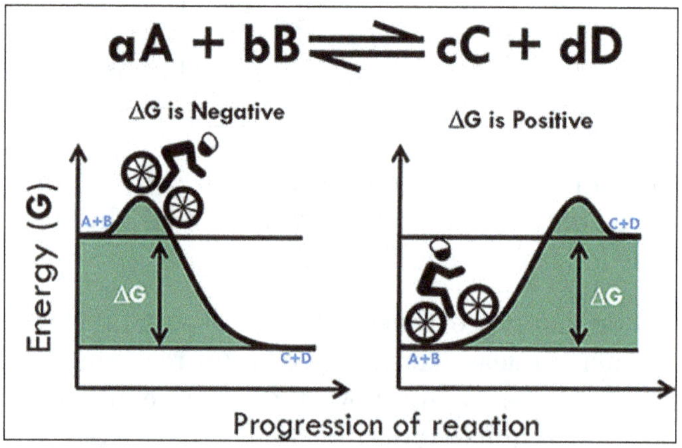

The effect of changes in ΔG on the reaction A+B and C+D.

If we put this into practical terms to consider the second reaction in glycolysis catalysed by glucose-6-phosphate (G6P) isomerase (G6PI) (equation below):

$$\text{Glucose-6-phosphate} \overset{G6PI}{\leftrightharpoons} \text{Fructose-6-phosphate}$$

The ΔG°′ for this reaction is +1.7 kJ.mol^{-1}, which is close to equilibrium, would be endogenic and therefore not spontaneous. However ΔG°′ describes the free energy available under standard state conditions. If we look at a more physiological condition, i.e. at 37°C, then if the concentration of G6P increases, this decreases the ΔG into negative and drives the forward reaction. The counter is true, so if fructose-6-phosphate (F6P) accumulates the forward reaction is inhibited (due to ΔG now becoming positive) then the reverse reaction is favored (as the ΔG for this is negative). This is an example of one of the freely reversible reactions in glycolysis.

Another thing to take into account is that some reactions will not be reversible under physiological conditions. For example, if the ΔG is very negative, then changing the amount of product or substrate will not affect, especially as the temperature can very rarely be altered to any real degree. One example of where this is particularly seen is in one of the key regulatory steps in glycolysis – the reaction catalyzed by phosphofructokinase 1 (PFK1), shown in (equation below). The ΔG°′ is −14.2 kJ.mol^{-1}, meaning that under physiological conditions the ΔG is very negative, and therefore even a build-up of fructose-1-phosphate is not sufficient to reverse the reaction. This reaction is effectively irreversible under physiological conditions; we will discuss this in more detail later.

$$Fructose-6-phosphate + ATP \underline{PFK1} Fructose-1,6-bisphosphate + ADP$$

One important caveat about ΔG and reactions is that ΔG does not predict the rate of reaction. The rate of reaction is governed by the activation energy and is catalyzed by enzymes. If ΔG is negative, then the reaction should occur spontaneously, but there is still a need to overcome the initial activation energy. Reactions that need catalysis can be controlled by the activity of the enzyme. For example, whilst the ΔG for is, PFK1 is regulated by feedback inhibition and activation. Increasing concentrations of either phosphoenolpyruvate (PEP; the penultimate intermediate in glycolysis) or ATP inhibit PFK1, as they act as a signal for sufficient energy demand. Conversely, an increase in the concentration of adenosine monophosphate (AMP) Signals an Energy Deficit in the Cell and Therefore Activates PFK1.

Coupled Reactions and Pathways

There is an intrinsic value to coupling reactions together to form complex pathways. As we can see from Table, several reactions have a positive ΔG and so in theory, should not be spontaneous nor go in reverse. However, by linking the reactions together, the product of one reaction becomes the substrate for the next. Thus, by coupling reactions together, substrates can build up and products can be rapidly removed, changing the equilibrium seesaw making the ΔG negative (for reversible reactions).

Table: Standard free energy and Gibbs free energy of each step in glycolysis.

Enzyme	$\Delta G^{o\prime}$ in $kJ.mol^{-1}$	ΔG in $kJ.mol^{-1}$
Hexokinase	−16.7	−33.5
G6PI	+1.7	−2.5
PFK	−14.2	−22.2
Aldolase	+23.8	−1.3
Triose phosphate isomerase	+7.5	+2.5
Glyceraldehyde 3 phosphate dehydrogenase	+6.3	−1.7
Phosphoglycerate kinase	−18.8	+1.3
Phosphoglycerate mutase	+4.6	+0.8
Enolase	+1.7	−3.3
Pyruvate kinase	−31.4	−16.7

Finally, ATP can be used to drive reactions forward by coupling the release of energy from its hydrolysis to overcome a positive ΔG. As we have previously seen, ATP stores a large amount of energy in the phosphate bonds and cleavage of one phosphate liberates −31 $kJ.mol^{-1}$. This can be used to overcome endergonic reactions such as the first reaction in glycolysis:

$$Glucose + Pi + H^{+} \rightarrow Glucose - 6 - phosphate + H_2O \ (endergonic) \ \Delta G^{o\prime} = +14 \ kJ.mol^{-1}$$

$$Glucose + Pi + H^{+} \rightarrow Glucose - 6 - phosphate + H_2O \ (endergonic) \ \Delta G^{o\prime} = +14 kj.mol^{-1}$$

$$Glucose + ATP \rightarrow G - 6 - phosphate + ADP \ (exergonic) \ \Delta G^{o\prime} = -17 \ kJ.mol^{-1}$$

The native reaction, the addition of inorganic phosphate, is endergonic and requires +14 $kJ.mol^{-1}$. However, because the hydrolysis of ATP liberates Pi and Pi is a substrate for the first reaction, these reactions can be coupled and the energy of hydrolysis of ATP is used to drive the reaction. Note this coupling only works because Pi is a shared intermediate in both reactions.

Reactions are favored when they have a negative ΔG. Coupling reactions together in pathways enable unfavorable reactions to occur. Metabolism is organized into distinct metabolic pathways to maximize the capture of energy and prevent uncontrolled combustion.

Metabolism from Thin Air

Organisms fall into two distinct groups, either heterotrophs or autotrophs. Autotrophs derived from the words 'auto' for 'self' and '-trophs' for 'food' are a set of organisms that can obtain energy from sunlight and inorganic nutrients from their environment to fix carbon into complex macromolecules. Heterotrophs represent a much larger group of organisms, including most microorganisms and animal cells. In contrast with autotrophs, they are not able to fix their own carbon, but instead use the carbon fixed by autotrophs to grow.

Autotrophs can generally be split into two groups, chemoautotrophs, and photoautotrophs. These organisms use the energy from inorganic chemical reactions (chemoautotrophs, i.e. Nitrobacter) or light (photoautotrophs, i.e. plants) to fix carbon dioxide as a usable energy source. One example of a key group of chemoautotrophs is the diazotrophs, which are nitrogen-fixing bacteria and archaea. They are responsible for fixing nitrogen from the air (N_2) into inorganic nitrogen compounds such as ammonia (NH_4^+), nitrite (NO_2^-), and nitrate (NO_3^-). They are also capable of using these compounds to obtain energy for biosynthesis of macromolecules. Ammonia-oxidizing (AOB) and nitrite-oxidizing (NOB) bacteria take advantage of processes similar to the eukaryotic ETC (in mitochondria) to generate their ATP. Oxidation of inorganic nitrogen compounds, such as NH_4^+ and NO_2^- release electrons that are captured and transferred by copper-containing enzymes to a final electron acceptor. The transfer of these electrons drive protons (H^+) to be pumped out to generate a proton gradient, also known as the proton motive force. These protons then flow back down their concentration gradient through an ATP synthase, to generate ATP. This ATP can then be utilized in biosynthetic processes such as the Calvin cycle, which captures CO_2 and converts it into carbohydrates.

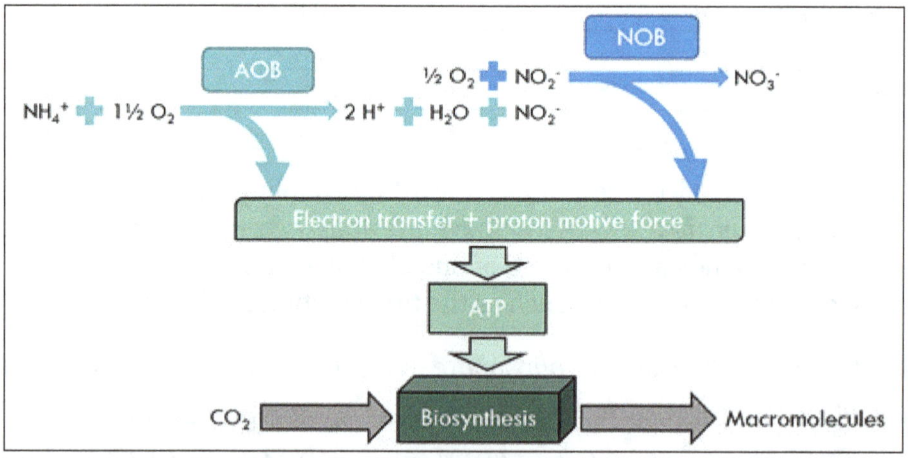

The role of nitrogen in generating macromolecules in chemoautotrophs – AOB, NOB, NH_4^+, NO_2^-, and NO_3^-.

Photoautotrophs represent the major class of autotrophs, which utilise light as an energy source to fix carbon dioxide. Multicellular plants and unicellular organisms (algae and cyanobacteria) are perhaps the best examples of this class, using the process of photosynthesis to turn sunlight into ATP and NADPH. These organisms use photons of light to obtain electrons from water and produce oxygen as a by-product. These electrons are again used to drive ATP formation, which in turn is used by pathways, such as the Calvin cycle to generate carbohydrates. In eukaryotic organisms such as green plants and algae, this process of capturing photons is achieved by the chloroplast. Cyanobacteria lack these organelles, instead photosynthesizing using folded membranes. Genome analysis data suggest that an ancient cyan bacterium might have provided the ancestor of the chloroplast in eukaryotic cells describes their work to identify the early photosynthetic eukaryotes).

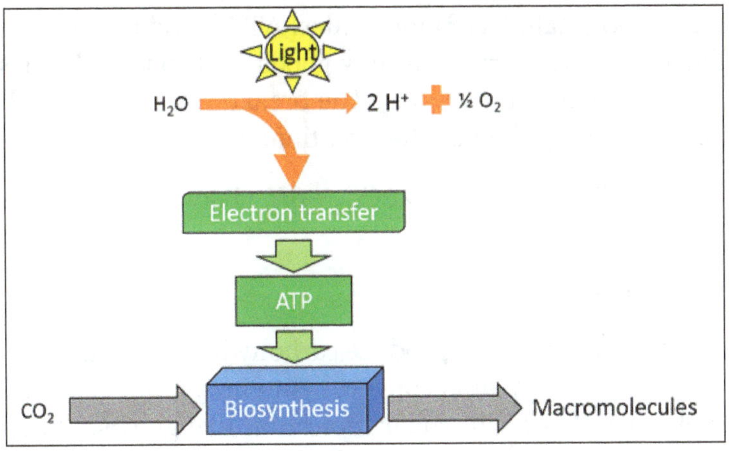

Metabolism of Carbohydrates

Organisms break down carbohydrates to produce energy for cellular processes, and photosynthetic plants produce carbohydrates. Carbohydrates are one of the major forms of energy for animals and plants. Plants build carbohydrates using light energy from the sun (during the process of photosynthesis), while animals eat plants or other animals to obtain carbohydrates. Plants store carbohydrates in long polysaccharides chains called starch, while animals store carbohydrates as the molecule glycogen. These large polysaccharides contain many chemical bonds and therefore store a lot of chemical energy. When these molecules are broken down during metabolism, the energy in the chemical bonds is released and can be harnessed for cellular processes.

All living things use carbohydrates as a form of energy: Plants, like this oak tree and acorn, use energy from sunlight to make sugar and other organic molecules. Both plants and animals (like this squirrel) use cellular respiration to derive energy from the organic molecules originally produced by plants.

Energy Production from Carbohydrates

The metabolism of any monosaccharide (simple sugar) can produce energy for the cell to use. Excess carbohydrates are stored as starch in plants and as glycogen in animals, ready for metabolism if the energy demands of the organism suddenly increase. When those energy demands increase, carbohydrates are broken down into constituent monosaccharaides, which are then distributed to all the living cells of an organism. Glucose ($C_6H_{12}O_6$) is a common example of the monosaccharaides used for energy production.

Inside the cell, each sugar molecule is broken down through a complex series of chemical reactions. As chemical energy is released from the bonds in the monosaccharide, it is harnessed to synthesize high-energy adenosine triphosphate (ATP) molecules. ATP is the primary energy currency of all cells. Just as the dollar is used as currency to buy goods, cells use molecules of ATP to perform immediate work and power chemical reactions. The breakdown of glucose during metabolism is call cellular respiration can be described by the equation:

$$C_6H_{12}O_6 + 6O_2 \rightarrow 6CO_2 + 6H_2O + energy$$

Producing Carbohydrates

Plants and some other types of organisms produce carbohydrates through the process called photosynthesis. During photosynthesis, plants convert light energy into chemical energy by building carbon dioxide gas molecules (CO_2) into sugar molecules like glucose. Because this process involves building bonds to synthesize a large molecule, it requires an input of energy (light) to proceed. The synthesis of glucose by photosynthesis is described by this equation (notice that it is the reverse of the previous equation):

$$6CO_2 + 6H_2O + energy \rightarrow C_6H_{12}O_6 + 6O_2$$

As part of plants' chemical processes, glucose molecules can be combined with and converted into other types of sugars. In plants, glucose is stored in the form of starch, which can be broken down back into glucose via cellular respiration in order to supply ATP.

References

- Biochemistry-is-the-application-of-the-chemistry-of-living-systems, biochemistry: mcgill.caRetrieved 12, May 2020
- Applications-of-biochemistry: azchemistry.com, Retrieved 20, April 2020
- General-concepts-of-plant-biochemistry: researchgate.net, Retrieved 04, Feb 2020
- Introduction-cell-biology, cell-and-molecular-biology-experiences: di.uq.edu.au, Retrieved 24, June 2020
- Energy-and-metabolism, boundless-biology: courses.lumenlearning.com, Retrieved 10, August 2020

2

Carbohydrates Biochemistry

Carbohydrates are the biomolecules which are made up of carbon, hydrogen and oxygen atoms. They are the primary sources of energy in all the living organisms. This chapter has been carefully written to provide an easy understanding of the varied facets of carbohydrate biochemistry.

Carbohydrates are one of the three major macronutrients in our diet representing a wide group of substances which include the sugars, starches, glycogen, gums and celluloses. Carbohydrates are known as the important energy source for living things. They comes from the process of photosynthesis, an endothermic reaction in which condensation of carbon dioxide takes place and which requires light energy and the pigment chlorophyll. The chemical component of carbohydrates is carbon, hydrogen, and oxygen.

Generally, but not always the hydrogen and oxygen in carbohydrates are present in the proportion of two hydrogen atoms to one oxygen atom as in H_2O, from which fact the term "carbohydrates" was derived. The carbohydrates are actually or potentially hydroxy aldehydes or ketones, and in most cases they are polyhydroxy aldehyde or ketones compounds that can be hydrolyzed to them. It also means "hydrate of carbon".

Plant biomass is made up of three fourths of carbohydrates but in animal body it is present in small quantities. In animal it is stored in the form of glycogen, sugar and their derivatives. All living organism contains derived monosaccharides such as the sugar acids, amino sugars and the deoxysugars. Derivatives of the carbohydrates also contain nitrogen, phosphates and sulfur compounds. Carbohydrates can combine with lipid and protein to form glycolipids and glycoproteins. Carbohydrates and their derivatives are present in all animal tissues and tissue fluids, blood and milk. The glucosides are classes of carbohydrates are derivatives which are frequent constituents of plants.

The major components of living matter are formed from carbohydrates, proteins and fats. This complex form of carbohydrates is required for the maintenance of cell structure and functions. Presence of functional group like alcohol on almost every carbon makes the chemistry of carbohydrates a complex structure.

Carbohydrate may be existing in either a straight chain or a closed ring structure. The general reaction between the aldehyde and ketone part of the carbohydrates and alcohol groups forms the derivative called hemiacetals and hemiketals leads to the formation of ring structure. The most common basic structure of carbohydrates is closed rings Glucose is an example of a single closed ring carbohydrate.

Functions of Carbohydrates

There are five primary functions of carbohydrates in the human body. They are energy production, energy storage, building macromolecules, sparing protein, and assisting in lipid metabolism.

Energy Production

The primary role of carbohydrates is to supply energy to all cells in the body. Many cells prefer glucose as a source of energy versus other compounds like fatty acids. Some cells, such as red blood cells, are only able to produce cellular energy from glucose. The brain is also highly sensitive to low blood-glucose levels because it uses only glucose to produce energy and function (unless under extreme starvation conditions). About 70 percent of the glucose entering the body from digestion is redistributed (by the liver) back into the blood for use by other tissues. Cells that require energy remove the glucose from the blood with a transport protein in their membranes. The energy from glucose comes from the chemical bonds between the carbon atoms. Sunlight energy was required to produce these high-energy bonds in the process of photosynthesis. Cells in our bodies break these bonds and capture the energy to perform cellular respiration. Cellular respiration is basically a controlled burning of glucose versus an uncontrolled burning. A cell uses many chemical reactions in multiple enzymatic steps to slow the release of energy (no explosion) and more efficiently capture the energy held within the chemical bonds in glucose.

The first stage in the breakdown of glucose is called glycolysis. Glycolysis, or the splitting of glucose, occurs in an intricate series of ten enzymatic-reaction steps. The second stage of glucose breakdown occurs in the energy factory organelles, called mitochondria. One carbon atom and two oxygen atoms are removed, yielding more energy. The energy from these carbon bonds is carried to another area of the mitochondria, making the cellular energy available in a form cells can use.

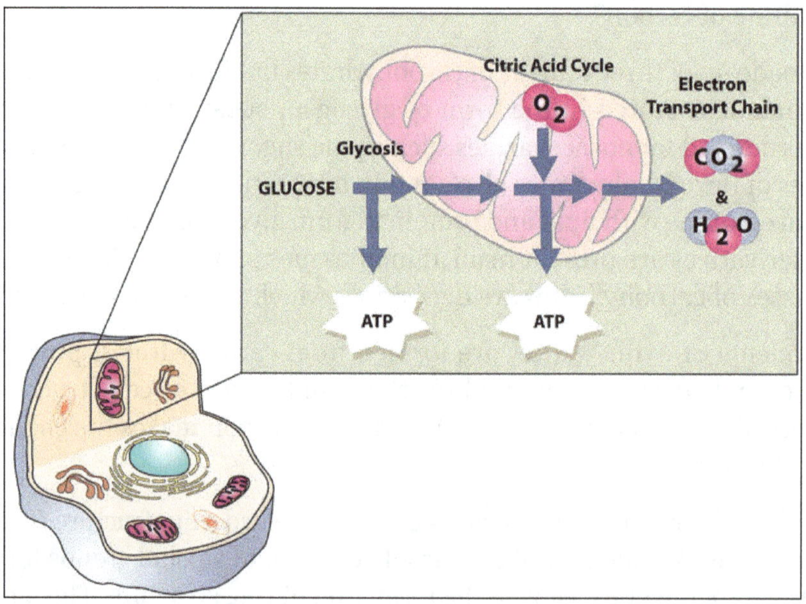

Cellular respiration is the process by which energy is captured from glucose.

Energy Storage

If the body already has enough energy to support its functions, the excess glucose is stored as glycogen (the majority of which is stored in the muscles and liver). A molecule of glycogen may contain in excess of fifty thousand single glucose units and is highly branched, allowing for the rapid dissemination of glucose when it is needed to make cellular energy.

The amount of glycogen in the body at any one time is equivalent to about 4,000 kilocalories—3,000 in muscle tissue and 1,000 in the liver. Prolonged muscle use (such as exercise for longer than a few hours) can deplete the glycogen energy reserve. Remember that this is referred to as "hitting the wall" or "bonking" and is characterized by fatigue and a decrease in exercise performance. The weakening of muscles sets in because it takes longer to transform the chemical energy in fatty acids and proteins to usable energy than glucose. After prolonged exercise, glycogen is gone and muscles must rely more on lipids and proteins as an energy source. Athletes can increase their glycogen reserve modestly by reducing training intensity and increasing their carbohydrate intake to between 60 and 70 percent of total calories three to five days prior to an event. People who are not hardcore training and choose to run 5-kilometer races for fun do not need to consume a big plate of pasta prior to a race since without long-term intense training the adaptation of increased muscle glycogen will not happen.

The liver, like muscle, can store glucose energy as a glycogen, but in contrast to muscle tissue it will sacrifice its stored glucose energy to other tissues in the body when blood glucose is low. Approximately one-quarter of total body glycogen content is in the liver (which is equivalent to about a four-hour supply of glucose) but this is highly dependent on activity level. The liver uses this glycogen reserve as a way to keep blood-glucose levels within a narrow range between meal times. When the liver's glycogen supply is exhausted, glucose is made from amino acids obtained from the destruction of proteins in order to maintain metabolic homeostasis.

Building Macromolecules

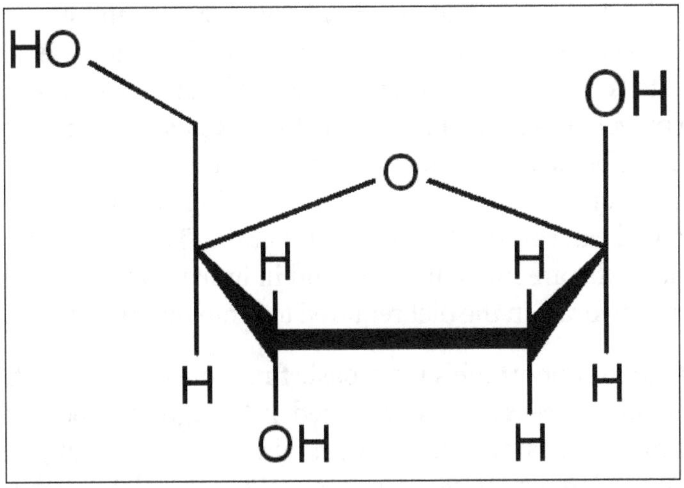

Chemical Structure of Deoxyribose: The sugar molecule deoxyribose is used to build the backbone of DNA.

Although most absorbed glucose is used to make energy, some glucose is converted to ribose and deoxyribose, which are essential building blocks of important macromolecules, such as RNA, DNA, and ATP. Glucose is additionally utilized to make the molecule NADPH, which is important for protection against oxidative stress and is used in many other chemical reactions in the body. If all of the energy, glycogen-storing capacity, and building needs of the body are met, excess glucose can be used to make fat. This is why a diet too high in carbohydrates and calories can add on the fat pounds.

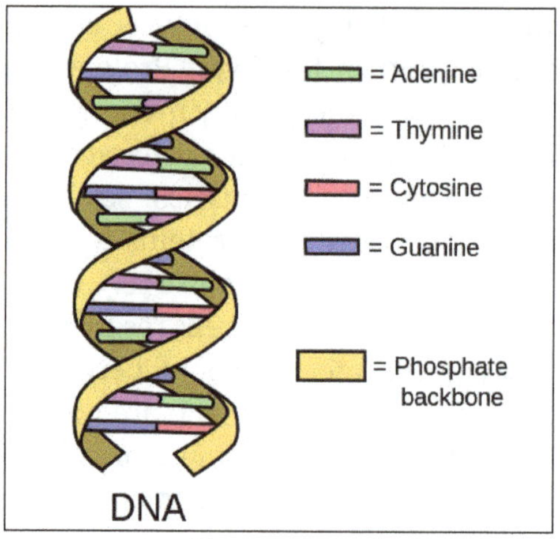

Double-stranded DNA.

Sparing Protein

In a situation where there is not enough glucose to meet the body's needs, glucose is synthesized from amino acids. Because there is no storage molecule of amino acids, this process requires the destruction of proteins, primarily from muscle tissue. The presence of adequate glucose basically spares the breakdown of proteins from being used to make glucose needed by the body.

Lipid Metabolism

As blood-glucose levels rise, the use of lipids as an energy source is inhibited. Thus, glucose additionally has a "fat-sparing" effect. This is because an increase in blood glucose stimulates release of the hormone insulin, which tells cells to use glucose (instead of lipids) to make energy. Adequate glucose levels in the blood also prevent the development of ketosis. Ketosis is a metabolic condition resulting from an elevation of ketone bodies in the blood. Ketone bodies are an alternative energy source that cells can use when glucose supply is insufficient, such as during fasting. Ketone bodies are acidic and high elevations in the blood can cause it to become too acidic. This is rare in healthy adults, but can occur in alcoholics, people who are malnourished, and in individuals who have Type 1 diabetes. The minimum amount of carbohydrate in the diet required to inhibit ketosis in adults is 50 grams per day.

Carbohydrates are critical to support life's most basic function—the production of energy. Without energy none of the other life processes are performed. Although our bodies can synthesize glucose it comes at the cost of protein destruction. As with all nutrients though, carbohydrates are to be consumed in moderation as having too much or too little in the diet may lead to health problems.

Glycolysis

Glycolysis is the first of the main metabolic pathways of cellular respiration to produce energy in the form of ATP. Through two distinct phases, the six-carbon ring of glucose is cleaved into two three-carbon sugars of pyruvate through a series of enzymatic reactions. The first phase of glycolysis requires energy, while the second phase completes the conversion to pyruvate and produces ATP and NADH for the cell to use for energy. Overall, the process of glycolysis produces a net gain of two pyruvate molecules, two ATP molecules, and two NADH molecules for the cell to use for energy. Following the conversion of glucose to pyruvate, the glycolytic pathway is linked to the Krebs cycle, where further ATP will be produced for the cell's energy needs.

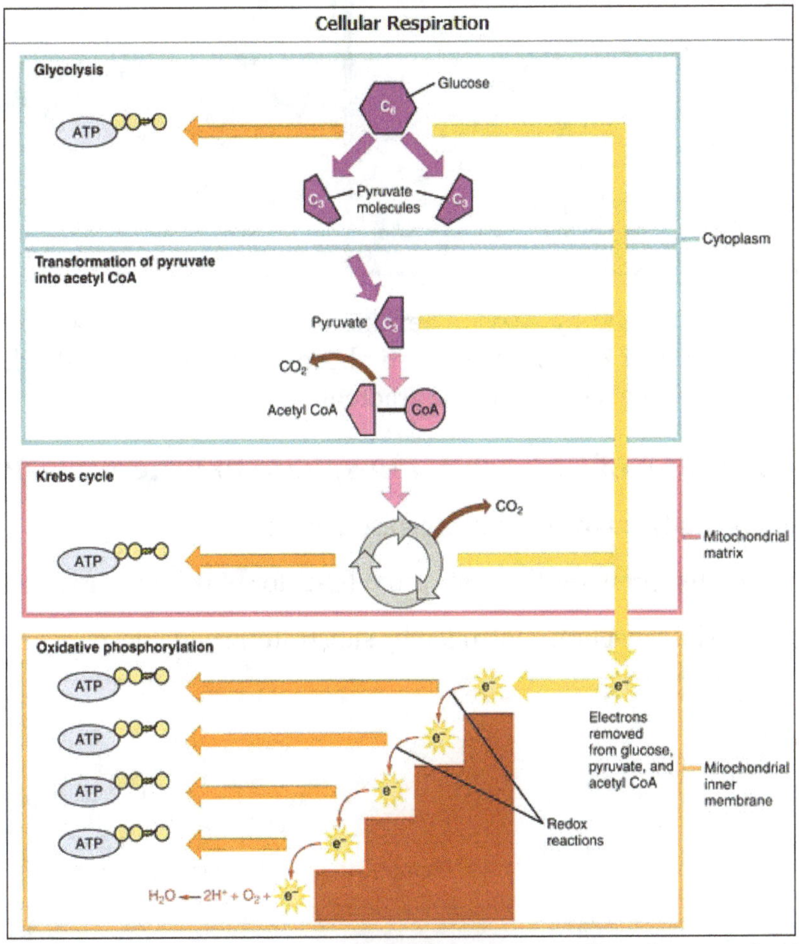

Glycolysis is the first pathway of cellular respiration that oxidizes glucose molecules. It is followed by the Krebs cycle and oxidative phosphorylation to produce ATP.

The Reactions of the Glycolytic Pathway

Stage – 1

Reaction 1: Synthesis of Glucose-6-Phosphate

- This reaction is a kinase reaction , added a phosphate to glucose immediately when glucose enter in the cell, due this phosphorylation glucose transport out of cell prevents and reactivity of oxygen of is also increase.
- The phosphorylation of glucose in all cell in body is catalyzed by several enzymes called the hexokinases, ATP is complexes with Mg^{+2} which a co- substrate in this reaction is.
- The reaction is essentially irreversible, and glucose is efficiently trapped inside the cell, as phosphorylated intermediates do not readily pass through cell membrane.

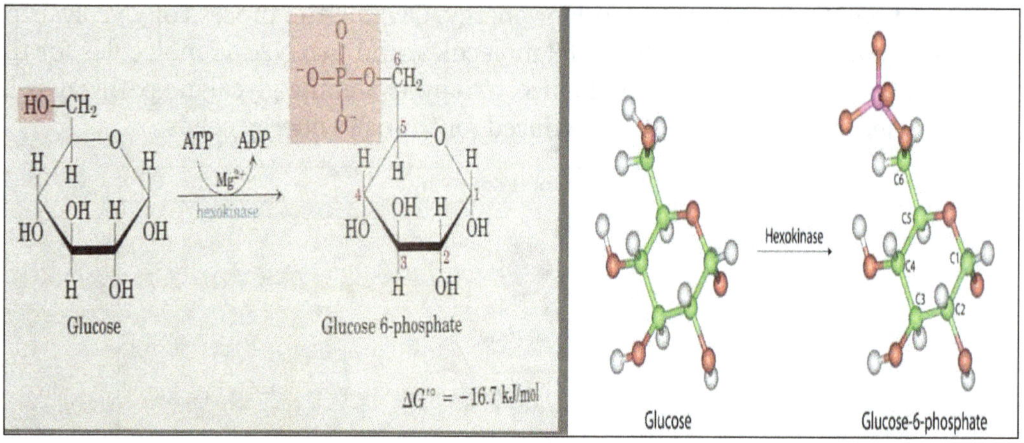

- This reaction is catalyzed by enzyme hexokinase, present in virtually extra hepatic cells, has a high affinity (low K_m) for glucose, so phosphorylates essentially all the glucose that enters cell, maintaining a large glucose gradient.

Reaction 2: Conversion of Glucose-6-Phosphate to Fructose-6-Phosphate

- The reaction is catalysed by phosphoglucoisomarase.
- This enzymatic step prepares the first carom (C-1) for phosphorylation.
- It is freely reversible reaction controlled by substrate-product levels.

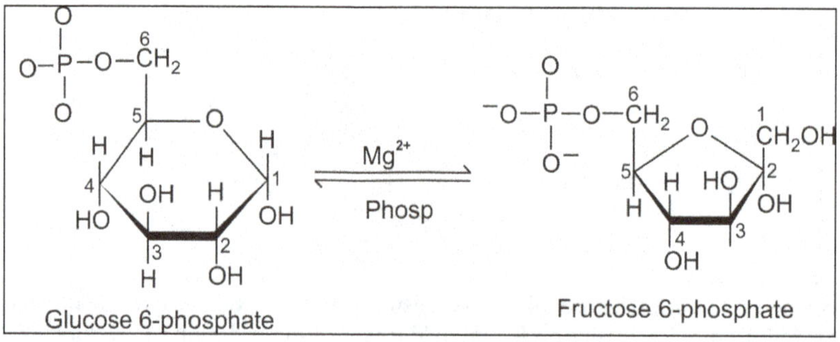

- Small change in standard free energy, the reaction proceeds readily in either direction, and require Mg^{2+}.

Reaction 3: Fructose-6- Phosphate to 1, 6 Diphosphate

- The reaction is catalyzed by phosphofructokinase.

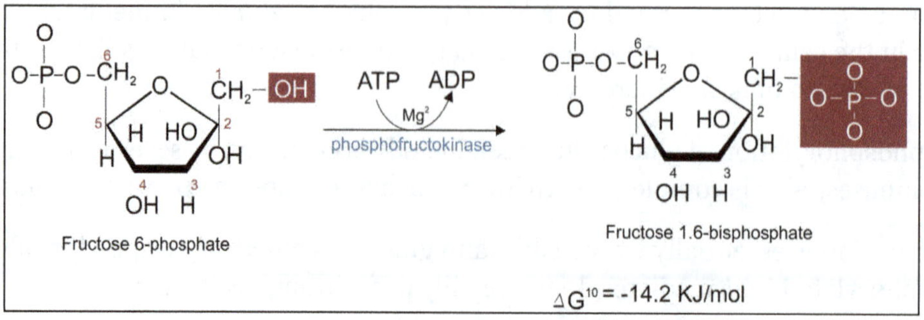

- The reaction is essentially irreversible.
- In this reaction of Glycolysis phosphofructokinase -1 catalyzes the transfer of phosphor group from ATP to fructose 6-phosphate to yield fructose 1, 6- bis phosphates.
- PFK is the rate limiting enzyme of Glycolysis in most tissues. It is the major regulatory enzyme of the glycolytic pathway.

Reaction 4: Fructose-1, 6-Bisphosphate to Dihydroxyacetone Phosphate and Glyceraldehyde's-3-Phosphate

- This reaction completes the first stages of Glycolysis. It is catalyzed by aldose.
- The enzyme fructose-1, 6-bisphosphate aldolase, often called simply aldolase, catalyzes a reversible aldol condensation.
- Fructose-1, 6-bisphosphate is cleaved to yield two different triose phosphates, glyceraldehyde -3-phosphate and an aldose and Dihydroxyacetone phosphate, a ketoses'.
- It is an energetically unfavorable reaction in the direction written, with standard free energy change ΔG° of+5.73 kcal, but the rapid conversion of G3P to pyruvates drives the reaction.
- All of the carbon of glucose can end up as pyruvate because of the equilibrium between DHAP and G3P, catalyzed by triose phosphate isomarase.
- As G3P is utilized by subsequent reaction of Glycolysis, carbon is drawn from DHAP to form G3P.

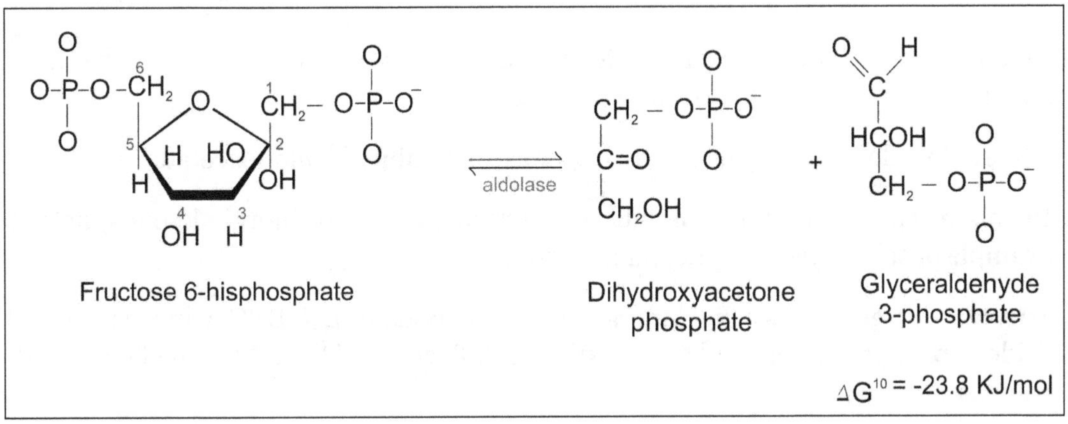

Reaction 5: The Inter Conversion of Triose Phosphate

- Only one of the two triose phosphates formed by aldoses, glyceraldehyde 3- Phosphate, can be directly degraded in the subsequent step of Glycolysis.
- The other product DHAP, is rapidly converted to, glyceraldehyde 3- Phosphate by fifth enzyme triose phosphate isomarase.

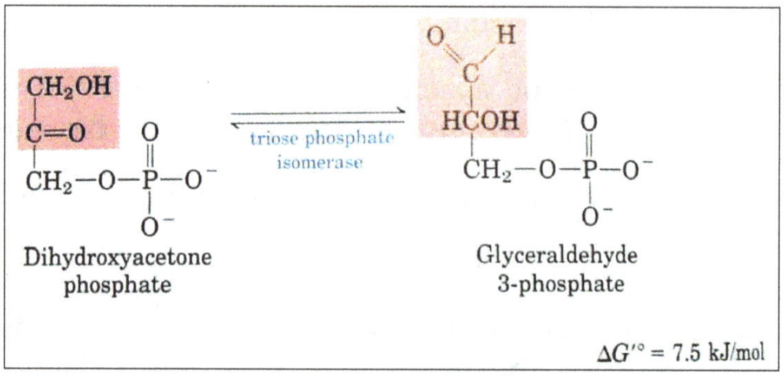

Stage -2

Reaction 6: Oxidation of Glyceraldehyde-3-Phosphate

- The reaction catalyzed by Glyceraldehyde -3-posphate dehydrogenase, requires nicotinamide adenine dinucleotide (NAD^+) as an electron carrier. In its oxidized form NAD^+ binds tightly to the enzyme.

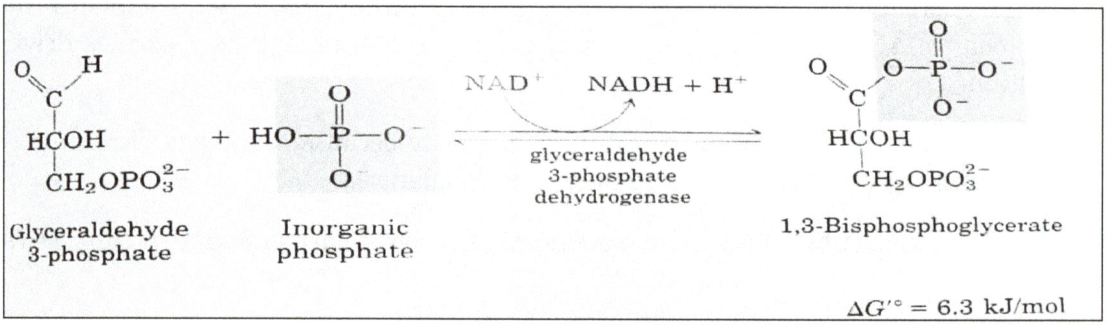

- Aldehyde group is dehydrogenated to an acyl phosphate, as aldehyde group has a very high standard free energy of hydrolysis (-49.3 kJ/mol).
- Glyceradehyde -3 -phosphate dehydrogenase is inhibited by iodoacetate.
- In this reaction, the phosphorylation occurs at expense of inorganic phosphate. It is an example of substrate-level phosphorylation.
- This reaction generates a high energy phosphate bond in 1, 3, DPG, which is a mixed anhydride of phosphoric acid and a carboxylic acid. Because of this, 1, 3 –DPG has a high group transfer potential.

Reaction 7: Phosphoryl Group Transfer

- The reaction is catalyzed by phosphoglycerate kinase.
- This is the first step in the Glycolysis that generates ATP. It is another of substrate level phosphorylation.
    - In the prior step, two molecules of 1, 3,-DPG were formed from each molecules of glucose. Therefore, two ATP molecules are now formed per original molecules of glucose.

- Because, up to triose formation (reaction-4), two molecules of ATP have been utilized per molecule of glucose consumed, the balance sheet for ATP utilization and formation is even at this step.

- The molecular structure of phosphoglycerate kinase is similar to hexokinase in that it has two lobes (jaws) that each bind one of the substrates (ADP-Mg^{2+} or 1,3-bisphosphoglycerate) leading to a large conformational change in the enzyme that brings the substrates close together and excludes H_2O from the active site.

- The bioenergetics of reaction-7 emphasizes two important concepts. First, reaction-6 and reaction-7 are coupled reactions in that the large change in standard free energy of reaction 7 (ΔG°' = -18.9 kJ/mol) pulls the less favorable reaction 6 (ΔG°' = +6.3 kJ/mol) to the right through the shared intermediate 1,3- bisphosphoglcerate as shown below:

  (Rxn 6) Glyceraldehyde-3-P + Pi + NAD+↔ 1, 3-bisphosphoglycerate + NADH + H^+

  ΔG°' = +6.3 kJ/mol ΔG = -1.3 kJ/mol

  (Rxn 7) 1, 3-bisphosphoglycerate + ADP ↔ 3-phosphoglycerate + ATP

  ΔG°' = -18.9 kJ/mol ΔG = +0.1 kJ/mol

  Glyceraldehyde-3-P + Pi + ADP + NAD^+↔3-phosphoglycerate + ATP+ NADH + H^+

  ΔG°' = -12.6 kJ/mol ΔG = -1.2 kJ/mol

- Second, the actual change in free energy for each of these two reactions is very close to zero (ΔG = -1.3 kJ/mol, ΔG = +0.1 kJ/mol), and therefore both reactions are in fact reversible insidethe cell. Again, this difference in ΔG°' and ΔG is due to the mass action ratio which takes into account the actual concentrations of substrates and products that exist in the cell.

- The reversibility of two reaction is important because when flux through gluconeogenesis is high, this two glycolytic reactions can be reversed and thus quickly respond to changing condition in the cell.

Reaction 8: The Interconversion of 3-Phosphoglycerate and 2-Phosphoglycerate

- The reaction is catalyzed by enzyme phosphoglycerate mutase.
- This reversible reaction has a ΔG°' of +1.1.

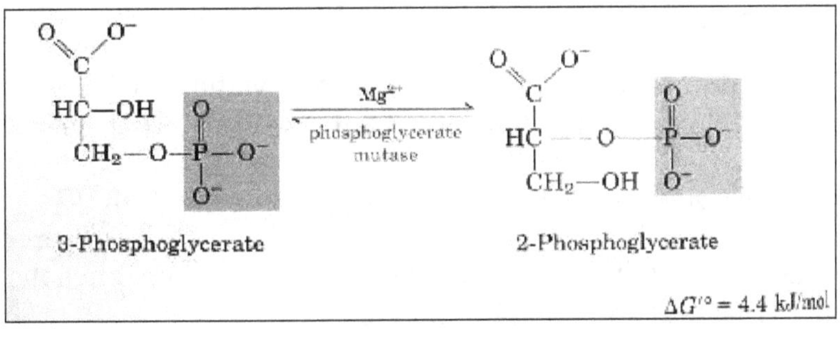

- This reaction is to generate a compound, 2-phosphoglycerate that can be converted to phosphoenol pyruvate in the next reaction, in preparation for a second substrate level phosphorylation that generates ATP earning in step.

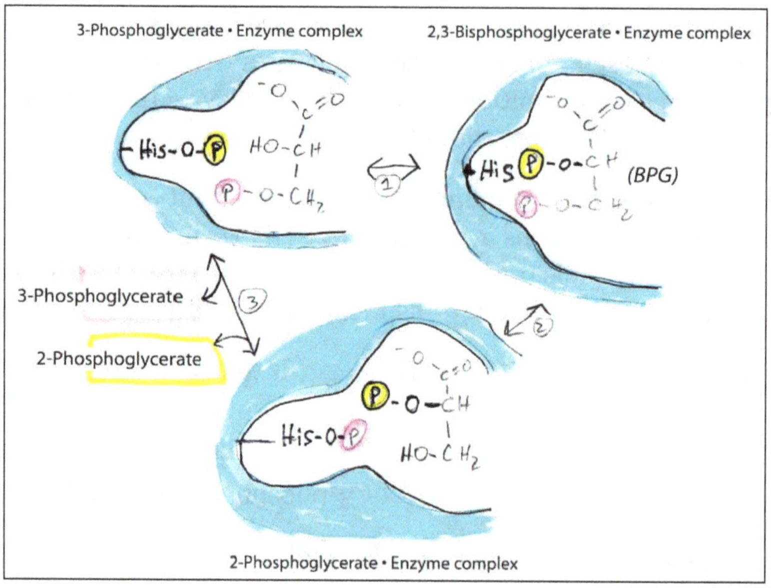

Mechanism of highly reversible reaction: It can be seen to require a phosphoryl transfer from a phosphorylated histidine residue (His-P) located in the enzyme active site which is phosphorylated by transfer of phosphoryl group.

- In step 1, the substrate 3-phosphoglycerate binds to the enzyme active site and is phosphorylated in the C2 position by a transfer reaction involving the His-P group.
- This type of substrate interaction with the enzyme is non-covalent and referred to as a substrate enzyme complex.
- The short-lived inter mediate 2,3bisphosphoglycerate (BPG) is created by Phosphoryl transfer from the histidine residue to the C2 atom of the 3-phospho glycerate.
- In the second step of the reaction, the C3 phosphate is transferred back to the histidine residue of the enzyme to regenerate His-P, leading to the release of 2- phosphoglycerate and binding of anew molecule of 3-phosphoglycerate in the third step.
- Note that the BPG formed in step 1 can diffuse out of the active site resulting in dephosphorylated enzyme, and you may remember that blood cells BPG has an important role in regulating oxygen binding to hemoglobin in red blood cells.
- When BPG leaves the active site without re-phosphorylating the His group, the enzyme can only be activated when trace amounts of BPG diffuse back into the active site.

Reaction 9: Dehydration of 2-Phosphoglycerate

- In this Reversible reaction water molecule is remove from phosphoglycerate to yield phosphoenol pyruvate is promoted by enolase.

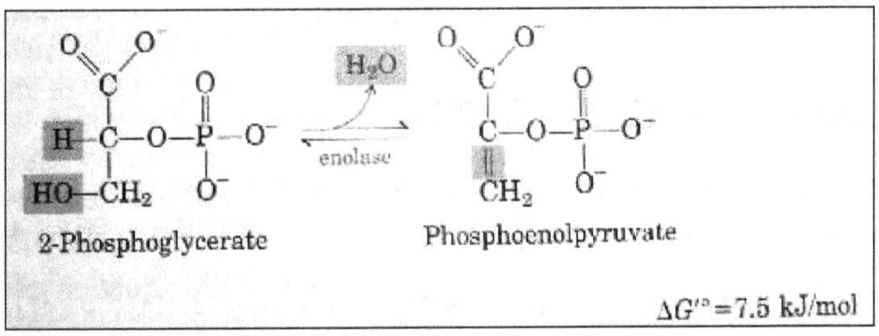

- In this step of glycolysis 2-phosphoglycerate is convert in to phosphoenol pyruvate by dehydration reaction, which is catalyzed by enzyme enolase.
- It is interesting that the change in standard free energy for this reaction is relatively small (ΔGº' = +1.7) kJ/mol), meaning that the overall metabolic energy available fromm2-phosphoglycerate and phosphoenol pyruvate is similar.
- However, when enolase converts 2-phosphoglycerate to phosphoenol pyruvate, it traps the phosphate group in an unstable enol form, resulting in a dramatic increase in the phosphoryl transfer potential of the triose sugar.
- The standard free energy change for phosphate hydrolysis in 2-phosphoglycerate is ΔGº' = -16kJ/mol, whereas for phosphoenol pyruvate it is an incredible ΔGº' = -62 kJ/mol.

Reaction 10: Synthesis of Pyruvate

- This last step in glycolysis is catalyzed by enzyme pyruvate kinase.

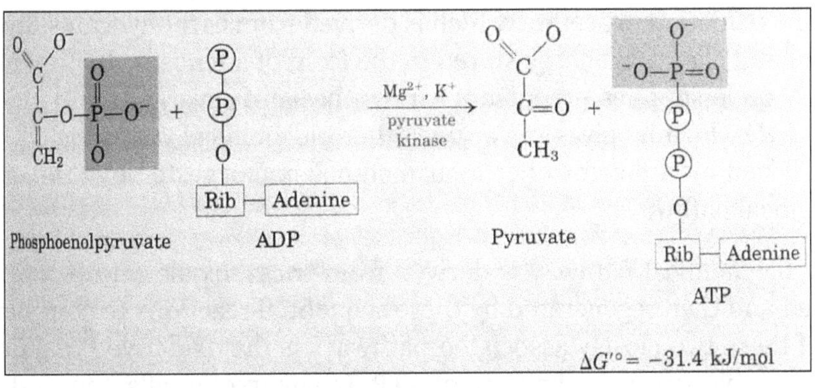

- The formation of ATP from ADP at the expense of the high energy phosphoenol bond of PEP.
- This step is also an important site of regulation.
- In this reaction, the high phosphoric transfer potential of PEP is used by the enzyme pyruvate kinase to generate pyruvate, the end product of.
- Glycolysis and 2 ATP are formed for every glucose molecule entering the pathway.
- This is the second of two substrate level phosphorylation reactions in glycolysis that couples energy released from phosphate hydrolysis (ΔGº' = -62 kJ/mol) to that of ATP synthesis (ΔGº' = +30.5 kJ/mol).

- Pyruvate is a stable compound in cells that is utilized by many other metabolic pathways.

Overall Balance Sheet - Net Gain of ATP

1Glucose + 2 ATP+ 2 NAD+ + 4 ADP + 2 Pi ⟶ 2 Pyruvate + 2 ADP + 2 NADH + 2 H^+ + 4 ATP + 2 H_2O

Glucose + 2 NAD^+ + 2 ADP + 2 Pi ⟶ 2 Pyruvate + 2 NADH + 2 H^+ + 2 ATP + 2 H_2O

- Under aerobic conditions, the two molecules of NADH are deoxidized to NAD^+ by transfer of their electrons to the respiratory chain in the mitochondrion.
    - 2 NADH + 2 H^+ + O_2 → 2 NAD^+ + 2 H_2O

During Glycolysis

- Carbon pathway - Glucose ⟶ 2x pyruvate.
- Phosphate pathway - 2 ADP + 2 Pi ⟶ 2 ATP.

Citric Acid Cycle

Citric acid cycle is also called Tricarboxylic acid (TCA) cycle or Krebs cycle is a sequence of biochemical reactions that occurs in all aerobic organisms for energy generation. Energy is generation is carried out by the oxidation of acetate, which is derived from carbohydrates, lipids and proteins converted into Co_2 and chemical energy stored in the form of adenosine triphosphate (ATP). Furthermore the TCA cycle supplies precursors for synthesis of several amino acids and reducing agent such as NADH, which involves in various other biochemical reactions. TCA cycle is one of the initially established mechanisms of cellular metabolism suggested by the central importance in various biochemical pathways.

The name of this biochemical pathway is derived from tricarboxylic acid (e.g. citric acid). Citric acid is first utilized and then regenerated by these sequential reactions to complete the cycle. The major function of these two closely associated pathways is the oxidative breakdown of nutrients into production of usable energy in the form of ATP. In eukaryotic cells, the Krebs cycle occurs in the mitochondrial matrix. In prokaryotic cells the TCA reaction occurs in the cytosol through the proton gradient for energy generation. In 1935 Albert Szent-Gyorgyi showed that:

Succinate ⟶ Fumarate ⟶ Malate ⟶ Oxaloacetate

Carl Martius and Franz Knoop showed:

Citrate ⟶ cis-aconitate ⟶ Isocitrate ⟶ α ketoglutarate ⟶ Succinate ⟶ Fumarate ⟶ Malate ⟶ Oxaloacetate

Overall reaction of the citric acid cycle is:

$3NAD^+$ + FAD + GDP + P_i + acetyl-CoA ⟶ 3NADH + $FADH_2$ + GTP + CoA + $2CO_2$

From glucose:

Glucose + 2NAD$^+$ + 2ADP + 2P$_i$ $\longrightarrow$ 2pyruvate + 2NADH + 2ATP

2pyruvate + 2NAD$^+$ + 2CoA $\longrightarrow$ 2acetyl-CoA + 2NADH + 2CO$_2$

2acetyl-CoA + 6NAD$^+$ + 2FAD + 2GDP + 2P$_i$ $\longrightarrow$ 6NADH + 2FADH$_2$ + 2GTP + 2CoA + 4CO$_2$

2GTP + 2ADP $\longrightarrow$ 2ATP + 2GDP

Glucose + 10NAD$^+$ + 4ADP + 4P$_i$ + 2FAD $\longrightarrow$ 10NADH + 2FADH$_2$ + 4ATP + 6CO$_2$ $\longrightarrow$ 30ATP + 4ATP + 4ATP = 38ATP

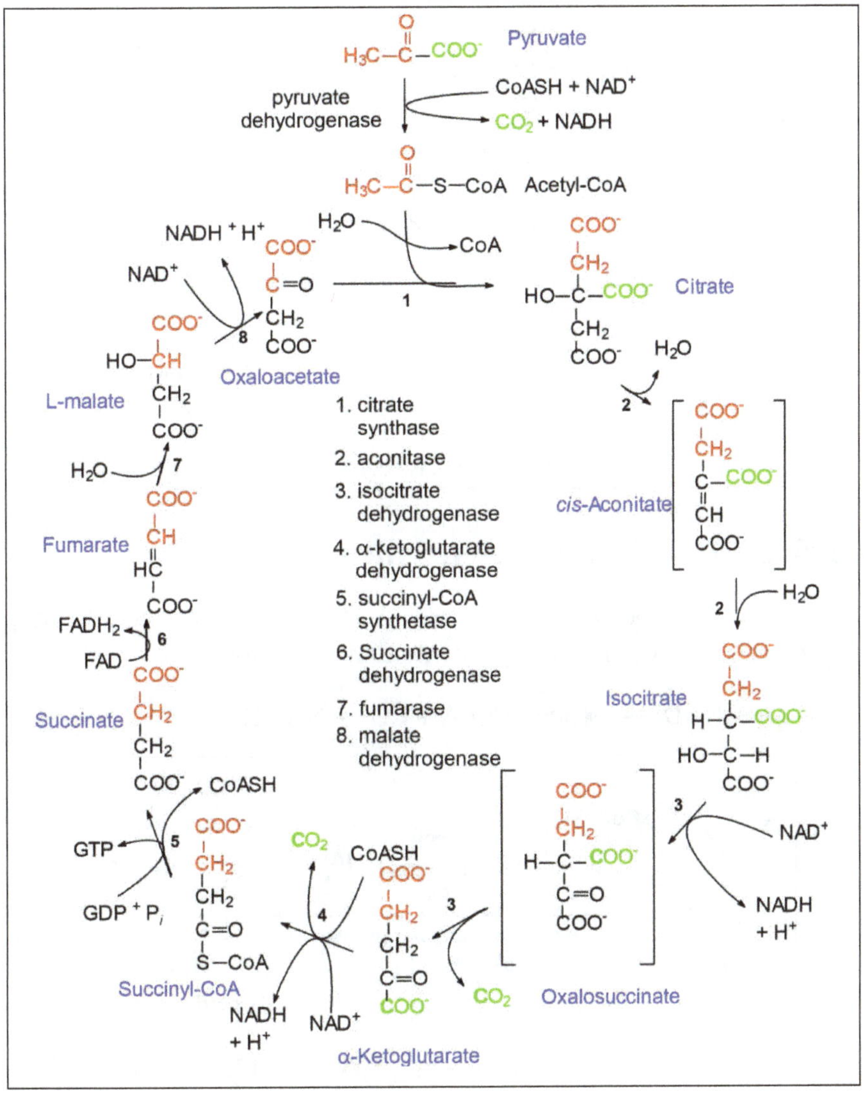

Overall reactions of citric acid cycle.

Conversion of Pyruvate to Activated Acetate by Pyruvate Dehydrogenase

- Pyruvate converts into the acetyl-CoA before enters into the TCA.

- The coenzyme A is act as a carrier for acetyl and other acyl group.
- Acetyl-CoA is a "high-energy" compound.

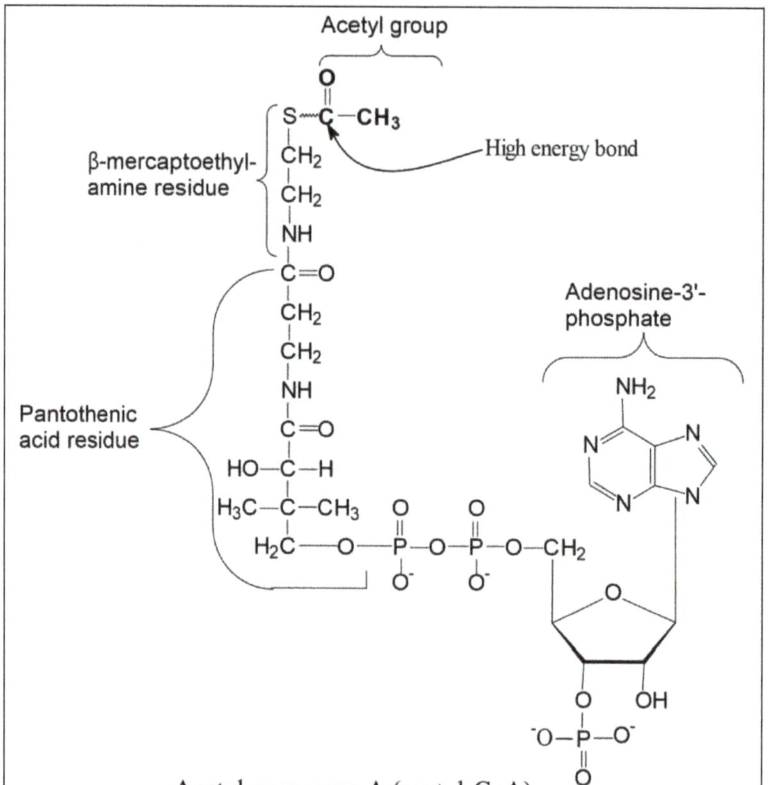

Acetyl-coenzyme A (acetyl-CoA)

Pyruvate Dehydrogenase is a Multi-enzyme Complex

- By the oxidative decarboxylation process Acetyl-CoA is formed from pyruvate using multi-enzyme complex named as a pyruvate dehydrogenase.

  Pyruvate + CoA + $NAD^+ \longrightarrow$ acetyl-CoA + CO_2 + NADH

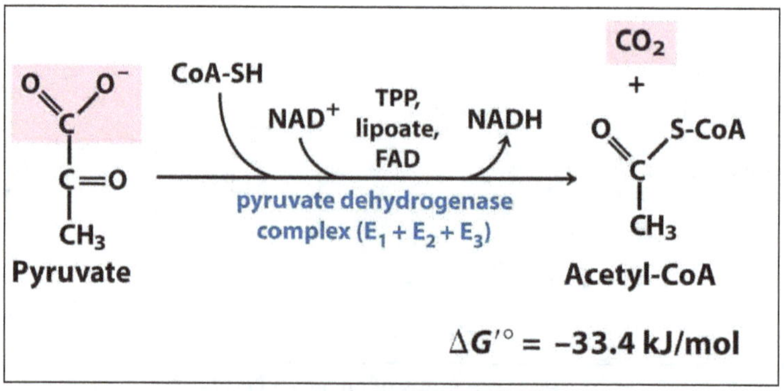

- Pyruvate dehydrogenase a multi-enzyme complex consists of:
  - Pyruvate dehydrogenase (E_1).

- Dihydrolipoyl transacetylase (E_2).
- Dihydrolipoyl dehydrogenase (E_3).

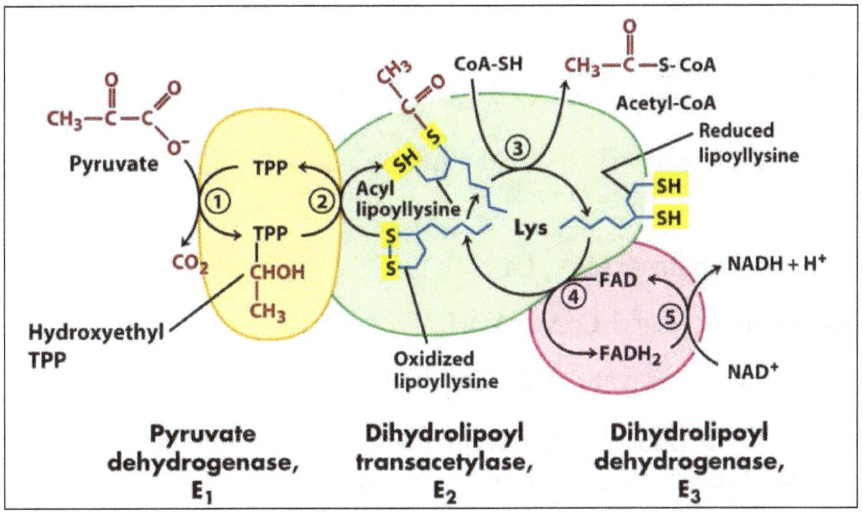

Conversion of Pyruvate to Acetyl-CoA by Pyruvate dehydrogenase multienzyme complex.

Control of Pyruvate Dehydrogenase

Product Inhibition

- When the relative concentrations of NADH and acetyl-CoA are high, the reversible reactions catalyzed by E_2 and E_3 are driven backwards. Therefore formation of acetyl-CoA is inhibited.

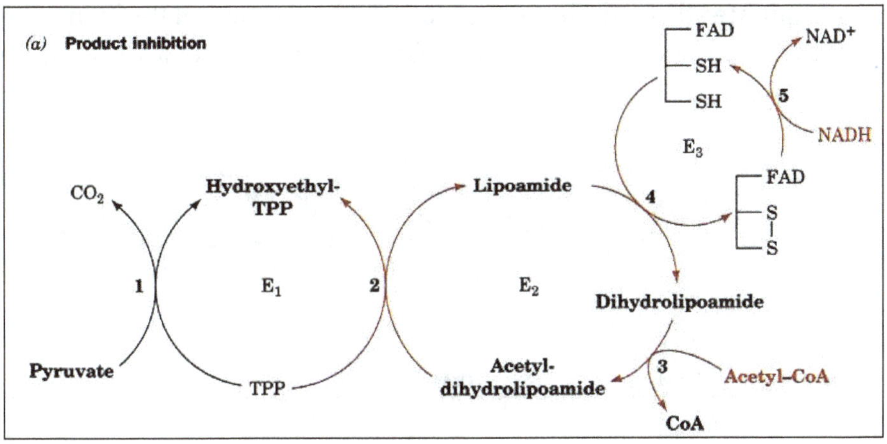

- Thus the E_2 and E_3 activities are controlled by product inhibition (acetyl-CoA for E_2 and NADH for E_3).

Covalent Modification

E1 is regulated by phosphorylation/dephosphorylating. When the Ser of E_1 is phosphorylated, the enzyme is inactivated.

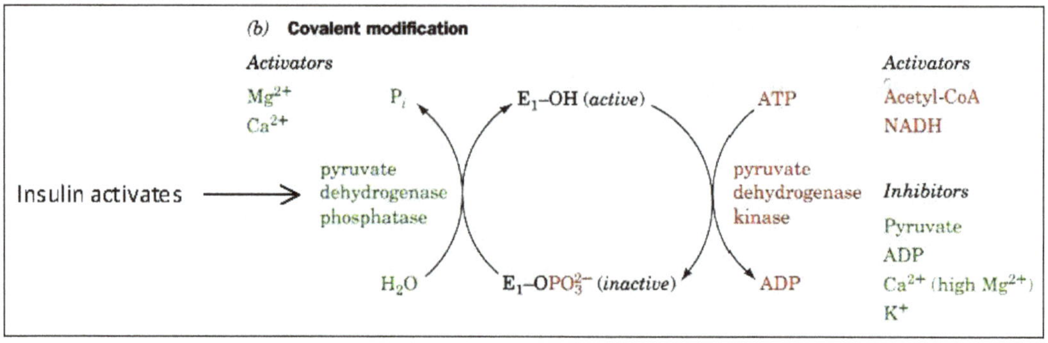

- Activators of phosphatase: Mg^{2+}, Ca^{2+}.
- Activators of kinase: Acetyl-CoA, NADH.
- Inhibitors of kinase: Pyruvate, ADP, Ca^{2+}, high Mg^{2+}, K^+.

Remember: Insulin inhibits phosphorylation and activates dephosphorylation in order to reduce the (glucose) in blood at the starting point of glycolysis.

- Now, insulin also works to reduce the end product of glycolysis, i.e., activates dephosphorylation of E_1 to convert pyruvate to acetyl-CoA.
- Acetyl-CoA is not only the fuel of citric acid cycle, but also the precursor of fatty acids.

Reactions of the Citric Acid Cycle

Citrate is formed from Oxaloacetate and Acetyl Coenzyme A by Citrate Synthase Enzyme

The citric acid cycle initiates through the condensation of an oxaloacetate (four-carbon unit), and the acetyl group of acetyl CoA (a two-carbon unit). Oxaloacetate reacts with acetyl CoA and H_2O to yield as citrate and CoA.

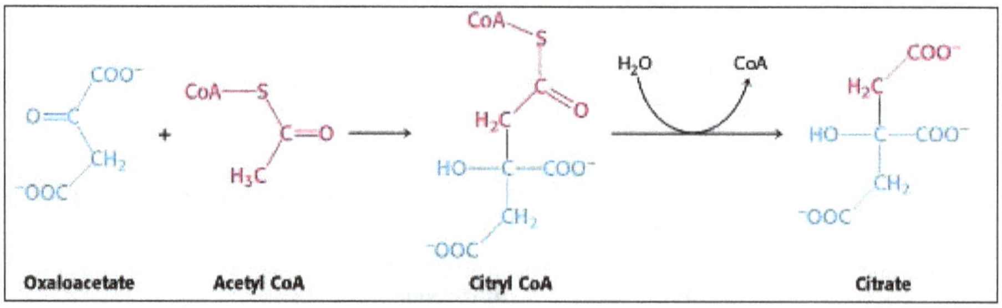

Isomerization of Citrate into Isocitrate

In the citrate molecule the tertiary hydroxyl group is not properly situated for the oxidative decarboxylations that follow. Therefore, isomerization occurs of citrate into isocitrate to allow the six-carbon component to undergo oxidative decarboxylation. The isomerization of citrate is accomplished by a dehydration reaction following a hydration reaction. The result is a substitution of a hydrogen atom and an OH^- group. Both the steps are catalyzed by the enzyme aconitase because cis-aconitate is an intermediate.

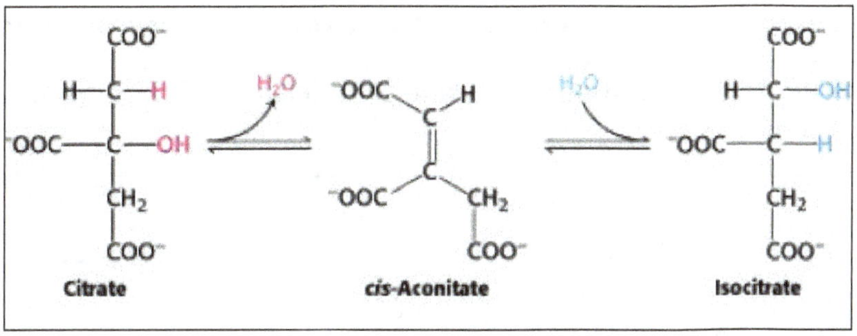

- $\Delta G^{o\prime}$=-13.3 kJ/mol

Fluorocitrate Inhibits Aconitase

Fluoroacetate, one of the most toxic small molecules (LD_{50} = 0.2 mg/kg), is converted to (2R, 3R)-fluorocitrate, which specifically inhibits aconitase since Ser-642 cannot remove the proton at C_2.

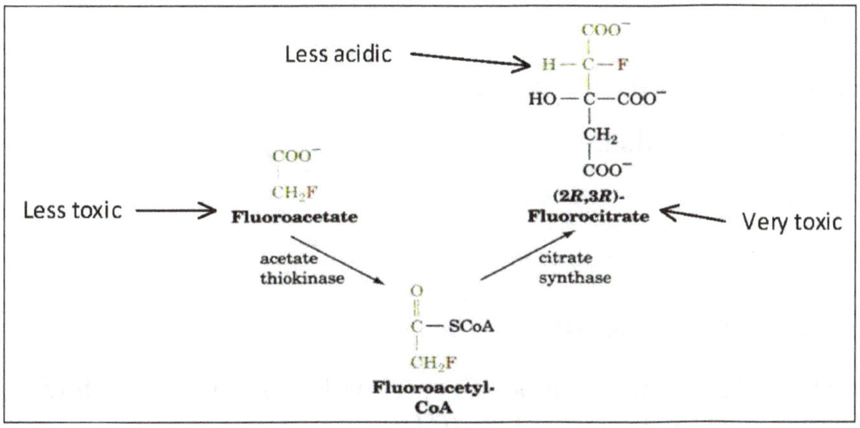

Oxidation and Decarboxylation of Isocitrate to a-Ketoglutarate

The isocitrate is oxidized and decarboxylated by enzyme isocitrate dehydrogenase. Oxalosuccinate act as an intermediate in this reaction.

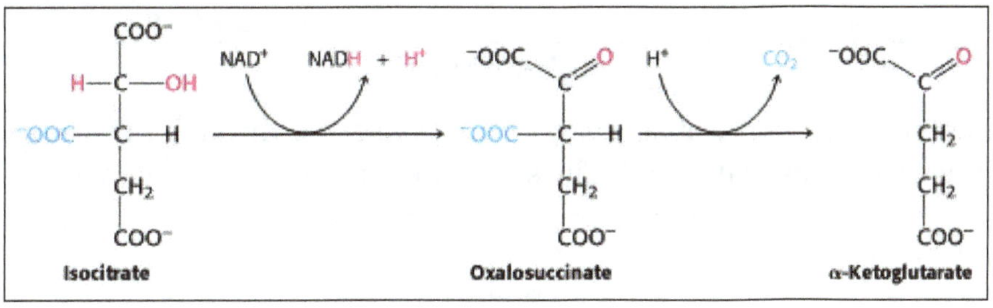

$\Delta G^{o\prime}$=-20.9 kJ/mol

- There are two isozymes in mammalian cells.
    - NAD^+ -dependent form is in mitochondria and requires Mn^{2+} or Mg^{2+}.
    - $NADP^+$ -dependent form is in both cytosol and mitochondria.

The Oxidative Decarboxylation of α-Ketoglutarate to Forms Succinyl CoA

Catalyzes the oxidative decarboxylation of a α-keto acid, releasing CO_2, forming succinyl-CoA and reducing NAD^+ to NADH.

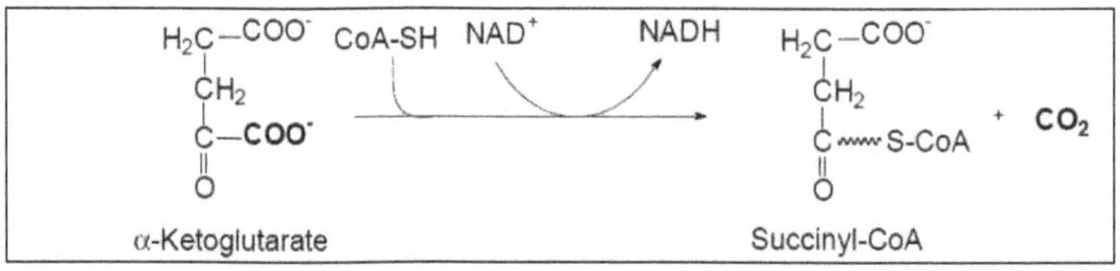

ΔGº'=-33.5 kJ/mol

- A α-Ketoglutarate dehydrogenase that consists of α-ketoglutarate dehydrogenase (E_1), dihydrolipoyl transsuccinylase (E_2), and dihydrolipoyl dehydrogenase (E_3).
- The overall reaction closely resembles that are catalyzed by the pyruvate dehydrogenase multienzyme complex, i.e.,
    - Decarboxylation -----------------------E_1.
    - Succinyl group transfer -----------E_2.
    - Succinyl-CoA formation. -------- E_2.
    - Oxidation of E_2. ------------------- E_3.
    - Reduction of NAD^+. ---------------E_3.

Succinate Formed from Succinyl-CoA

- Hydrolysis of "high-energy" compound succinyl-CoA is coupled with the production of a "high- energy" nucleosidetriphosphate (GTP).

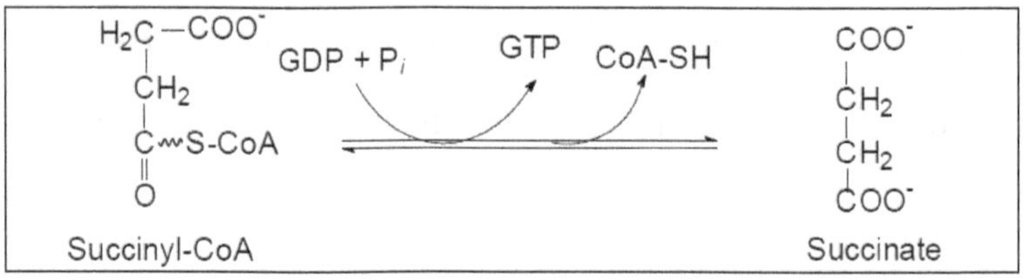

ΔGº'=-2.9 kJ/mol

- The thioester bond energy of succinyl-CoA is conserved through the formation of a series of "high-energy" phosphate (~P_i). The succinate formation is as follows:

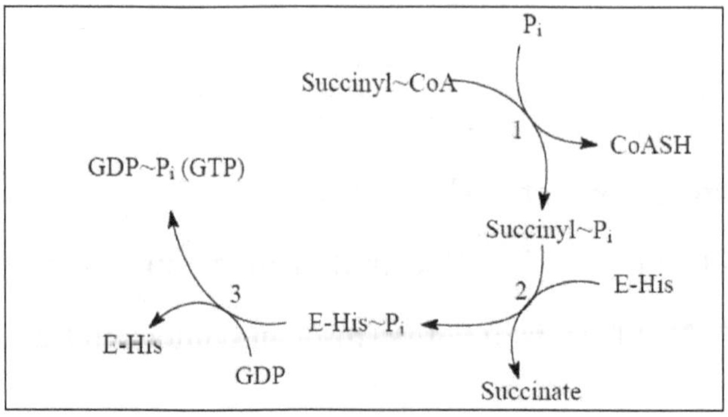

- GTP is converted into ATP by nucleoside diphosphate kinase.

  GTP + ADP ↔ GDP + ATP $\Delta G^{o'}$ = 0 kJ/mol

Fumarate is Formed from Succinate

- Stereospecific dehydrogenation occurs of succinate to fumarate and produces $FADH_2$.

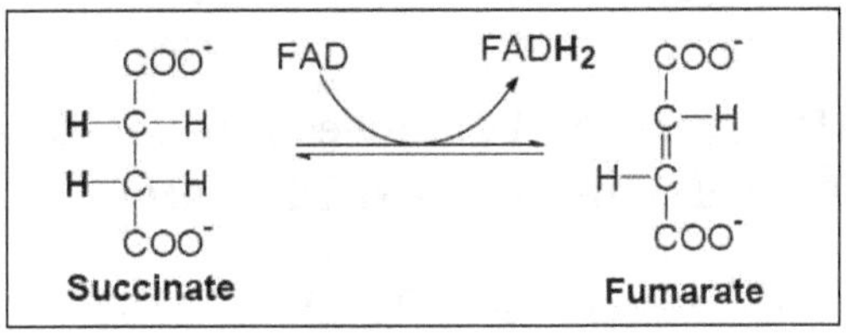

$\Delta G^{o'}$=-0 kJ/mol

- The FAD is covalently bound to the succinate dehydrogenase enzyme. Thus, $FADH_2$ cannot be oxidized as a cofactor. $FADH_2$ is oxidized by the electron transport chain reaction.
- For the reason, succinate dehydrogenase is the only membrane-bound enzyme of citric acid cycle. The others are dissolved in the mitochondrial matrix.
- The enzyme is sturdily inhibited by malonate (structural analog of succinate).

Malate is Formed from Fumarate by Hydrogenation

- Hydrogenation occurs of fumarate's double bond to form L-malate.

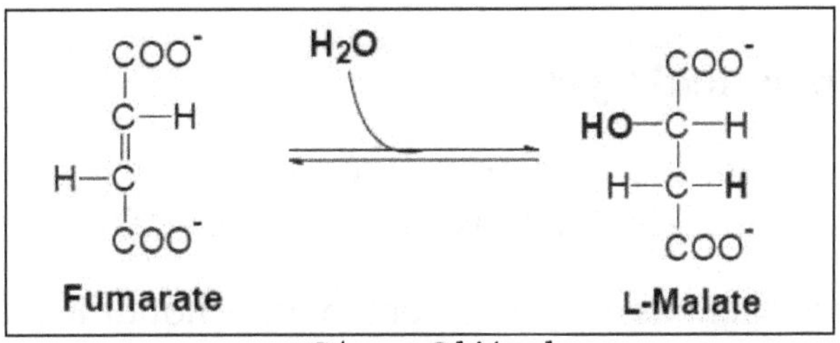

$\Delta G^{a'}$ = - 3.8 kj/mol

Oxaloacetate Regenerates from Malate

- Oxaloacetate regenerates by the oxidation of hydroxyl group of L-malate to ketone in a NAD^+ -dependent reaction.

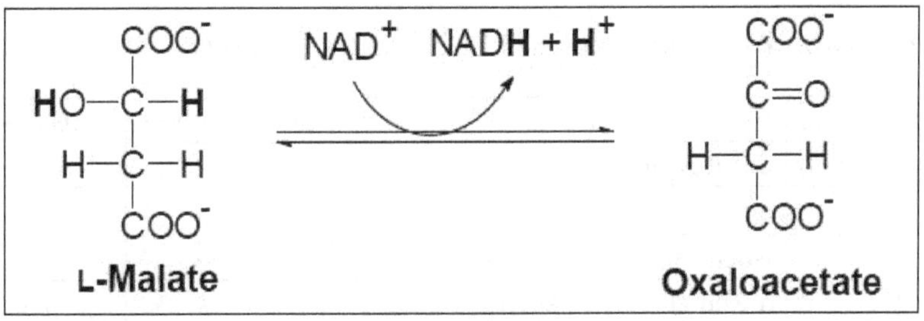

$\Delta G^{a'}$ = 29.7 kj/mol

- This reaction is relatively high endergonic reaction (ΔG>0).

Integration of the Citric Acid Cycle

Following chemical transformations occurs in Citric acid cycle:

- One acetyl group (-$COCH_3$) $\longrightarrow$ $2CO_2$ (4-electron pair process).

$$CoA{-}S{-}\overset{\overset{O}{\|}}{C}{-}CH_3 + 3H_2O \longrightarrow 2CO_2 + CoA{-}{-}SH + 8H^+ + 8e^-$$

- Reduction of three NAD^+ to three NADH (3-electron pairs process) and equivalent to 9ATP generation, i.e., $3NAD^+ + 6H^+ + 6e^- \longrightarrow 3NADH + 3H^+$
- Reduction of one FAD to $FADH_2$ (1-electron pairs process) and equivalent to 2ATP generation, i.e., $FAD + 2H^+ + 2e^- \longrightarrow FADH_2$
- Generation of one GTP (ATP):
    - Four electron pairs generated by one acetyl group oxidation are carried by 3NADH and $FADH_2$ to the oxidative phosphorylation pathway to generate 11ATP.
    - Thus, citric acid cycle generates 12ATP from one acetyl group and sends 4-electron pairs (8 electrons) to electron-transport chain, where they reduce two molecules of O_2 to $4H_2O$, i.e.,

      $O_2 + 8H^+ + 8e^- \longrightarrow 4H_2O$

Regulation of the Citric Acid Cycle

- Rate-limiting enzymes of the citric acid cycle are Citrate synthase, isocitrate dehydrogenase and α-ketoglutarate dehydrogenase because those ΔG are negative.
- The citric acid cycle reactions are carried out in mitochondria, but most of the metabolites of citric acid cycle are present in both mitochondria and cytosol. Therefore it is difficult to establish the rate-determining steps.
- However, three of the eight steps have significantly negative physiological free energy changes. The enzymes involved in those steps are likely to function distant from equilibrium under physiological conditions.

Table: Standard ($\Delta G^{o\prime}$) and Physiological (ΔG) free Energy Changes.

Reaction	Enzyme	$\Delta G^{o\prime}$ (kJ/mol)	ΔG (kJ/mol)
1	Citrate synthase	-32.2	Negative
2	Aconitase	+13.3	~0
3	Isocitrate dehydrogenase	-20.9	Negative
4	α-Ketoglutarate dehydrogenase	-33.5	Negative
5	Succinyl-CoA synthetase	-2.9	~0

6	Succinate dehydrogenase	0.0	~0
7	Fumarase	-3.8	~0
8	Malate dehydrogenase	+29.7	~0

The citric acid cycle is mainly regulated by:

- Substrate availability (rate of diffusion of substrate into mitochondria).
- Product inhibition. (NADH, ATP, citrate).
- Competitive feedback inhibition by intermediates further along the cycle.

Products and NADH are Involved in Feedback Inhibition

- ADP and ATP are allosteric regulators of isocitrate dehydrogenase. High [ADP] activates the enzyme whereas high [ATP] inhibits the enzyme.
- Pyruvate dehydrogenase, isocitrate dehydrogenase and α-ketoglutarate dehydrogenase enzymes are activates by Ca^{2+} ion.

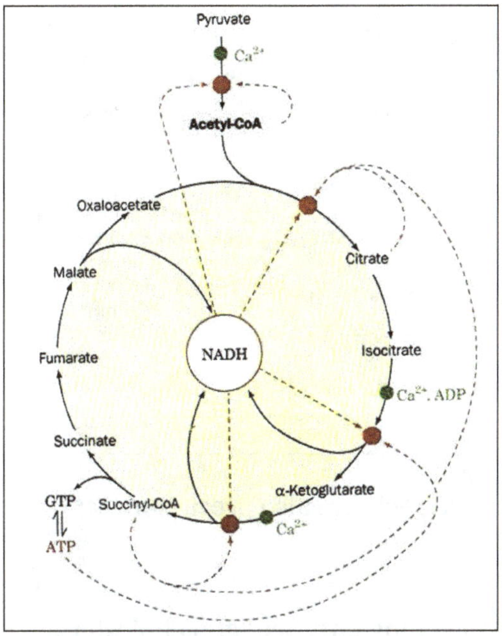

A diagram of the citric acid cycle and the pyruvate dehydrogenase reaction, indicating their points of inhibition (red octagons) and the pathway intermediates that function as inhibitors (dashed red arrows). ADP and Ca^{2+} (green dots) are activators.

The Amphibolic Nature of the Citric Acid Cycle

- In the muscle, the citric acid cycle works mainly degradation of acetyl-CoA to produce bioenergies (ATP).
- In the liver, the citric acid cycle is amphibolic.

Amphibolic = both anabolic and catabolic processes.

Anabolism:	Amino acids Sugars Fatty acids, etc.	⇒	Proteins Nucleic acids Lipids, etc.
Catabolism:	Energy yielding materials, such as proteins	⇒	Energy poor end products, such as CO_2, NH_3, H_2O

Intermediates of Citric Acid Cycle are also Various Precursors

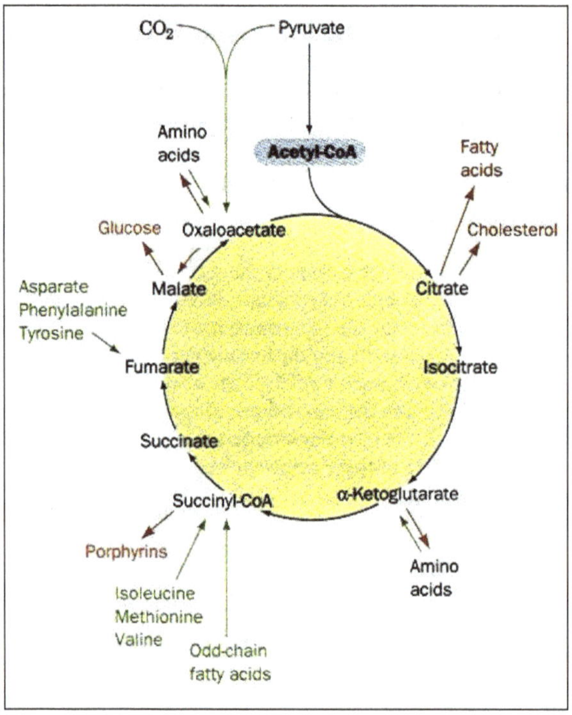

- Intermediates of citric acid cycle are also precursors of:
  - Glucose biosynthesis.
  - Lipid biosynthesis including fatty acid and cholesterol.

Lipid biosynthesis is taken place in cytosol, but the mitochondrial acetyl-CoA (processor) cannot be transported across the inner mitochondrial membrane. Thus, acetylCoA is converted to citrate by ATP-citrate lyase since citrate can cross the membrane. Why citrate synthase is not used? Because no ATP is produced. ADP + P_i + oxaloacetate + acetyl-CoA ↔ ATP + citrate + CoA.

- Amino acid biosynthesis:

  α-ketoglutarate + NAD(P)H + NH_4^+ ↔ Glu + $NAD(P)^+$ + H_2O

  α-ketoglutarate + Ala ↔ Glu + pyruvate

  Oxaloacetate + Ala ↔ Asp + pyruvate

- Porphyrin biosynthesis.
- Succinyl-CoA Utilize as a starting material.

When the citric acid cycle intermediates are transported too much as precursors, the concentration of oxaloacetate is very low. In this case, it is necessary to replenish citric acid cycle intermediates. The main reaction is:

$$\text{Pyruvate} + CO_2 + ATP + H_2O \leftrightarrow \text{oxaloacetate} + ADP + P_i$$

The Citric Acid Cycle is the Center of Metabolism

- Reduced products: NADH and $FADH_2$ are reoxidized to produce ATP.
- The citric acid intermediates are utilized in the biosynthesis of many vital cellular constituents.

Oxidative Phosphorylation

The NADH and $FADH_2$ formed in glycolysis, fatty acid oxidation, and the citric acid cycle are energy-rich molecules because each contains a pair of electrons having a high transfer potential. When these electrons are used to reduce molecular oxygen to water, a large amount of free energy is liberated, which can be used to generate ATP. Oxidative phosphorylation is the process in which ATP is formed as a result of the transfer of electrons from NADH or $FADH_2$ to O_2 by a series of electron carriers. This process, which takes place in mitochondria, is the major source of ATP in aerobic organisms. For example, oxidative phosphorylation generates 26 of the 30 molecules of ATP that are formed when glucose is completely oxidized to CO_2 and H_2O.

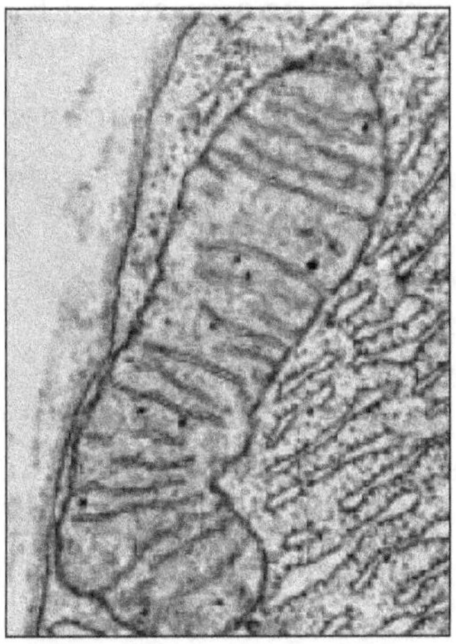

Electron Micrograph of a Mitochondrion.

Oxidative phosphorylation is conceptually simple and mechanistically complex. Indeed, the unraveling of the mechanism of oxidative phosphorylation has been one of the most challenging problems of biochemistry. The flow of electrons from NADH or $FADH_2$ to O_2 through protein complexes located in the mitochondrial inner membrane leads to the pumping of protons out of the mitochondrial matrix. The resulting uneven distribution of protons generates a pH gradient and a Trans-membrane electrical potential that creates a proton-motive force. ATP is synthesized when protons flow back to the mitochondrial matrix through an enzyme complex. Thus, the oxidation of fuels and the phosphorylation of ADP are coupled by a proton gradient across the inner mitochondrial membrane.

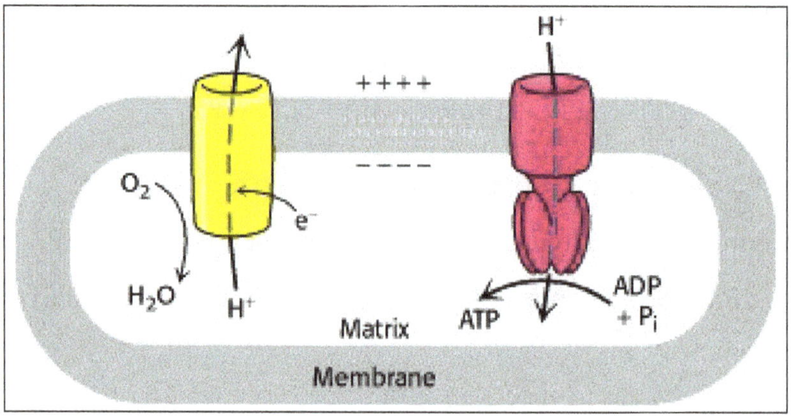

Essence of Oxidative Phosphorylation. Oxidation and ATP synthesis are coupled by transmembrane proton fluxes.

Oxidative phosphorylation is the culmination of a series of energy transformations that are called cellular respiration or simply respiration in their entirety. First, carbon fuels are oxidized in the citric acid cycle to yield electrons with high transfer potential. Then, this electro-motive force is converted into a proton-motive force and, finally, the proton-motive force is converted into phosphoryl transfer potential. The conversion of electro-motive force into proton-motive force is carried out by three electron-driven proton pumps—NADH-Q oxidoreductase, Q-cytochrome c oxidoreductase, and cytochrome c oxidase. These large Trans-membrane complexes contain multiple oxidation-reduction centers, including quinones, flavins, iron-sulfur clusters, hemes, and copper ions. The final phase of oxidative phosphorylation is carried out by ATP synthase, an ATP-synthesizing assembly that is driven by the flow of protons back into the mitochondrial matrix. Components of this remarkable enzyme rotate as part of its catalytic mechanism. Oxidative phosphorylation vividly shows that proton gradients are an interconvertible currency of free energy in biological systems.

Electron Transport Chain

The electron transport chain is the last component of aerobic respiration and is the only part of glucose metabolism that uses atmospheric oxygen. Oxygen continuously diffuses into plants; in animals, it enters the body through the respiratory system. Electron transport is a series of redox reactions that resemble a relay race or bucket brigade in that electrons are passed rapidly from one component to the next, to the endpoint of the chain where the electrons reduce molecular oxygen, producing water. There are four complexes composed of proteins, labeled I through IV in Figure, and the aggregation of these four complexes, together with associated mobile, accessory electron carriers, is called the electron transport chain. The electron transport chain is present in multiple copies in the inner mitochondrial membrane of eukaryotes and the plasma membrane of prokaryotes. Note, however, that the electron transport chain of prokaryotes may not require oxygen as some live in anaerobic conditions. The common feature of all electron transport chains is the presence of a proton pump to create a proton gradient across a membrane.

Complex I

To start, two electrons are carried to the first complex aboard NADH. This complex, labeled I, is composed of flavin mononucleotide (FMN) and an iron-sulfur (Fe-S)-containing protein. FMN, which is derived from vitamin B_2, also called riboflavin, is one of several prosthetic groups or co-factors in the electron transport chain. A prosthetic group is a non-protein molecule required for the activity of a protein. Prosthetic groups are organic or inorganic, non-peptide molecules bound to a protein that facilitates its function; prosthetic groups include co-enzymes, which are the prosthetic groups of enzymes. The enzyme in complex I is NADH dehydrogenase and is a very large protein, containing 45 amino acid chains. Complex I can pump four hydrogen ions across the membrane from the matrix into the intermembrane space, and it is in this way that the hydrogen ion gradient is established and maintained between the two compartments separated by the inner mitochondrial membrane.

Q and Complex II

Complex II directly receives $FADH_2$, which does not pass through complex I. The compound connecting the first and second complexes to the third is ubiquinone (Q). The Q molecule is lipid soluble and freely moves through the hydrophobic core of the membrane. Once it is reduced, (QH_2), ubiquinone delivers its electrons to the next complex in the electron transport chain. Q receives the electrons derived from NADH from complex I and the electrons derived from $FADH_2$ from complex II, including succinate dehydrogenase. This enzyme and $FADH_2$ form a small complex that delivers electrons directly to the electron transport chain, bypassing the first complex. Since these electrons bypass and thus do not energize the proton pump in the first complex, fewer ATP molecules are made from the $FADH_2$ electrons. The number of ATP molecules ultimately obtained is directly proportional to the number of protons pumped across the inner mitochondrial membrane.

Complex III

The third complex is composed of cytochrome b, another Fe-S protein, Rieske center (2Fe-2S center), and cytochrome c proteins; this complex is also called cytochrome oxidoreductase. Cytochrome proteins have a prosthetic group of heme. The heme molecule is similar to the heme in hemoglobin, but it carries electrons, not oxygen. As a result, the iron ion at its core is reduced and oxidized as it passes the electrons, fluctuating between different oxidation states: Fe^{++} (reduced) and Fe^{+++} (oxidized). The heme molecules in the cytochromes have slightly different characteristics due to the effects of the different proteins binding them, giving slightly different characteristics to each complex. Complex III pumps protons through the membrane and passes its electrons to cytochrome c for transport to the fourth complex of proteins and enzymes (cytochrome c is the acceptor of electrons from Q; however, whereas Q carries pairs of electrons, cytochrome c can accept only one at a time).

Complex IV

The fourth complex is composed of cytochrome proteins c, a, and a_3. This complex contains two heme groups (one in each of the two cytochromes, a, and a_3) and three copper ions (a pair of Cu_A and one Cu_B in cytochrome a_3). The cytochromes hold an oxygen molecule very tightly between the iron and copper ions until the oxygen is completely reduced. The reduced oxygen then picks up two hydrogen ions from the surrounding medium to make water (H_2O). The removal of the hydrogen ions from the system contributes to the ion gradient used in the process of chemiosmosis.

Chemiosmosis

In chemiosmosis, the free energy from the series of redox reactions just described is used to pump hydrogen ions (protons) across the membrane. The uneven distribution of H^+ ions across the membrane establishes both concentration and electrical gradients (thus, an electrochemical gradient), owing to the hydrogen ions' positive charge and their aggregation on one side of the membrane. If the membrane were open to diffusion by the hydrogen ions, the ions would tend to diffuse back across into the matrix, driven by their electrochemical gradient. Recall that many ions cannot diffuse through the nonpolar regions of phospholipid membranes without the aid of ion channels. Similarly, hydrogen ions in the matrix space can only pass through the inner mitochondrial membrane through an integral membrane protein called ATP synthase. This complex protein acts as a tiny generator, turned by the force of the hydrogen ions diffusing through it, down their electrochemical gradient. The turning of parts of this molecular machine facilitates the addition of a phosphate to ADP, forming ATP, using the potential energy of the hydrogen ion gradient.

Chemiosmosis is used to generate 90 percent of the ATP made during aerobic glucose catabolism; it is also the method used in the light reactions of photosynthesis to harness the energy of sunlight in the process of photophosphorylation. Recall that the production of ATP using the process of chemiosmosis in mitochondria is called oxidative phosphorylation. The overall result of these re-actions is the production of ATP from the energy of the electrons removed from hydrogen atoms. These atoms were originally part of a glucose molecule. At the end of the pathway, the electrons are used to reduce an oxygen molecule to oxygen ions. The extra electrons on the oxygen attract hydrogen ions (protons) from the surrounding medium, and water is formed.

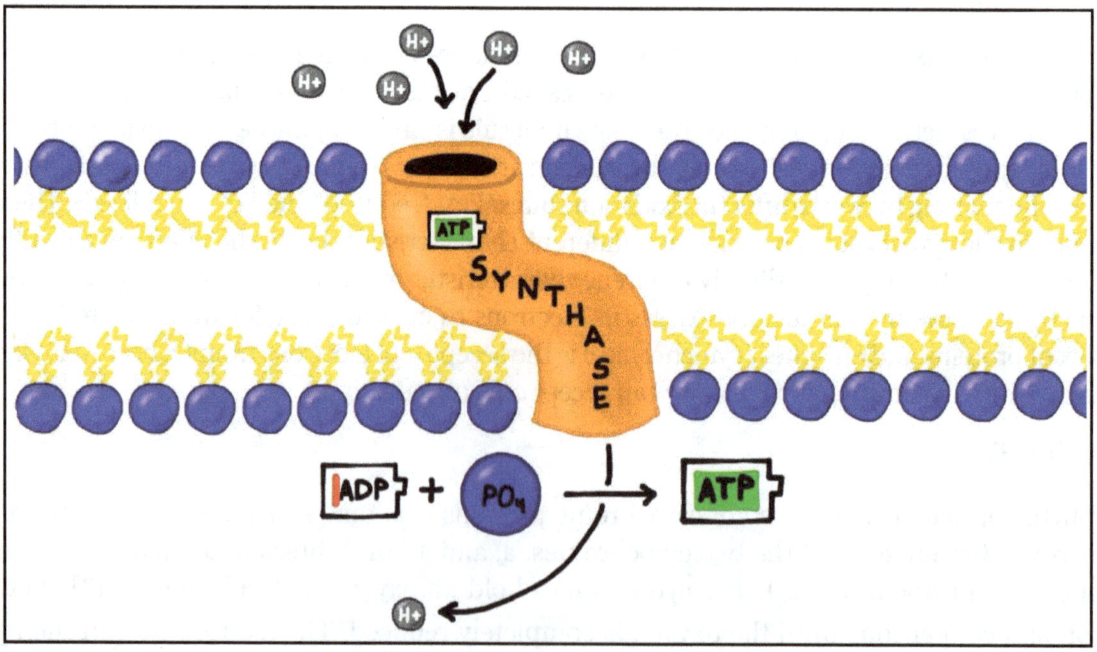

ATP Synthesis and Chemiosmosis in Photosynthesis.

ATP Yield

The number of ATP molecules generated from the catabolism of glucose varies. For example, the number of hydrogen ions that the electron transport chain complexes can pump through the membrane varies between species. Another source of variance stems from the shuttle of electrons across the membranes of the mitochondria. (The NADH generated from glycolysis cannot easily enter mitochondria.) Thus, electrons are picked up on the inside of mitochondria by either NAD+ or FAD^+. As you have learned earlier, these FAD^+ molecules can transport fewer ions; consequently, fewer ATP molecules are generated when FAD^+ acts as a carrier. NAD^+ is used as the electron transporter in the liver and FAD^+ acts in the brain.

Another factor that affects the yield of ATP molecules generated from glucose is the fact that intermediate compounds in these pathways are used for other purposes. Glucose catabolism connects with the pathways that build or break down all other biochemical compounds in cells, and the result is somewhat messier than the ideal situations described thus far. For example, sugars other than glucose are fed into the glycolytic pathway for energy extraction. Moreover, the five-carbon sugars that form nucleic acids are made from intermediates in glycolysis. Certain nonessential amino acids can be made from intermediates of both glycolysis and the citric acid cycle. Lipids, such as cholesterol and triglycerides, are also made from intermediates in these pathways, and both amino acids and triglycerides are broken down for energy through these pathways. Overall, in living systems, these pathways of glucose catabolism extract about 34 percent of the energy contained in glucose.

3

Protein Biochemistry

Proteins are the macromolecules which are made up of one or more long chains of amino acid residues. Proteins are crucial for providing structure to the cells. This chapter closely examines the key concepts of protein biochemistry to provide an extensive understanding of the subject.

Proteins are polymers of amino acids. Each amino acid contains a central carbon, hydrogen, a carboxyl group, an amino group, and a variable R group. The R group specifies which class of amino acids it belongs to: electrically charged hydrophilic side chains, polar but uncharged side chains, nonpolar hydrophobic side chains, and special cases. Proteins have different "layers" of structure: primary, secondary, tertiary, quaternary.

Proteins have a variety of function in cells. Major functions include acting as enzymes, receptors, transport molecules, regulatory proteins for gene expression, and so on. Enzymes are biological catalysts that speed up a chemical reaction without being permanently altered. They have "active sites" where the substrate/reactant binds, and they can be either activated or inhibited (competitive and/or noncompetitive inhibitors).

Amino Acids

Proteins are one of the most abundant organic molecules in living systems and have the most diverse range of functions of all macromolecules. Proteins may be structural, regulatory, contractile, or protective; they may serve in transport, storage, or membranes; or they may be toxins or enzymes. Each cell in a living system may contain thousands of proteins, each with a unique function. Their structures, like their functions, vary greatly. They are all, however, polymers of amino acids, arranged in a linear sequence.

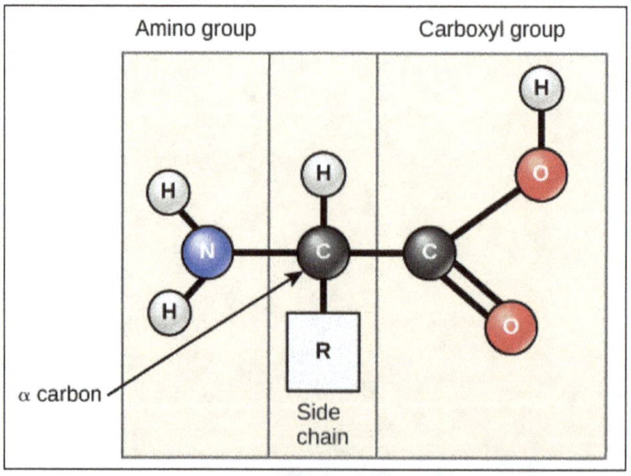

Amino acids have a central asymmetric carbon to which an amino group, a carboxyl group, a hydrogen atom, and a side chain (R group) are attached.

Amino acids are the monomers that make up proteins. Each amino acid has the same fundamental structure, which consists of a central carbon atom, also known as the alpha (α) carbon, bonded to an amino group (NH_2), a carboxyl group (COOH), and to a hydrogen atom. Every amino acid also has another atom or group of atoms bonded to the central atom known as the R group.

The name "amino acid" is derived from the fact that they contain both amino group and carboxyl-acid-group in their basic structure. There are 20 amino acids present in proteins. Ten of these are considered essential amino acids in humans because the human body cannot produce them and they are obtained from the diet. For each amino acid, the R group (or side chain) is different.

The chemical nature of the side chain determines the nature of the amino acid (that is, whether it is acidic, basic, polar, or nonpolar). For example, the amino acid glycine has a hydrogen atom as the R group. Amino acids such as valine, methionine, and alanine are nonpolar or hydrophobic in nature, while amino acids such as serine, threonine, and cysteine are polar and have hydrophilic side chains. The side chains of lysine and arginine are positively charged, and therefore these amino acids are also known as basic amino acids. Proline has an R group that is linked to the amino group, forming a ring-like structure. Proline is an exception to the standard structure of an animo acid since its amino group is not separate from the side chain.

Amino acids are represented by a single upper case letter or a three-letter abbreviation. For example, valine is known by the letter V or the three-letter symbol val. Just as some fatty acids are essential to a diet, some amino acids are necessary as well. They are known as essential amino acids, and in humans they include isoleucine, leucine, and cysteine. Essential amino acids refer to those necessary for construction of proteins in the body, although not produced by the body; which amino acids are essential varies from organism to organism.

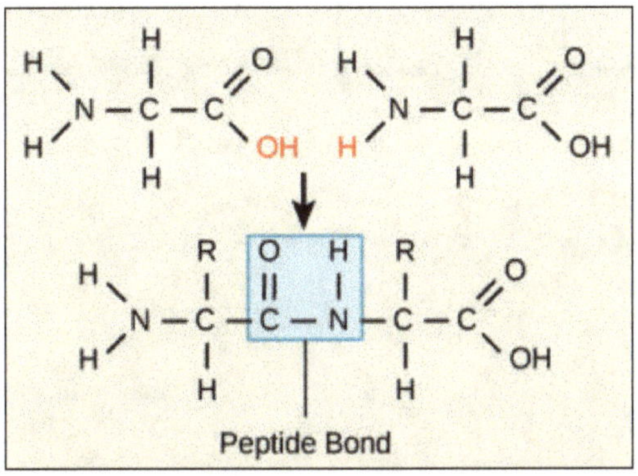

Peptide bond formation is a dehydration synthesis reaction. The carboxyl group of one amino acid is linked to the amino group of the incoming amino acid. In the process, a molecule of water is released.

The sequence and the number of amino acids ultimately determine the protein's shape, size, and function. Each amino acid is attached to another amino acid by a covalent bond, known as a peptide bond, which is formed by a dehydration reaction. The carboxyl group of one amino acid and the amino group of the incoming amino acid combine, releasing a molecule of water. The resulting bond is the peptide bond.

The products formed by such linkages are called peptides. As more amino acids join to this growing chain, the resulting chain is known as a polypeptide. Each polypeptide has a free amino group at one end. This end is called the N terminal, or the amino terminal, and the other end has a free carboxyl group, also known as the C or carboxyl terminal. While the terms polypeptide and protein are sometimes used interchangeably, a polypeptide is technically a polymer of amino acids, whereas the term protein is used for a polypeptide or polypeptides that have combined together, often have bound non-peptide prosthetic groups, have a distinct shape, and have a unique function. After protein synthesis (translation), most proteins are modified. These are known as post-translational modifications. They may undergo cleavage, phosphorylation, or may require the addition of other chemical groups. Only after these modifications is the protein completely functional.

Protein Structure

The shape of a protein is critical to its function. For example, an enzyme can bind to a specific substrate at a site known as the active site. If this active site is altered because of local changes or changes in overall protein structure, the enzyme may be unable to bind to the substrate. To understand how the protein gets its final shape or conformation, we need to understand the four levels of protein structure: primary, secondary, tertiary, and quaternary.

Primary Structure

The unique sequence of amino acids in a polypeptide chain is its primary structure. For example, the pancreatic hormone insulin has two polypeptide chains, A and B, and they are linked together by disulfide bonds. The N terminal amino acid of the A chain is glycine, whereas the C terminal amino acid is asparagine. The sequences of amino acids in the A and B chains are unique to insulin.

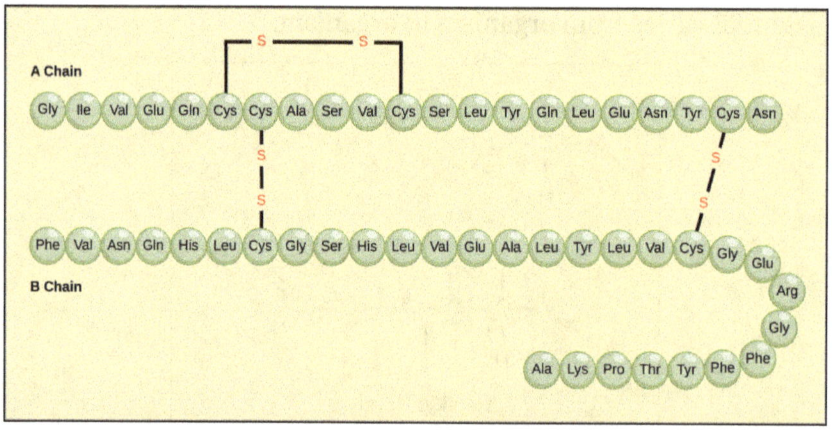

Bovine serum insulin is a protein hormone made of two peptide chains, A (21 amino acids long) and B (30 amino acids long). In each chain, primary structure is indicated by three-letter abbreviations that represent the names of the amino acids in the order they are present. The amino acid cysteine (cys) has a sulfhydryl (SH) group as a side chain. Two sulfhydryl groups can react in the presence of oxygen to form a disulfide (S-S) bond. Two disulfide bonds connect the A and B chains together, and a third helps the A chain fold into the correct shape. Note that all disulfide bonds are the same length, but are drawn different sizes for clarity.

The unique sequence for every protein is ultimately determined by the gene encoding the protein. A change in nucleotide sequence of the gene's coding region may lead to a different amino acid being added to the growing polypeptide chain, causing a change in protein structure and function.

In sickle cell anemia, the hemoglobin β chain (a small portion of which is shown in Figure) has a single amino acid substitution, causing a change in protein structure and function.

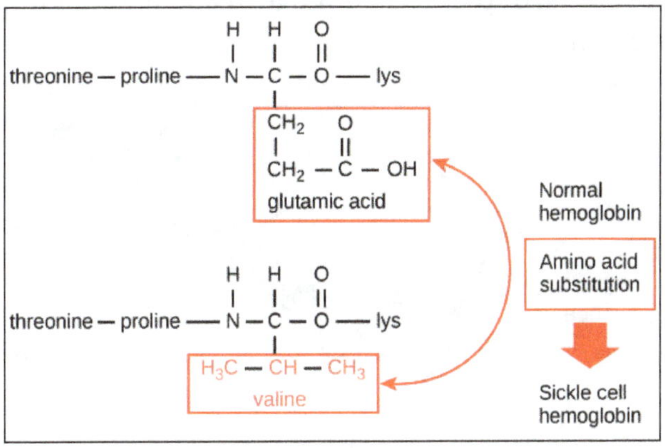

The beta chain of hemoglobin is 147 residues in length, yet a single amino acid substitution leads to sickle cell anemia. In normal hemoglobin, the amino acid at position seven is glutamate. In sickle cell hemoglobin, this glutamate is replaced by a valine.

Specifically, the amino acid glutamic acid is substituted by valine in the β chain. What is most remarkable to consider is that a hemoglobin molecule is made up of two alpha chains and two beta chains that each consist of about 150 amino acids? The molecule, therefore, has about 600 amino acids. The structural difference between a normal hemoglobin molecule and a sickle cell molecule—which dramatically decreases life expectancy—is a single amino acid of the 600. What is even more remarkable is that those 600 amino acids are encoded by three nucleotides each, and the mutation is caused by a single base change (point mutation), 1 in 1800 bases.

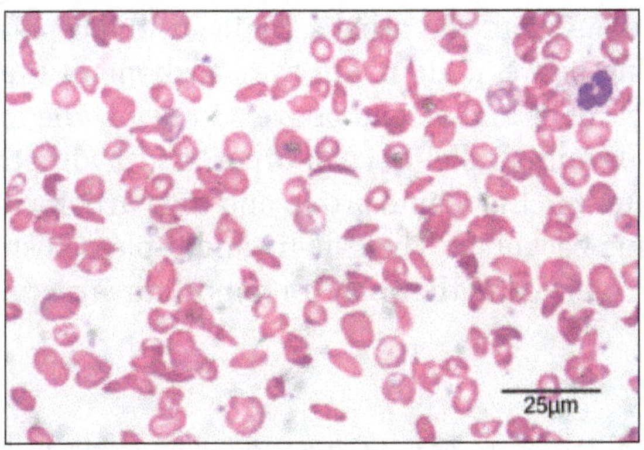

In this blood smear, visualized at 535x magnification using bright field microscopy, sickle cells are crescent shaped, while normal cells are disc-shaped.

Because of this change of one amino acid in the chain, hemoglobin molecules form long fibers that distort the biconcave, or disc-shaped, red blood cells and assume a crescent or "sickle" shape, which clogs arteries. This can lead to myriad serious health problems such as breathlessness, dizziness, headaches, and abdominal pain for those affected by this disease.

Secondary Structure

The local folding of the polypeptide in some regions gives rise to the secondary structure of the protein. The most common are the α-helix and β-pleated sheet structures. Both structures are the α-helix structure—the helix held in shape by hydrogen bonds. The hydrogen bonds form between the oxygen atom in the carbonyl group in one amino acid and another amino acid that is four amino acids farther along the chain.

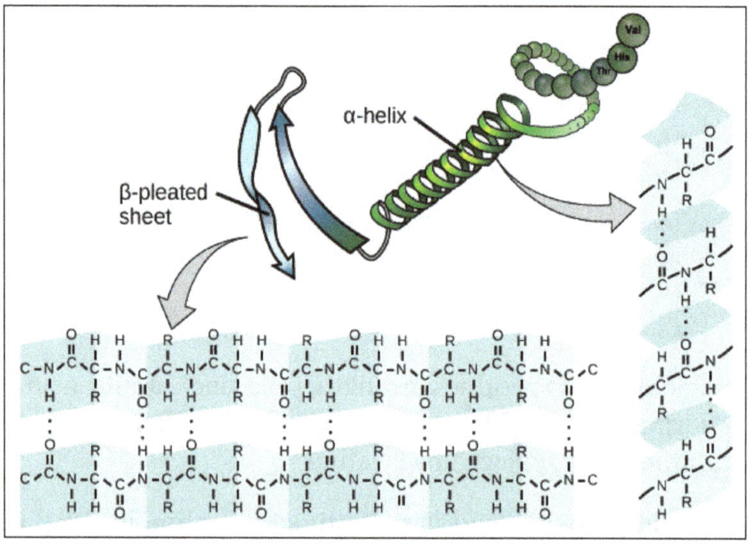

The α-helix and β-pleated sheet are secondary structures of proteins that form because of hydrogen bonding between carbonyl and amino groups in the peptide backbone. Certain amino acids have a propensity to form an α-helix, while others have a propensity to form a β-pleated sheet.

Every helical turn in an alpha helix has 3.6 amino acid residues. The R groups (the variant groups) of the polypeptide protrude out from the α-helix chain. In the β-pleated sheet, the "pleats" are formed by hydrogen bonding between atoms on the backbone of the polypeptide chain. The R groups are attached to the carbons and extend above and below the folds of the pleat. The pleated segments align parallel or antiparallel to each other, and hydrogen bonds form between the partially positive nitrogen atom in the amino group and the partially negative oxygen atom in the carbonyl group of the peptide backbone. The α-helix and β-pleated sheet structures are found in most globular and fibrous proteins and they play an important structural role.

Tertiary Structure

The unique three-dimensional structure of a polypeptide is its tertiary structure. This structure is in part due to chemical interactions at work on the polypeptide chain. Primarily, the interaction among R groups creates the complex three-dimensional tertiary structure of a protein. The nature of the R groups found in the amino acids involved can counteract the formation of the hydrogen bonds described for standard secondary structures. For example, R groups with like charges are repelled by each other and those with unlike charges are attracted to each other (ionic bonds). When protein folding takes place, the hydrophobic R groups of nonpolar amino acids lay in the interior of the protein, whereas the hydrophilic R groups lay on the outside. The former types of interactions are also known as hydrophobic interactions. Interaction between cysteine side chains forms disulfide linkages in the presence of oxygen, the only covalent bond forming during protein folding.

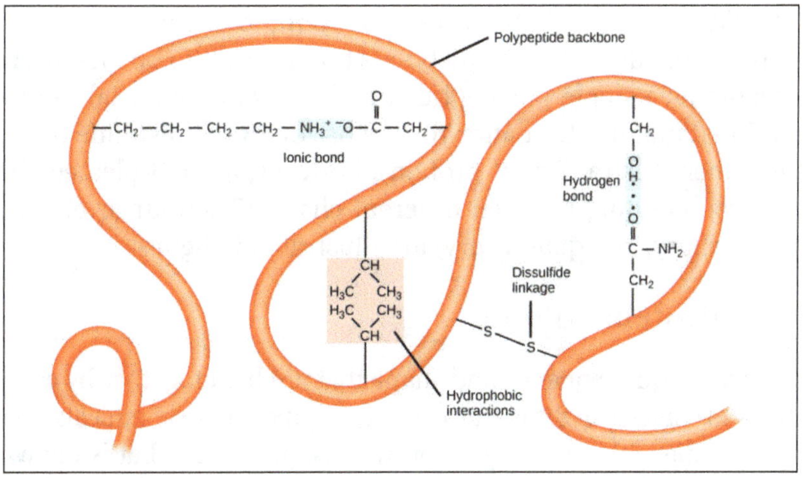

The tertiary structure of proteins is determined by a variety of chemical interactions. These include hydrophobic interactions, ionic bonding, hydrogen bonding and disulfide linkages.

All of these interactions, weak and strong, determine the final three-dimensional shape of the protein. When a protein loses its three-dimensional shape, it may no longer be functional.

Quaternary Structure

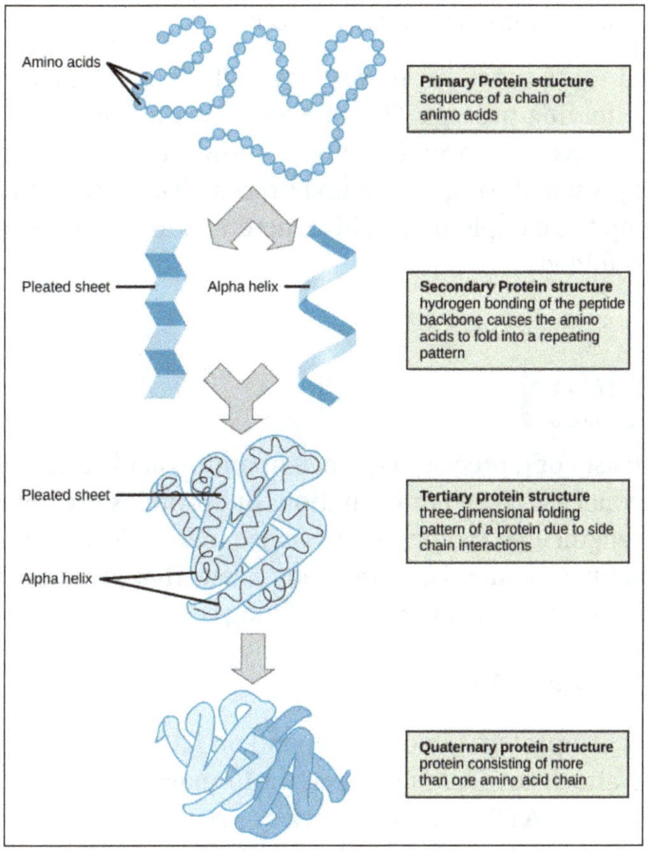

The four levels of protein structure can be observed in these illustrations.

In nature, some proteins are formed from several polypeptides, also known as subunits, and the interaction of these subunits forms the quaternary structure. Weak interactions between the subunits help to stabilize the overall structure. For example, insulin (a globular protein) has a combination of hydrogen bonds and disulfide bonds that cause it to be mostly clumped into a ball shape. Insulin starts out as a single polypeptide and loses some internal sequences in the presence of post-translational modification after the formation of the disulfide linkages that hold the remaining chains together. Silk (a fibrous protein), however, has a β-pleated sheet structure that is the result of hydrogen bonding between different chains. The four levels of protein structure (primary, secondary, tertiary, and quaternary) are illustrated in figure.

Denaturation and Protein Folding

Each protein has its own unique sequence and shape that are held together by chemical interactions. If the protein is subject to changes in temperature, pH, or exposure to chemicals, the protein structure may change, losing its shape without losing its primary sequence in what is known as denaturation. Denaturation is often reversible because the primary structure of the polypeptide is conserved in the process if the denaturing agent is removed, allowing the protein to resume its function. Sometimes denaturation is irreversible, leading to loss of function. One example of irreversible protein denaturation is when an egg is fried. The albumin protein in the liquid egg white is denatured when placed in a hot pan. Not all proteins are denatured at high temperatures; for instance, bacteria that survive in hot springs have proteins that function at temperatures close to boiling. The stomach is also very acidic, has a low pH, and denatures proteins as part of the digestion process; however, the digestive enzymes of the stomach retain their activity under these conditions.

Protein folding is critical to its function. It was originally thought that the proteins themselves were responsible for the folding process. Only recently was it found that often they receive assistance in the folding process from protein helpers known as chaperones (or chaperonins) that associate with the target protein during the folding process. They act by preventing aggregation of polypeptides that make up the complete protein structure, and they disassociate from the protein once the target protein is folded.

Protein Function

Each type of protein consists of a precise sequence of amino acids that allows it to fold up into a particular three-dimensional shape, or conformation. But proteins are not rigid lumps of material. They can have precisely engineered moving parts whose mechanical actions are coupled to chemical events. It is this coupling of chemistry and movement that gives proteins the extraordinary capabilities that underlie the dynamic processes in living cells.

All Proteins Bind to other Molecules

The biological properties of a protein molecule depend on its physical interaction with other molecules. Thus, antibodies attach to viruses or bacteria to mark them for destruction, the enzyme hexokinase binds glucose and ATP so as to catalyze a reaction between them, actin molecules bind to each other to assemble into actin filaments, and so on. Indeed, all proteins stick, or bind, to other molecules. In some cases, this binding is very tight; in others, it is weak and short-lived. But the binding always shows great specificity, in the sense that each protein molecule can usually bind just one or a few molecules out of the many thousands of different types it encounters. The substance that is bound by the protein—no matter whether it is an ion, a small molecule, or a macromolecule— is referred to as a ligand for that protein.

The ability of a protein to bind selectively and with high affinity to a ligand depends on the formation of a set of weak, non-covalent bonds—hydrogen bonds, ionic bonds, and van der Waals attractions—plus favorable hydrophobic interactions. Because each individual bond is weak, an effective binding interaction requires that many weak bonds be formed simultaneously. This is possible only if the surface contours of the ligand molecule fit very closely to the protein, matching it like a hand in a glove.

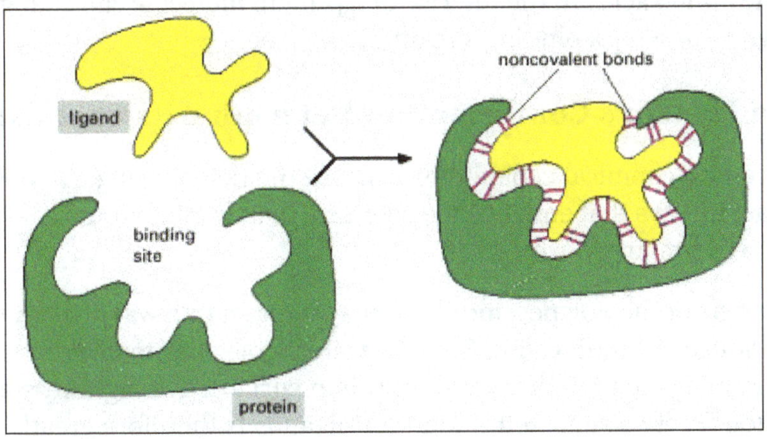

Figure shows the selective binding of a protein to another molecule. Many weak bonds are needed to enable a protein to bind tightly to a second molecule, which is called a ligand for the protein. A ligand must therefore fit precisely into a protein's binding site, like a hand into a glove, so that a large number of non-covalent bonds can be formed between the protein and the ligand.

The region of a protein that associates with a ligand, known as the ligand's binding site, usually consists of a cavity in the protein surface formed by a particular arrangement of amino acids. These amino acids can belong to different portions of the polypeptide chain that are brought together when the protein folds. Separate regions of the protein surface generally provide binding sites for different ligands, allowing the protein's activity to be regulated. And other parts of the protein can serve as a handle to place the protein in a particular location in the cell—an example is the SH2 domain, which is often used to move a protein containing it to sites in the plasma membrane in response to particular signals.

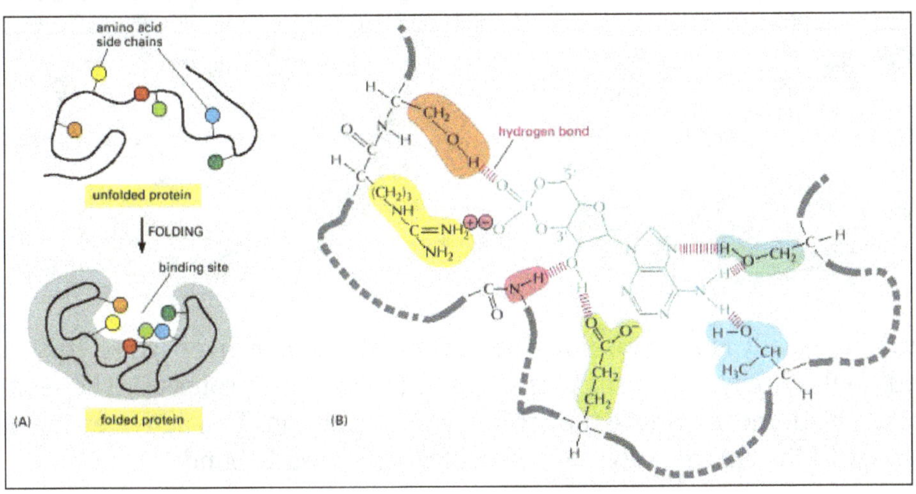

Figure shows the binding site of a protein. (A) The folding of the polypeptide chain typically creates a crevice or cavity on the protein surface. This crevice contains a set of amino acid side chains disposed in such a way that they can make non-covalent bonds only with certain ligands. (B) A close-up of an actual binding site showing the hydrogen bonds and ionic interactions formed between a protein and its ligand (in this example, cyclic AMP is the bound ligand).

Although the atoms buried in the interior of the protein have no direct contact with the ligand, they provide an essential scaffold that gives the surface its contours and chemical properties. Even small changes to the amino acids in the interior of a protein molecule can change its three-dimensional shape enough to destroy a binding site on the surface.

The Details of a Protein's Conformation Determine its Chemistry

Proteins have impressive chemical capabilities because the neighboring chemical groups on their surface often interact in ways that enhance the chemical reactivity of amino acid side chains. These interactions fall into two main categories.

First, neighboring parts of the polypeptide chain may interact in a way that restricts the access of water molecules to a ligand binding site. Because water molecules tend to form hydrogen bonds, they can compete with ligands for sites on the protein surface. The tightness of hydrogen bonds (and ionic interactions) between proteins and their ligands is therefore greatly increased if water molecules are excluded. Initially, it is hard to imagine a mechanism that would exclude a molecule as small as water from a protein surface without affecting the access of the ligand itself. Because of the strong tendency of water molecules to form water–water hydrogen bonds, however, water molecules exist in a large hydrogen-bonded network. In effect, a ligand binding site can be kept dry because it is energetically unfavorable for individual water molecules to break away from this network, as they must do to reach into a crevice on a protein's surface.

Second, the clustering of neighboring polar amino acid side chains can alter their reactivity. If a number of negatively charged side chains are forced together against their mutual repulsion by the way the protein folds, for example, the affinity of the site for a positively charged ion is greatly increased. In addition, when amino acid side chains interact with one another through hydrogen bonds, normally unreactive side groups (such as the $-CH_2OH$ on the serine shown in Figure) can become reactive, enabling them to enter into reactions that make or break selected covalent bonds.

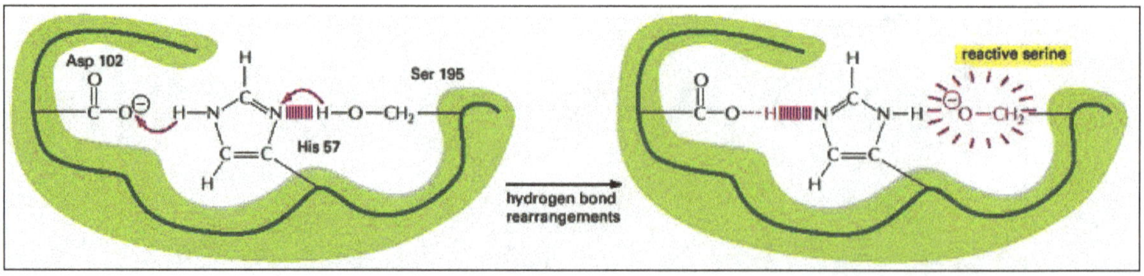

Figure shows an unusually reactive amino acid at the active site of an enzyme. This example is the "catalytic triad" found in chymotrypsin, elastase, and other serine proteases. The aspartic acid side chain induces the histidine to remove the proton from serine 195. This activates the serine to form a covalent bond with the enzyme substrate, hydrolyzing a peptide bond.

The surface of each protein molecule therefore has a unique chemical reactivity that depends not only on which amino acid side chains are exposed, but also on their exact orientation relative to one another. For this reason, even two slightly different conformations of the same protein molecule may differ greatly in their chemistry.

Sequence Comparisons between Protein Family Members Highlight Crucial Ligand Binding Sites

Many of the domains in proteins can be grouped into families that show clear evidence of their evolution from a common ancestor, and genome sequences reveal large numbers of proteins that contain one or more common domains. The three-dimensional structures of the members of the same domain family are remarkably similar. For example, even when the amino acid sequence identity falls to 25%, the backbone atoms in a domain have been found to follow a common protein fold within 0.2 nanometers (2 Å).

These facts allow a method called "evolutionary tracing" to be used to identify those sites in a protein domain that are the most crucial to the domain's function. For this purpose, those amino acids that are unchanged, or nearly unchanged, in all of the known protein family members are mapped onto a structural model of the three-dimensional structure of one family member. When this is done, the most invariant positions often form one or more clusters on the protein surface, as illustrated in Figure for the SH2 domain described previously. These clusters generally correspond to ligand binding sites.

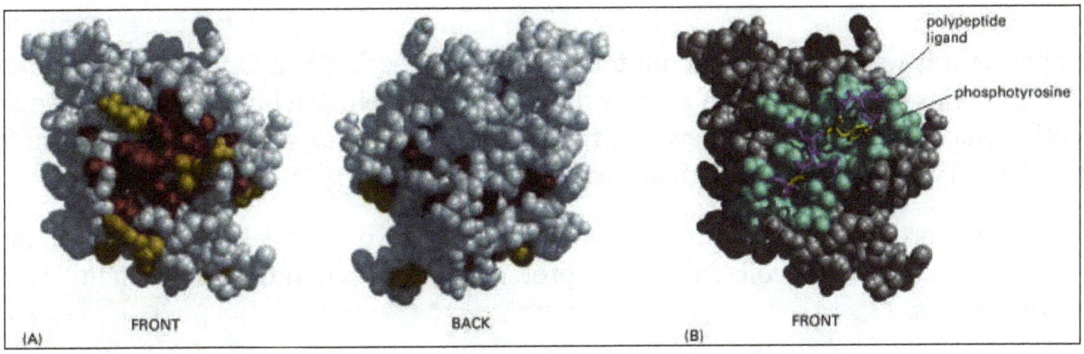

Figure shows the evolutionary trace method applied to the SH2 domain. (A) Front and back views of a space-filling model of the SH2 domain, with evolutionarily conserved amino acids on the protein surface colored yellow, and those more toward the protein interior colored red. (B) The structure of the SH2 domain with its bound polypeptide. Here, those amino acids located within 0.4 nm of the bound ligand are colored blue. The two key amino acids of the ligand are yellow, and the others are purple.

The SH2 domain is a module that functions in protein–protein interactions. It binds the protein containing it to a second protein that contains a phosphorylated tyrosine side chain in a specific amino acid sequence context, as shown in Figure. The amino acids located at the binding site for the phosphorylated polypeptide have been the slowest to change during the long evolutionary process that produced the large SH2 family of peptide recognition domains. Because mutation is a random process, this result is attributed to the preferential elimination during evolution of all organisms whose SH2 domains became altered in a way that inactivated the SH2-binding site, thereby destroying the function of the SH2 domain.

In this era of extensive genome sequencing, many new protein families have been discovered whose functions are unknown. By identifying the critical binding sites on a three-dimensional structure determined for one family member, the above method of evolutionary tracing is being used to help determine the functions of such proteins.

Proteins Bind to other Proteins through Several Types of Interfaces

Proteins can bind to other proteins in at least three ways. In many cases, a portion of the surface of one protein contacts an extended loop of polypeptide chain (a "string") on a second protein. Such a surface–string interaction, for example, allows the SH2 domain to recognize a phosphorylated polypeptide as a loop on a second protein, as just described, and it also enables a protein kinase to recognize the proteins that it will phosphorylate.

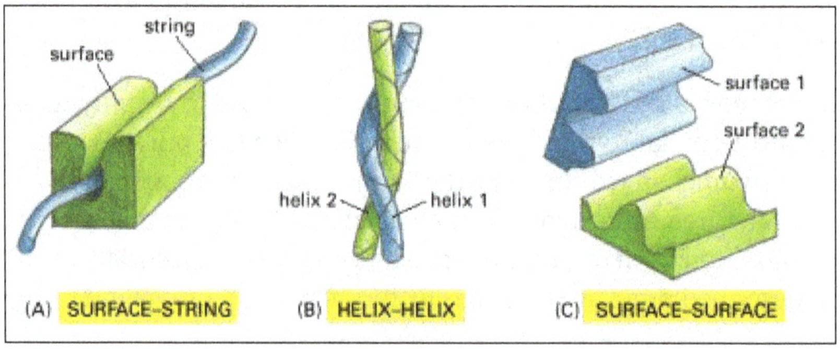

Figure shows three ways in which two proteins can bind to each other. Only the interacting parts of the two proteins are shown. (A) A rigid surface on one protein can bind to an extended loop of polypeptide chain (a "string") on a second protein. (B) Two α helices can bind together to form a coiled-coil. (C) Two complementary rigid surfaces often link two proteins together.

A second type of protein–protein interface is formed when two α helices, one from each protein, pair together to form a coiled-coil. This type of protein interface is found in several families of gene regulatory proteins.

The most common way for proteins to interact, however, is by the precise matching of one rigid surface with that of another. Such interactions can be very tight, since a large number of weak bonds can form between two surfaces that match well. For the same reason, such surface–surface interactions can be extremely specific, enabling a protein to select just one partner from the many thousands of different proteins found in a cell.

The Binding Sites of Antibodies are especially Versatile

All proteins must bind to particular ligands to carry out their various functions. This capacity for tight selective binding is displayed to an extraordinary degree by the antibody family. Antibodies, or immunoglobulin, are proteins produced by the immune system in response to foreign molecules, such as those on the surface of an invading microorganism. Each antibody binds to a particular target molecule extremely tightly, thereby either inactivating the target directly or marking it for destruction. An antibody recognizes its target (called an antigen) with remarkable specificity. Because there are potentially billions of different antigens we might encounter, we have to be able to produce billions of different antibodies.

Antibodies are Y-shaped molecules with two identical binding sites that are complementary to a small portion of the surface of the antigen molecule. A detailed examination of the antigen-binding sites of antibodies reveals that they are formed from several loops of polypeptide chain that protrude from the ends of a pair of closely juxtaposed protein domains. The enormous diversity of antigen-binding sites possessed by different antibodies is generated by changing only the length and amino acid sequence of these loops, without altering the basic protein structure.

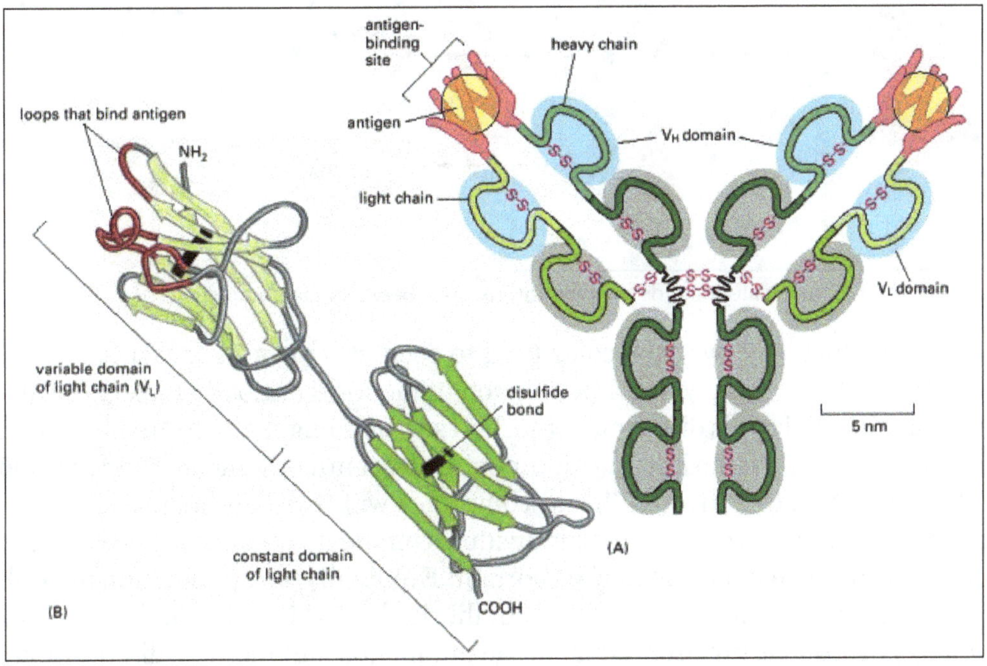

Figure shows an antibody molecule. (A) A typical antibody molecule is Y-shaped and has two identical binding sites for its antigen, one on each arm of the Y. The protein is composed of four polypeptide chains (two identical heavy chains and two identical and smaller light chains) held together by disulfide bonds. Each chain is made up of several different immunoglobulin domains, here shaded either blue or gray. The antigen-binding site is formed where a heavy-chain variable domain (V_H) and a light-chain variable domain (V_L) come close together. These are the domains that differ most in their sequence and structure in different antibodies. (B) This ribbon model of a light chain shows the parts of the V_L domain most closely involved in binding to the antigen in red. They contribute half of the fingerlike loops that fold around each of the antigen molecules in (A).

Loops of this kind are ideal for grasping other molecules. They allow a large number of chemical groups to surround a ligand so that the protein can link to it with many weak bonds. For this reason, loops are often used to form the ligand-binding sites in proteins.

Binding Strength is Measured by the Equilibrium Constant

Molecules in the cell encounter each other very frequently because of their continual random thermal movements. When colliding molecules have poorly matching surfaces, few non-covalent bonds form, and the two molecules dissociate as rapidly as they come together. At the other extreme, when many non-covalent bonds form, the association can persist for a very long time. Strong interactions occur in cells whenever a biological function requires that molecules remain associated for a long time—for example, when a group of RNA and protein molecules come together to make a subcellular structure such as a ribosome.

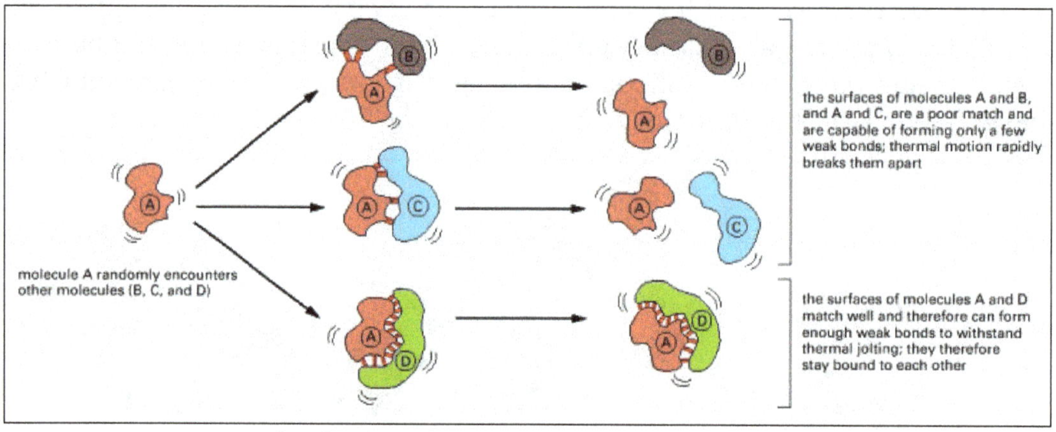

Non-covalent bonds mediate interactions between macromolecules.

The strength with which any two molecules bind to each other can be measured directly. As an example, imagine a situation in which a population of identical antibody molecules suddenly encounters a population of ligands diffusing in the fluid surrounding them. At frequent intervals, one of the ligand molecules will bump into the binding site of an antibody and form an antibody–ligand complex. The population of antibody–ligand complexes will therefore increase, but not without limit: over time, a second process, in which individual complexes break apart because of thermally induced motion, will become increasingly important. Eventually, any population of antibody molecules and ligands will reach a steady state, or equilibrium, in which the number of binding (association) events per second is precisely equal to the number of "unbinding" (dissociation) events.

From the concentrations of the ligand, antibody, and antibody–ligand complex at equilibrium, one can calculate a convenient measure—termed the equilibrium constant (K)—of the strength of binding. The equilibrium constant is greater the greater the binding strength and it is a direct measure of the free-energy difference between the bound and Free states. Even a change of a few non-covalent bonds can have a striking effect on a binding interaction, as shown by the example in Figure.

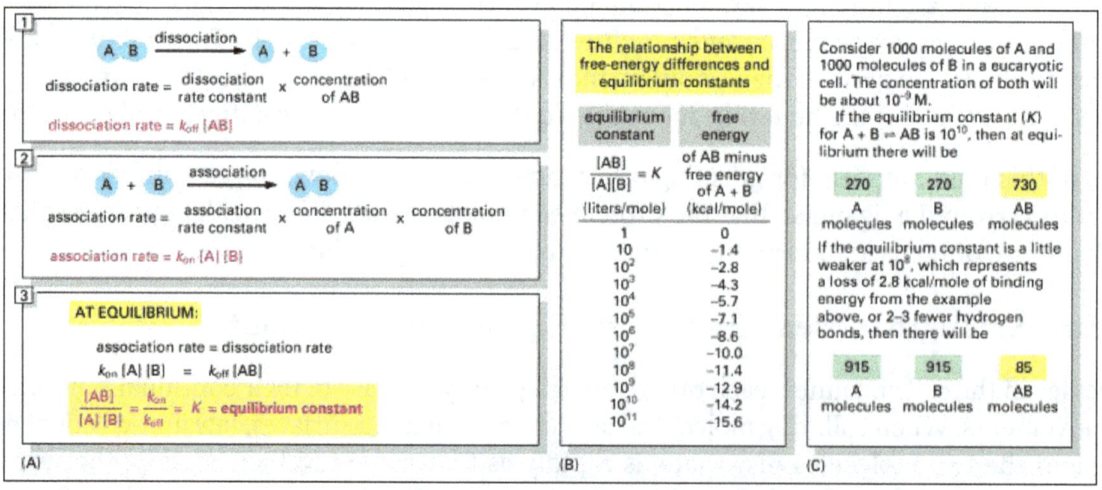

Figure shows relating binding energies to the equilibrium constant. (A) The equilibrium between molecules A and B and the complex AB is maintained by a balance between the two opposing reactions shown in panels 1 and 2. Molecules A and B must collide if they are to react, and the association rate is therefore proportional to the product of their individual concentrations [A] × [B]. (Square brackets indicate concentration.) As shown in panel 3, the ratio of the rate constants for the association and the dissociation reactions is equal to the equilibrium constant (K) for the reaction. (B) The equilibrium constant in panel 3 is that for the association reaction $A + B \leftrightarrow AB$, and the larger its value, the stronger the binding between A and B. Note that, for every 1.4 kcal/mole of free-energy drop, the equilibrium constant increases by a factor of 10. (C) An example of the dramatic effect that the presence or absence of a few weak bonds can have in a biological context.

The equilibrium constant here has units of liters/mole: for simple binding interactions it is also called the affinity constant or association constant, denoted K_a. The reciprocal of K_a is called the dissociation constant, K_d (in units of moles/liter).

Enzymes are Powerful and Highly Specific Catalysts

Many proteins can perform their function simply by binding to another molecule. An actin molecule, for example, need only associate with other actin molecules to form a filament. There are other proteins, however, for which ligand binding is only a necessary first step in their function. This is the case for the large and very important class of proteins called enzymes. Enzymes are remarkable molecules that determine all the chemical transformations that make and break covalent bonds in cells. They bind to one or more ligands, called substrates, and convert them into one or more chemically modified products, doing this over and over again with amazing rapidity. Enzymes speed up reactions, often by a factor of a million or more, without themselves being changed—that is, they act as catalysts that permit cells to make or break covalent bonds in a controlled way. It is the catalysis of organized sets of chemical reactions by enzymes that creates and maintains the cell, making life possible.

Enzymes can be grouped into functional classes that perform similar chemical reactions. Each type of enzyme within such a class is highly specific, catalyzing only a single type of reaction. Thus, hexokinase adds a phosphate group to d-glucose but ignores its optical isomer l-glucose; the blood-clotting enzyme thrombin cuts one type of blood protein between a particular arginine and its adjacent glycine and nowhere else, and so on. Enzymes work in teams, with the product of one enzyme becoming the substrate for the next. The result is an elaborate network of metabolic pathways that provides the cell with energy and generates the many large and small molecules that the cell needs.

Table: Some Common Types of Enzymes.

Enzyme	Reaction Catalyzed
Hydrolases	General term for enzymes that catalyze a hydrolytic cleavage reaction.
Nucleases	Break down nucleic acids by hydrolyzing bonds between nucleotides.
Proteases	Break down proteins by hydrolyzing bonds between amino acids.
Synthases	General name used for enzymes that synthesize molecules in anabolic reactions by condensing two smaller molecules together.
Isomerases	Catalyze the rearrangement of bonds within a single molecule.
Polymerases	Catalyze polymerization reactions such as the synthesis of DNA and RNA.

Kinases	Catalyze the addition of phosphate groups to molecules. Protein kinases are an important group of kinases that attach phosphate groups to proteins.
Phosphatases	Catalyze the hydrolytic removal of a phosphate group from a molecule.
Oxido-Reductases	General name for enzymes that catalyze reactions in which one molecule is oxidized while the other is reduced. Enzymes of this type are often called oxidases, reductases, and dehydrogenases.
ATPases	Hydrolyze ATP. Many proteins with a wide range of roles have an energy-harnessing ATPase activity as part of their function, for example, motor proteins such as myosin and membrane transport proteins such as the sodium–potassium pump.

Substrate Binding is the First Step in Enzyme Catalysis

For a protein that catalyzes a chemical reaction (an enzyme), the binding of each substrate molecule to the protein is an essential prelude. In the simplest case, if we denote the enzyme by E, the substrate by S, and the product by P, the basic reaction path is E + S → ES → EP → E + P. From this reaction path, we see that there is a limit to the amount of substrate that a single enzyme molecule can process in a given time. If the concentration of substrate is increased, the rate at which product is formed also increases, up to a maximum value. At that point the enzyme molecule is saturated with substrate, and the rate of reaction (V_{max}) depends only on how rapidly the enzyme can process the substrate molecule. This maximum rate divided by the enzyme concentration is called the turnover number. The turnover number is often about 1000 substrate molecules processed per second per enzyme molecule, although turnover numbers between 1 and 10,000 are known.

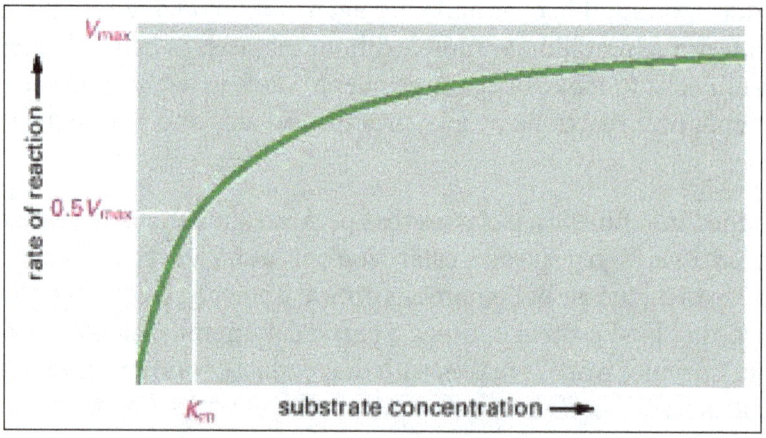

Figure shows enzyme kinetics. The rate of an enzyme reaction (V) increases as the substrate concentration increases until a maximum value (V_{max}) is reached. At this point all substrate-binding sites on the enzyme molecules are fully occupied, and the rate of reaction is limited by the rate of the catalytic process on the enzyme surface. For most enzymes, the concentration of substrate at which the reaction rate is half-maximal (K_m) is a measure of how tightly the substrate is bound, with a large value of K_m corresponding to weak binding.

The other kinetic parameter frequently used to characterize an enzyme is its K_m, the concentration of substrate that allows the reaction to proceed at one-half its maximum rate (0.5 V_{max}). A low K_m value means that the enzyme reaches its maximum catalytic rate at a low concentration of substrate and generally indicates that the enzyme binds to its substrate very tightly, whereas a high K_m value corresponds to weak binding.

Enzymes Speed Reactions by Selectively Stabilizing Transition States

Extremely high rates of chemical reaction are achieved by enzymes—far higher than for any synthetic catalysts. This efficiency is attributable to several factors. The enzyme serves, first, to increase the local concentration of substrate molecules at the catalytic site and to hold all the appropriate atoms in the correct orientation for the reaction that is to follow. More importantly, however, some of the binding energy contributes directly to the catalysis. Substrate molecules must pass through a series of intermediate states of altered geometry and electron distribution before they form the ultimate products of the reaction. The free energy required to attain the most unstable transition state is called the activation energy for the reaction, and it is the major determinant of the reaction rate. Enzymes have a much higher affinity for the transition state of the substrate than they have for the stable form. Because this tight binding greatly lowers the energies of the transition state, the enzyme greatly accelerates a particular reaction by lowering the activation energy that is required.

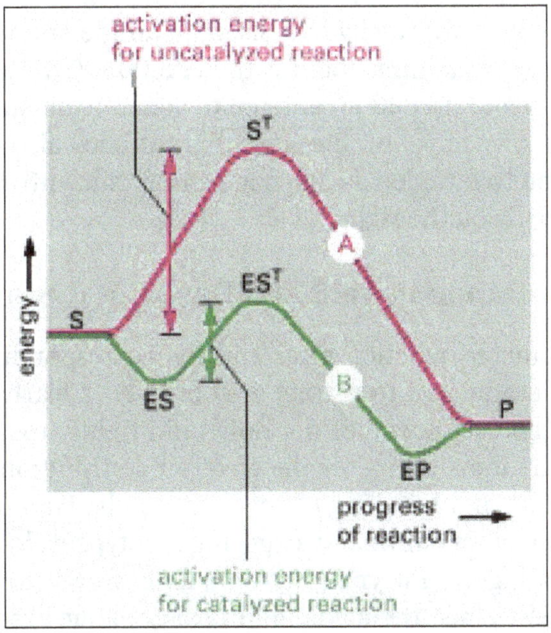

Figure shows enzymatic acceleration of chemical reactions by decreasing the activation energy. Often both the uncatalyzed reaction (A) and the enzyme-catalyzed reaction (B) can go through several transition states. It is the transition state with the highest energy (S^T and ES^T) that determines the activation energy and limits the rate of the reaction. (S = substrate; P = product of the reaction).

A dramatic proof that stabilizing a transition state can greatly increase a reaction rate is provided by the intentional production of antibodies that act like enzymes. Consider, for example, the hydrolysis of an amide bond, which is similar to the peptide bond that joins two adjacent amino acids in a protein. In an aqueous solution, an amide bond hydrolyzes very slowly by the mechanism shown in Figure. In the central intermediate, or transition state, the carbonyl carbon is bonded to four atoms arranged at the corners of a tetrahedron. By generating monoclonal antibodies that bind tightly to a stable analog of this very unstable tetrahedral intermediate, an antibody that functions like an enzyme can be obtained. Because this catalytic antibody binds to and stabilizes the tetrahedral intermediate, it increases the spontaneous rate of amide-bond hydrolysis by more than 10,000-fold.

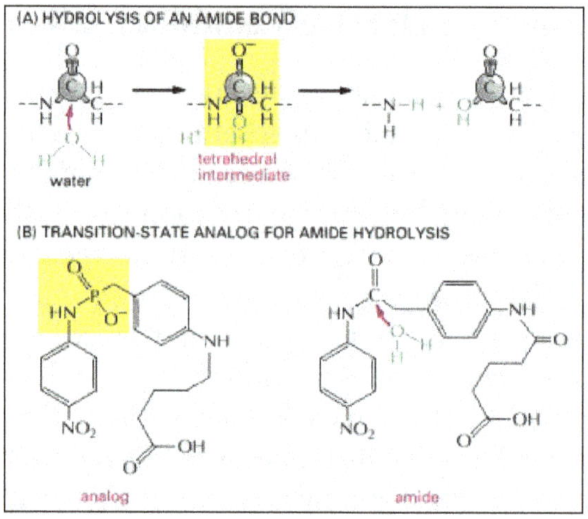

Figure shows catalytic antibodies. The stabilization of a transition state by an antibody creates an enzyme. (A) The reaction path for the hydrolysis of an amide bond goes through a tetrahedral intermediate, the high-energy transition state for the reaction. (B) The molecule on the left was covalently linked to a protein and used as an antigen to generate an antibody that binds tightly to the region of the molecule shown in yellow. Because this antibody also bound tightly to the transition state in (A), it was found to function as an enzyme that efficiently catalyzed the hydrolysis of the amide bond in the molecule on the right.

Enzymes can use Simultaneous Acid and Base Catalysis

Figure compares the spontaneous reaction rates and the corresponding enzyme-catalyzed rates for five enzymes. Rate accelerations of 109–1023 are observed. Clearly, enzymes are much better catalysts than catalytic antibodies. Enzymes not only bind tightly to a transition state, they also contain precisely positioned atoms that alter the electron distributions in those atoms that participate directly in the making and breaking of covalent bonds. Peptide bonds, for example, can be hydrolyzed in the absence of an enzyme by exposing a polypeptide to either a strong acid or a strong base, as explained in Figure. Enzymes are unique, however, in being able to use acid and base catalysis simultaneously, since the acidic and basic residues required are prevented from combining with each other (as they would do in solution) by being tied to the rigid framework of the protein itself.

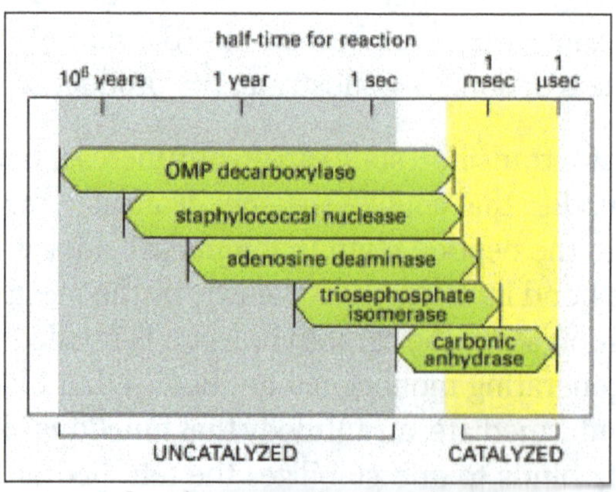

The rate accelerations caused by five different enzymes.

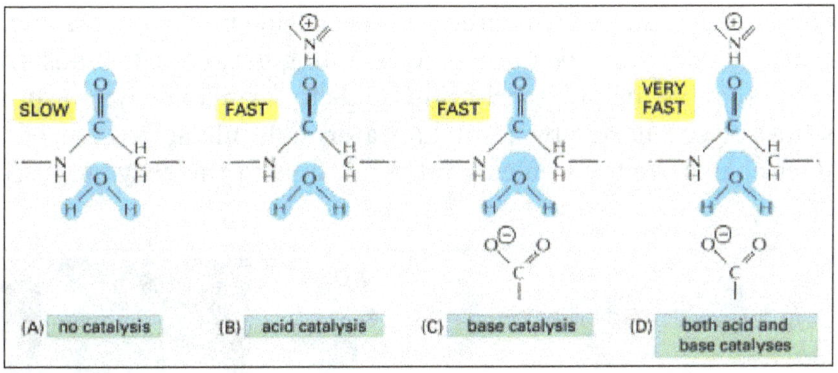

Figure shows acid catalysis and base catalysis. (A) The start of the uncatalyzed reaction shown in Figure is diagrammed, with blue indicating electron distribution in the water and carbonyl bonds. (B) An acid likes to donate a proton ($H^{\pm}$) to other atoms. By pairing with the carbonyl oxygen, an acid causes electrons to move away from the carbonyl carbon, making this atom much more attractive to the electronegative oxygen of an attacking water molecule. (C) A base likes to take up H+. By pairing with hydrogen of the attacking water molecule, a base causes electrons to move toward the water oxygen, making it a better attacking group for the carbonyl carbon. (D) By having appropriately positioned atoms on its surface, an enzyme can perform both acid catalysis and base catalysis at the same time. The fit between an enzyme and its substrate needs to be precise. A small change introduced by genetic engineering in the active site of an enzyme can have a profound effect. Replacing a glutamic acid with an aspartic acid in one enzyme, for example, shifts the position of the catalytic carboxylate ion by only 1 Å (about the radius of a hydrogen atom); yet this is enough to decrease the activity of the enzyme a thousand fold.

Lysozyme Illustrates how an Enzyme Works

To demonstrate how enzymes catalyze chemical reactions, we shall use the example of an enzyme that acts as a natural antibiotic in egg white, saliva, tears, and other secretions. Lysozyme is an enzyme that catalyzes the cutting of polysaccharide chains in the cell walls of bacteria. Because the bacterial cell is under pressure from osmotic forces, cutting even a small number of polysaccharide chains causes the cell wall to rupture and the cell to burst. Lysozyme is a relatively small and stable protein that can be easily isolated in large quantities. For these reasons, it has been intensively studied, and it was the first enzyme to have its structure worked out in atomic detail by x-ray crystallography.

The reaction catalyzed by lysozyme is a hydrolysis: a molecule of water is added to a single bond between two adjacent sugar groups in the polysaccharide chain, thereby causing the bond to break. The reaction is energetically favorable because the free energy of the severed polysaccharide chain is lower than the free energy of the intact chain. However, the pure polysaccharide can sit for years in water without being hydrolyzed to any detectable degree. For a colliding water molecule to break a bond linking two sugars, the polysaccharide molecule has to be distorted into a particular shape—the transition state—in which the atoms around the bond have an altered geometry and electron distribution. Because of this distortion, large activation energy must be supplied through random collisions before the reaction can take place. In an aqueous solution at room temperature, the energy of collisions almost never exceeds the activation energy. Consequently, hydrolysis occurs extremely slowly, if at all.

This situation is drastically changed when the polysaccharide binds to lysozyme. The active site of lysozyme, because its substrate is a polymer, is a long groove that holds six linked sugars at the same time. As soon as the polysaccharide binds to form an enzyme–substrate complex, the enzyme cuts the polysaccharide by adding a water molecule across one of its sugar–sugar bonds. The product chains are then quickly released, freeing the enzyme for further cycles of reaction.

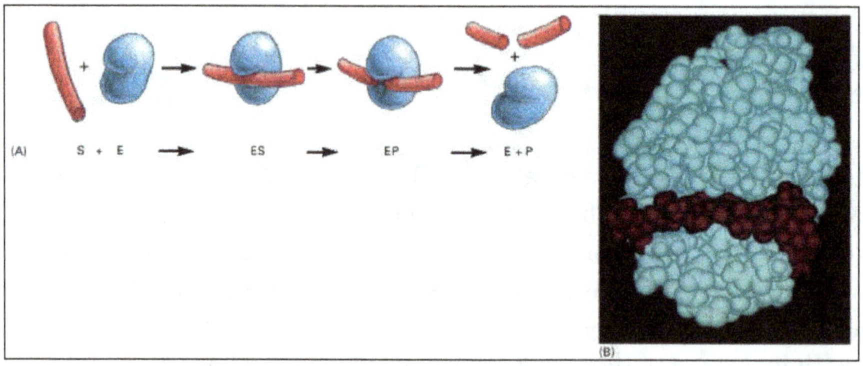

Figure shows the reaction catalyzed by lysozyme. (A) The enzyme lysozyme (denoted as E) catalyzes the cutting of a polysaccharide chain, which is its substrate (S). The enzyme first binds to the chain to form an enzyme–substrate complex (ES) and then catalyzes the cleavage of a specific covalent bond in the backbone of the polysaccharide, forming an enzyme–product complex (EP) that rapidly dissociates. Release of the severed chain (the products P) leaves the enzyme free to act on another substrate molecule. (B) A space-filling model of the lysozyme molecule bound to a short length of polysaccharide chain before cleavage.

The chemistry that underlies the binding of lysozyme to its substrate is the same as that for antibody binding to its antigen—the formation of multiple non-covalent bonds. However, lysozyme holds its polysaccharide substrate in a particular way, so that one of the two sugars involved in the bond to be broken is distorted from its normal, most stable conformation. The bond to be broken is also held close to two amino acids with acidic side chains (a glutamic acid and an aspartic acid) within the active site.

Conditions are thereby created in the microenvironment of the lysozyme active site that greatly reduces the activation energy necessary for the hydrolysis to take place. Figure shows three central steps in this enzymatically catalyzed reaction.

- The enzyme stresses its bound substrate by bending some critical chemical bonds in one sugar, so that the shape of this sugar more closely resembles the shape of high-energy transition states formed during the reaction.
- The negatively charged aspartic acid reacts with the C1 carbon atom on the distorted sugar, breaking the sugar–sugar bond and leaving the aspartic acid side chain covalently linked to the site of bond cleavage.
- Aided by the negatively charged glutamic acid, a water molecule reacts with the C1 carbon atom, displacing the aspartic acid side chain and completing the process of hydrolysis.

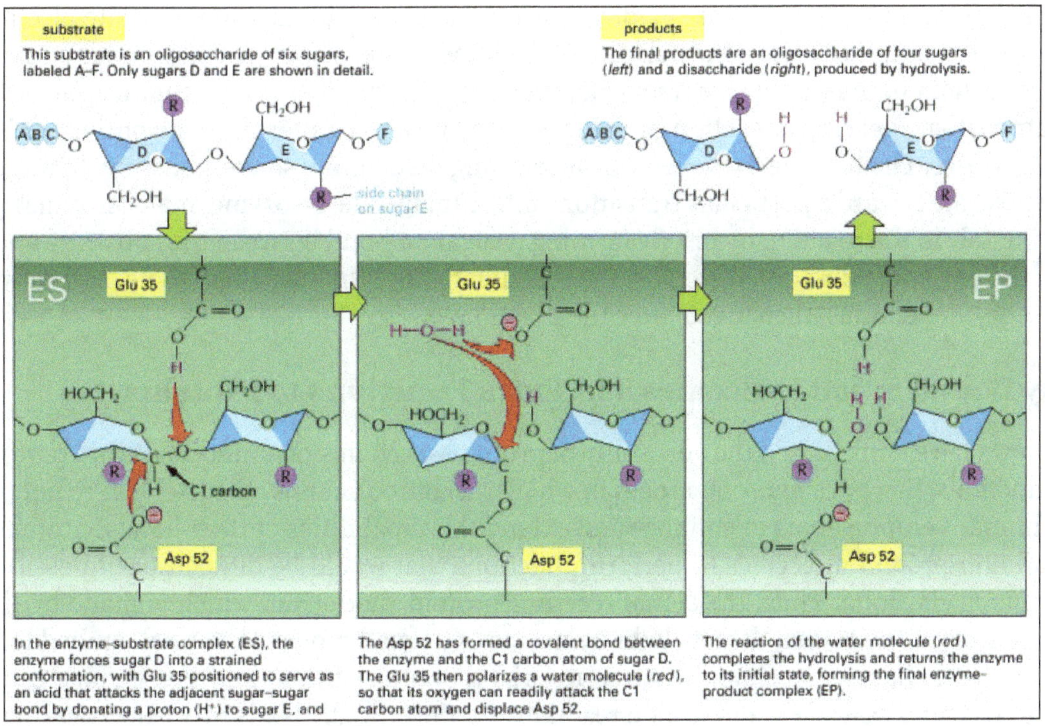

Figure shows events at the active site of lysozyme. The top left and top right drawings depict the free substrate and the free products, respectively, whereas the other three drawings depict the sequential events at the enzyme active site. Note the change in the conformation of sugar D in the enzyme–substrate complex; this shape change stabilizes the oxocarbenium ion-like transition states required for formation and hydrolysis of the covalent intermediate shown in the middle panel.

The overall chemical reaction, from the initial binding of the polysaccharide on the surface of the enzyme through the final release of the severed chains, occurs many millions of times faster than it would in the absence of enzyme.

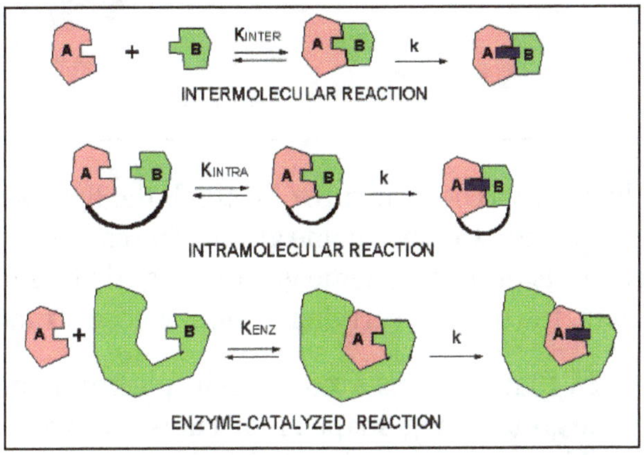

Some general strategies of enzyme catalysis: (A) Holding substrates together in a precise alignment. (B) Charge stabilization of reaction intermediates. (C) Altering bond angles in the substrate to increase the rate of a particular reaction.

Similar mechanisms are used by other enzymes to lower activation energies and speed up the reactions they catalyze. In reactions involving two or more reactants, the active site also acts like a template, or mold, that brings the substrates together in the proper orientation for a reaction to occur between them. As we saw for lysozyme, the active site of an enzyme contains precisely positioned atoms that speed up a reaction by using charged groups to alter the distribution of electrons in the substrates. The binding to the enzyme also changes substrate shapes, bending bonds so as to drive a substrate toward a particular transition state. Finally, like lysozyme, many enzymes participate intimately in the reaction by briefly forming a covalent bond between the substrate and a side chain of the enzyme. Subsequent steps in the reaction restore the side chain to its original state, so that the enzyme remains unchanged after the reaction.

Tightly Bound Small Molecules add Extra Functions to Proteins

Although we have emphasized the versatility of proteins as chains of amino acids that perform different functions, there are many instances in which the amino acids by themselves are not enough. Just as humans employ tools to enhance and extend the capabilities of their hands, proteins often use small non-protein molecules to perform functions that would be difficult or impossible to do with amino acids alone. Thus, the signal receptor protein rhodopsin, which is made by the photoreceptor cells in the retina, detects light by means of a small molecule, retinal, embedded in the protein. Retinal changes its shape when it absorbs a photon of light and this change causes the protein to trigger a cascade of enzymatic reactions that eventually leads to an electrical signal being carried to the brain.

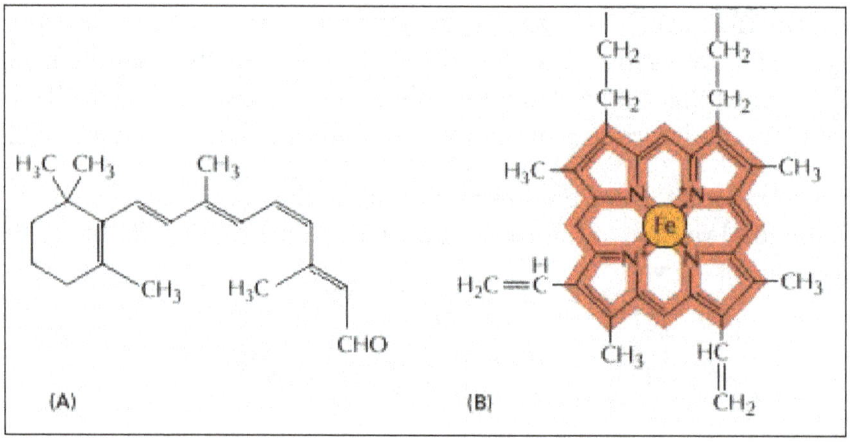

Figure shows retinal and heme. (A) The structure of retinal, the light-sensitive molecule attached to rhodopsin in the eye. (B) The structure of a heme group. The carbon-containing heme ring is red and the iron atom at its center is orange. A heme group is tightly bound to each of the four polypeptide chains in hemoglobin, the oxygen-carrying protein whose structure is shown in Figure.

Another example of a protein that contains a nonprotein portion is hemoglobin. A molecule of hemoglobin carries four heme groups, ring-shaped molecules each with a single central iron atom. Heme gives hemoglobin (and blood) its red color. By binding reversibly to oxygen gas through its iron atom, heme enables hemoglobin to pick up oxygen in the lungs and release it in the tissues.

Sometimes these small molecules are attached covalently and permanently to their protein, thereby becoming an integral part of the protein molecule itself. Those proteins are often anchored to cell membranes through covalently attached lipid molecules. And membrane proteins exposed on the surface of the cell, as well as proteins secreted outside the cell, are often modified by the covalent addition of sugars and oligosaccharides.

Enzymes frequently have a small molecule or metal atom tightly associated with their active site that assists with their catalytic function. Carboxypeptidase, for example, an enzyme that cuts polypeptide chains, carries a tightly bound zinc ion in its active site. During the cleavage of a peptide bond by carboxypeptidase, the zinc ion forms a transient bond with one of the substrate atoms, thereby assisting the hydrolysis reaction. In other enzymes, a small organic molecule serves a similar purpose. Such organic molecules are often referred to as coenzymes. An example is biotin, which is found in enzymes that transfer a carboxylate group ($-COO^-$) from one molecule to another. Biotin participates in these reactions by forming a transient covalent bond to the $-COO^-$ group to be transferred, being better suited to this function than any of the amino acids used to make proteins.

Table: Many Vitamins Provide Critical Coenzymes for Human Cells.

Vitamin	Coenzyme	Enzyme-Catalyzed Reactions Requiring These Coenzymes
Thiamine (vitamin B_1)	Thiamine pyrophosphate	Activation and transfer of aldehydes.
Riboflavin (vitamin B_2)	FADH	Oxidation–reduction.
Niacin	NADH, NADPH	Oxidation–reduction.
Pantothenic acid	Voenzyme A	Acyl group activation and transfer.
Pyridoxine	Pyridoxal phosphate	Reactions involving amino acid activation.
Biotin	Biotin	CO_2 activation and transfer.
Lipoic acid	Lipoamide	Acyl group activation; oxidation–reduction.
Folic acid	Tetrahydrofolate	Activation and transfer of single carbon groups.
Vitamin B_{12}	Cobalamin coenzymes	Isomerization and methyl group transfers.

Multienzyme Complexes Help to Increase the Rate of Cell Metabolism

The efficiency of enzymes in accelerating chemical reactions is crucial to the maintenance of life. Cells, in effect, must race against the unavoidable processes of decay, which—if left unattended—cause macromolecules to run downhill toward greater and greater disorder. If the rates of desirable reactions were not greater than the rates of competing side reactions, a cell would soon die. Some idea of the rate at which cell metabolism proceeds can be obtained by measuring the rate of ATP utilization. A typical mammalian cell "turns over" (i.e., hydrolyzes and restores by phosphorylation) its entire ATP pool once every 1 or 2 minutes. For each cell this turnover represents the utilization of roughly 10^7 molecules of ATP per second (or, for the human body, about 1 gram of ATP every minute).

The rates of reactions in cells are rapid because of the effectiveness of enzyme catalysis. Many important enzymes have become so efficient that there is no possibility of further useful improvement. The factor that limits the reaction rate is no longer the enzyme's intrinsic speed of action; rather, it is the frequency with which the enzyme collides with its substrate. Such a reaction is said to be diffusion-limited.

If an enzyme-catalyzed reaction is diffusion-limited, its rate depends on the concentration of both the enzyme and its substrate. If a sequence of reactions is to occur extremely rapidly, each metabolic intermediate and enzyme involved must be present in high concentration. However, given the enormous number of different reactions performed by a cell, there are limits to the concentrations of substrates that can be achieved. In fact, most metabolites are present in micro molar (10^{-6} M) concentrations and most enzyme concentrations are much lower. How is it possible, therefore, to maintain very fast metabolic rates?

The answer lies in the spatial organization of cell components. Reaction rates can be increased without raising substrate concentrations by bringing the various enzymes involved in a reaction sequence together to form a large protein assembly known as a multi enzyme complex. Because this allows the product of enzyme A to be passed directly to enzyme B, and so on, diffusion rates need not be limiting, even when the concentrations of the substrates in the cell as a whole are very low. It is perhaps not surprising, therefore, that such enzyme complexes are very common, and they are involved in nearly all aspects of metabolism—including the central genetic processes of DNA, RNA, and protein synthesis. In fact, few enzymes in eucaryotic cells may be left to diffuse freely in solution; instead, most seem to have evolved binding sites that concentrate them with other proteins of related function in particular regions of the cell, thereby increasing the rate and efficiency of the reactions that they catalyze.

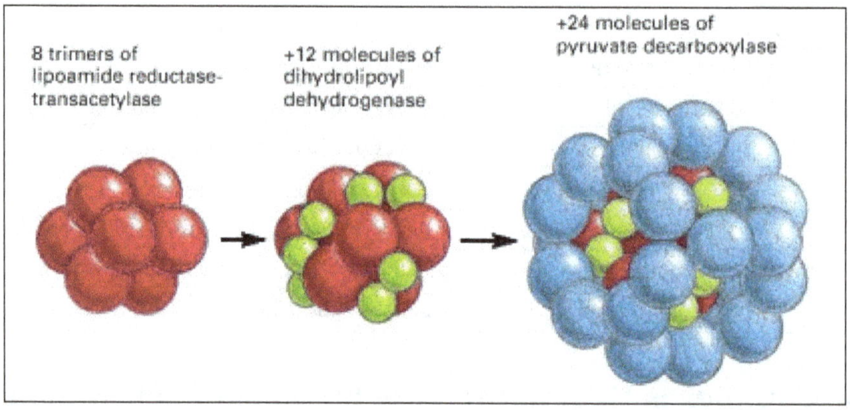

Figure shows the structure of pyruvate dehydrogenase. This enzyme complex catalyzes the conversion of pyruvate to acetyl CoA, as part of the pathway that oxidizes sugars to CO_2 and H_2O. It is an example of a large multi enzyme complex in which reaction intermediates are passed directly from one enzyme to another.

Eucaryotic cells have yet another way of increasing the rate of metabolic reactions, using their intracellular membrane systems. These membranes can segregate particular substrates and the enzymes that act on them into the same membrane-enclosed compartment, such as the endoplasmic reticulum or the cell nucleus. If, for example, a compartment occupies a total of 10% of the volume of the cell, the concentration of reactants in the compartment may be increased as much as 10 times compared with the same cell with no compartmentalization. Reactions that would otherwise be limited by the speed of diffusion can thereby be speeded up by a factor of 10.

The Catalytic Activities of Enzymes are Regulated

A living cell contains thousands of enzymes, many of which operate at the same time and in the same small volume of the cytosol. By their catalytic action, these enzymes generate a complex web of metabolic pathways, each composed of chains of chemical reactions in which the product of one enzyme becomes the substrate of the next. In this maze of pathways, there are many branch points where different enzymes compete for the same substrate. The system is so complex that elaborate controls are required to regulate when and how rapidly each reaction occurs.

Regulation occurs at many levels. At one level, the cell controls how many molecules of each enzyme it makes by regulating the expression of the gene that encodes that enzyme. The cell also controls enzymatic activities by confining sets of enzymes to particular subcellular compartments, enclosed by distinct membranes. The rate of protein destruction by targeted proteolysis represents yet another important regulatory mechanism. But the most rapid and general process that adjusts reaction rates operates through a direct, reversible change in the activity of an enzyme in response to specific molecules that it encounters.

The most common type of control occurs when a molecule other than one of the substrates binds to an enzyme at a special regulatory site outside the active site, thereby altering the rate at which the enzyme converts its substrates to products. In feedback inhibition, enzyme acting early in a reaction pathway is inhibited by a late product of that pathway. Thus, whenever large quantities of the final product begin to accumulate, this product binds to the first enzyme and slows down its catalytic action, thereby limiting the further entry of substrates into that reaction pathway. Where pathways branch or intersect, there are usually multiple points of control by different final products, each of which works to regulate its own synthesis. Feedback inhibition can work almost instantaneously and is rapidly reversed when the level of the product falls.

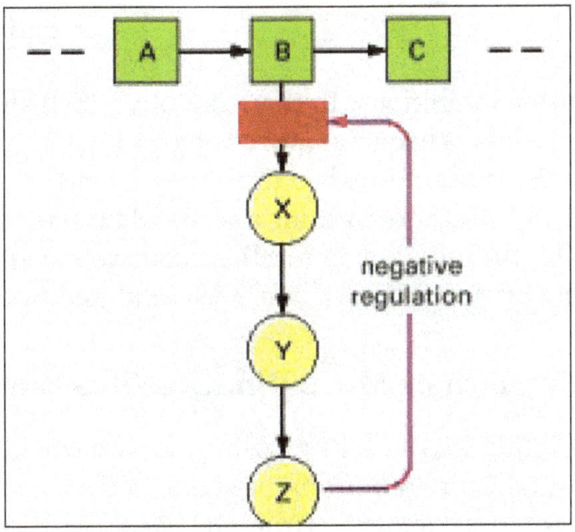

Feedback inhibition of a single biosynthetic pathway: The end-product Z inhibits the first enzyme that is unique to its synthesis and thereby controls its own level in the cell. This is an example of negative regulation.

Feedback inhibition is negative regulation: it prevents an enzyme from acting. Enzymes can also be subject to positive regulation, in which the enzyme's activity is stimulated by a regulatory molecule rather than being shut down. Positive regulation occurs when a product in one branch of the metabolic maze stimulates the activity of an enzyme in another pathway. As one example, the accumulation of ADP activates several enzymes involved in the oxidation of sugar molecules, thereby stimulating the cell to convert more ADP to ATP.

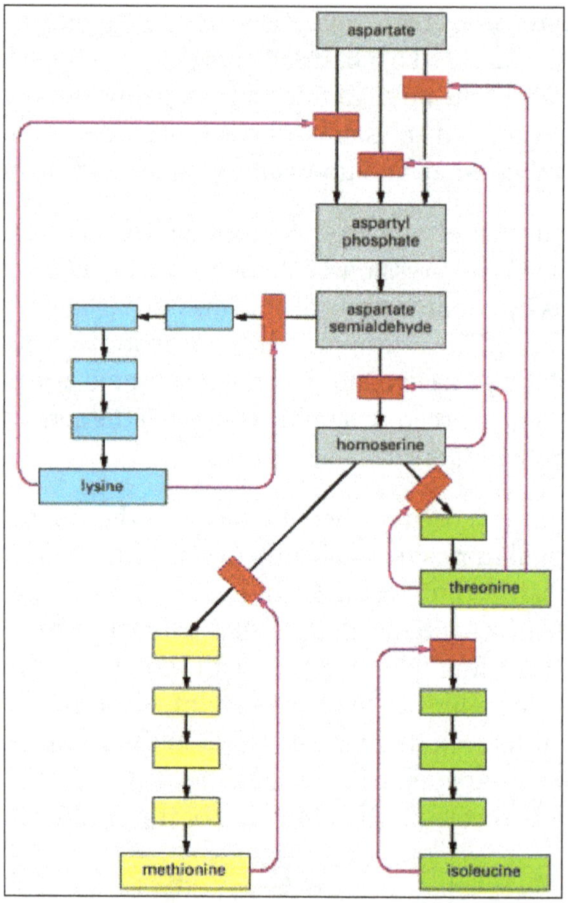

Figure shows multiple feedback inhibitions. In this example, which shows the biosynthetic pathways for four different amino acids in bacteria, the red arrows indicate positions at which products feedback to inhibit enzymes. Each amino acid controls the first enzyme specific to its own synthesis, thereby controlling its own levels and avoiding a wasteful buildup of intermediates. The products can also separately inhibit the initial set of reactions common to all the syntheses; in this case, three different enzymes catalyze the initial reaction, each inhibited by a different product.

Allosteric Enzymes have Two or More Binding Sites that Interact

One feature of feedback inhibition was initially puzzling to those who discovered it: the regulatory molecule often has a shape totally different from the shape of the substrate of the enzyme. This is why this form of regulation is termed allostery (from the Greek words allos, meaning "other," and stereos, meaning "solid" or "three-dimensional"). As more was learned about feedback inhibition, it was recognized that many enzymes must have at least two different binding sites on their surface—an active site that recognizes the substrates, and a regulatory site that recognizes a regulatory molecule. These two sites must somehow communicate in a way that allows the catalytic events at the active site to be influenced by the binding of the regulatory molecule at its separate site on the protein's surface.

The interaction between separated sites on a protein molecule is now known to depend on a conformational change in the protein: binding at one of the sites causes a shift from one folded shape to a slightly different folded shape. During feedback inhibition, for example, the binding of an inhibitor at one site on the protein causes the protein to shift to a conformation in which its active site—located elsewhere in the protein—becomes incapacitated.

It is thought that most protein molecules are allosteric. They can adopt two or more slightly different conformations, and a shift from one to another caused by the binding of a ligand can alter their activity. This is true not only for enzymes but also for many other proteins—including receptors, structural proteins, and motor proteins. There is nothing mysterious about the allosteric regulation of these proteins: each conformation of the protein has somewhat different surface contours, and the protein's binding sites for ligands are altered when the protein changes shape. Each ligand stabilizes the conformation that it binds to most strongly, and thus—at high enough concentrations—tends to "switch" the protein toward the conformation it prefers.

Two Ligands whose Binding Sites are Coupled Must Reciprocally Affect Each other's Binding

The effects of ligand binding on a protein follow from a fundamental chemical principle known as linkage. Suppose, for example, that a protein that binds glucose also binds another molecule, X, at a distant site on the protein's surface. If the binding site for X changes shape as part of the conformational change induced by glucose binding, the binding sites for X and for glucose are said to be coupled. Whenever two ligands prefer to bind to the same conformation of an allosteric protein, it follows from basic thermodynamic principles that each ligand must increase the affinity of the protein for the other. Thus, if the shift of the protein in Figure to the closed conformation that binds glucose best also causes the binding site for X to fit X better, then the protein will bind glucose more tightly when X is present than when X is absent.

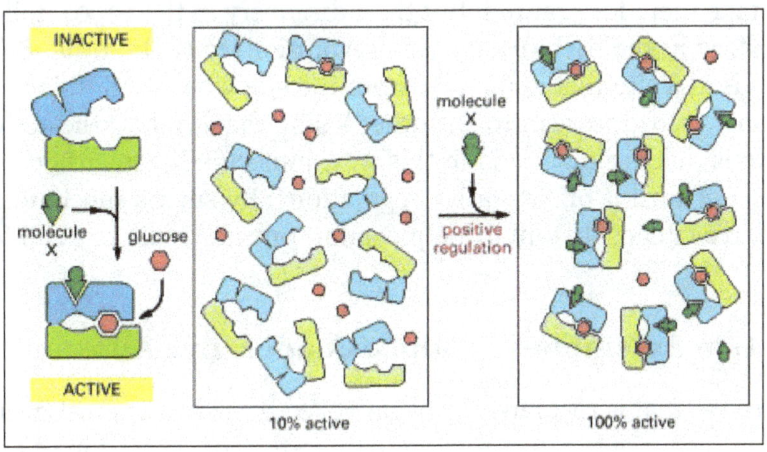

Figure shows positive regulation caused by conformational coupling between two distant binding sites. In this example, both glucose and molecule X bind best to the closed conformation of a protein with two domains. Because both glucose and molecule X drive the protein toward its closed conformation, each ligand helps the other to bind. Glucose and molecule X are therefore said to bind cooperatively to the protein.

Conversely, linkage operates in a negative way if two ligands prefer to bind to different conformations of the same protein. In this case, the binding of the first ligand discourages the binding of the second ligand. Thus, if a shape change caused by glucose binding decreases the affinity of a protein for molecule X, the binding of X must also decrease the protein's affinity for glucose. The linkage relationship is quantitatively reciprocal, so that, for example, if glucose has a very large effect on the binding of X, X has a very large effect on the binding of glucose.

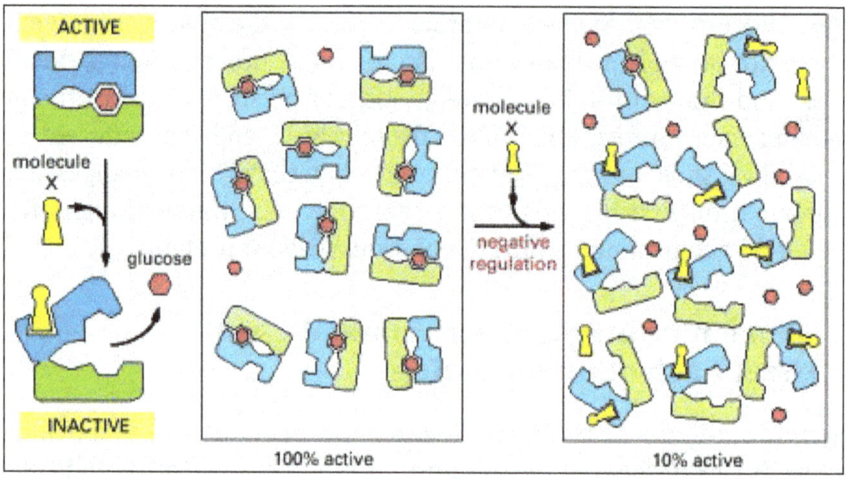

Figure shows negative regulation caused by conformational coupling between two distant binding sites. The scheme here resembles that in the previous figure, but here molecule X prefers the open conformation, while glucose prefers the closed conformation. Because glucose and molecule X drive the protein toward opposite conformations (closed and open, respectively), the presence of either ligand interferes with the binding of the other.

The relationships shown in Figures underlie all of cell biology. They seem so obvious in retrospect that we now take them for granted. But their discovery in the 1950s, followed by a general description of allostery in the early 1960s, was revolutionary at the time. Since molecule X in these examples binds at a site that is distinct from the site where catalysis occurs, it need have no chemical relationship to glucose or to any other ligand that binds at the active site. As we have just seen, for enzymes that are regulated in this way, molecule X can either turn the enzyme on (positive regulation) or turn it off (negative regulation). By such a mechanism, allosteric proteins serve as general switches that, in principle, allow one molecule in a cell to affect the fate of any other.

Symmetric Protein Assemblies Produce Cooperative Allosteric Transitions

A single subunit enzyme that is regulated by negative feedback can at most decrease from 90% to about 10% activity in response to a 100-fold increase in the concentration of an inhibitory ligand that it binds (red line). Responses of this type are apparently not sharp enough for optimal cell regulation, and most enzymes that are turned on or off by ligand binding consist of symmetric assemblies of identical subunits. With this arrangement, the binding of a molecule of ligand to a single site on one subunit can trigger an allosteric change in the subunit that can be transmitted to the neighboring subunits, helping them to bind the same ligand. As a result, a cooperative allosteric transition occurs (blue line), allowing a relatively small change in ligand concentration in the cell to switch the whole assembly from an almost fully active to an almost fully inactive conformation (or vice versa).

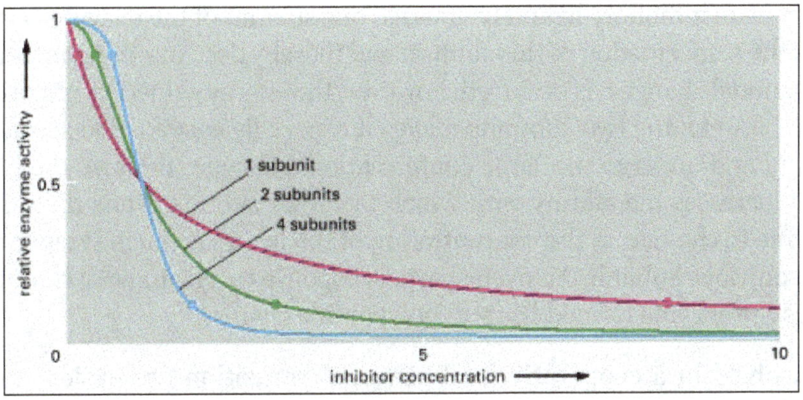

Figure shows enzyme activity versus the concentration of inhibitory ligand for single-subunit and multi-subunit allosteric enzymes. For an enzyme with a single subunit (red line), a drop from 90% enzyme activity to 10% activity (indicated by the two dots on the curve) requires a 100-fold increase in the concentration of inhibitor. The enzyme activity is calculated from the simple equilibrium relationship K = [I] [P]/ [IP], where P is active protein, I is inhibitor, and IP is the inactive protein bound to inhibitor. An identical curve applies to any simple binding interaction between two molecules, A and B. In contrast, a multi subunit allosteric enzyme can respond in a switch like manner to a change in ligand concentration: the steep response is caused by a cooperative binding of the ligand molecules, as explained in Figure. Here, the green line represents the idealized result expected for the cooperative binding of two inhibitory ligand molecules to an allosteric enzyme with two subunits, and the blue line shows the idealized response of an enzyme with four subunits. As indicated by the two dots on each of these curves, the more complex enzymes drop from 90% to 10% activity over a much narrower range of inhibitor concentration than does the enzyme composed of a single subunit.

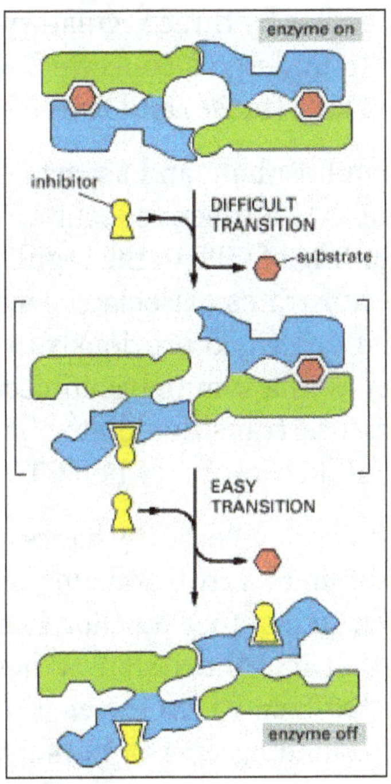

Figure shows a cooperative allosteric transition in an enzyme composed of two identical subunits. This diagram illustrates how the conformation of one subunit can influence that of its neighbor. The binding of a single molecule of an inhibitory ligand (yellow) to one subunit of the enzyme occurs with difficulty because it changes the conformation of this subunit and thereby destroys the symmetry of the enzyme. Once this conformational change has occurred, however, the energy gained by restoring the symmetric pairing interaction between the two subunits makes it especially easy for the second subunit to bind the inhibitory ligand and undergo the same conformational change. Because the binding of the first molecule of ligand increases the affinity with which the other subunit binds the same ligand, the response of the enzyme to changes in the concentration of the ligand is much steeper than the response of an enzyme with only one subunit. An even sharper response to a ligand can be obtained with larger assemblies, such as the enzyme formed from 12 polypeptide chains.

The principals involved in a cooperative "all-or-none" transition are easiest to visualize for an enzyme that forms a symmetric dimer. In the example shown in Figure, the first molecule of an inhibitory ligand binds with great difficulty since its binding destroys an energetically favorable interaction between the two identical monomers in the dimer. A second molecule of inhibitory ligand now binds more easily, however, because its binding restores the monomer–monomer contacts of a symmetric dimer (and also completely inactivates the enzyme).

The Allosteric Transition in Aspartate Transcarbamoylase is Understood in Atomic Detail. One enzyme used in the early studies of allosteric regulation was aspartate transcarbamoylase from E. coli. It catalyzes the important reaction that begins the synthesis of the pyrimidine ring of C, U, and T nucleotides: carbamoylphosphate + aspartate → N-carbamoylaspartate. One of the final products of this pathway, cytosine triphosphate (CTP), binds to the enzyme to turn it off whenever CTP is plentiful.

Aspartate transcarbamoylase is a large complex of six regulatory and six catalytic subunits. The catalytic subunits are present as two trimers, each arranged like an equilateral triangle; the two trimers face each other and are held together by three regulatory dimers that form a bridge between them. The entire molecule is poised to undergo a concerted, all-or-none, allosteric transition between two conformations, designated as T (tense) and R (relaxed) states.

The binding of substrates (carbamoylphosphate and aspartate) to the catalytic trimers drives aspartate transcarbamoylase into its catalytically active R state, from which the regulatory CTP molecules dissociate. By contrast, the binding of CTP to the regulatory dimers converts the enzyme to the inactive T state, from which the substrates dissociate. This tug-of-war between CTP and substrates is identical in principle to that described previously in Figure for a simpler allosteric protein. But because the tug-of-war occurs in a symmetric molecule with multiple binding sites, the enzyme undergoes a cooperative allosteric transition that will turn it on suddenly as substrates accumulate (forming the R state) or shut it off rapidly when CTP accumulates (forming the T state).

A combination of biochemistry and x-ray crystallography has revealed many fascinating details of this allosteric transition. Each regulatory subunit has two domains, and the binding of CTP causes the two domains to move relative to each other, so that they function like a lever that rotates the two catalytic trimmers and pulls them closer together into the T state. When this occurs, hydrogen bonds form between opposing catalytic subunits that help widen the cleft that forms the active site within each catalytic subunit, thereby destroying the binding sites for the substrates. Adding large amounts of substrate has the opposite effect, favoring the R state by binding in the cleft of each catalytic subunit and opposing the above conformational change. Conformations that are intermediate between R and T are unstable, so that the enzyme mostly clicks back and forth between its R and T forms, producing a mixture of these two species in proportions that depend on the relative concentrations of CTP and substrates.

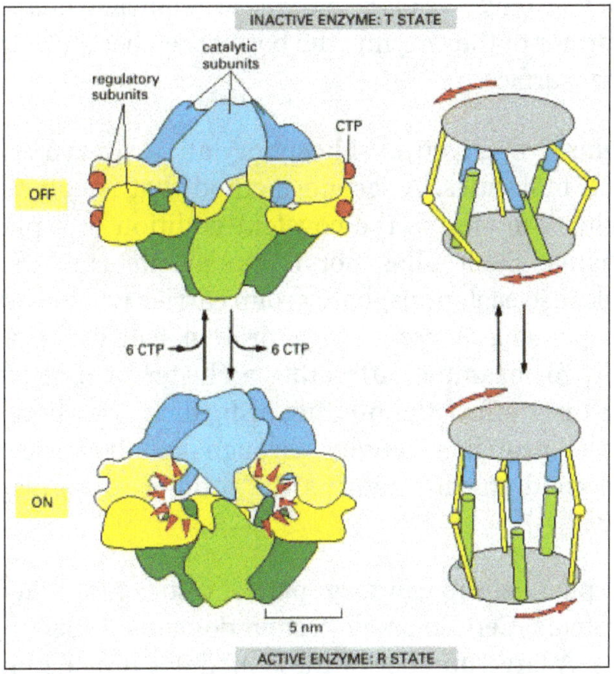

Figure shows the transition between R and T states in the enzyme aspartate transcarbamoylase. The enzyme consists of a complex of six catalytic subunits and six regulatory subunits, and the structures of its inactive (T state) and active (R state) forms have been determined by x-ray crystallography. The enzyme is turned off by feedback inhibition when CTP concentrations rise. Each regulatory subunit can bind one molecule of CTP, which is one of the final products in the pathway. By means of this negative feedback regulation, the pathway is prevented from producing more CTP than the cell needs.

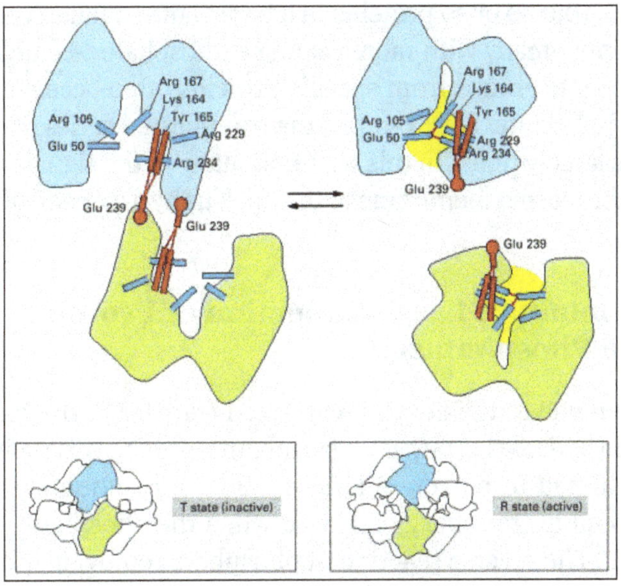

Figure shows part of the on–off switch in the catalytic subunits of aspartate transcarbamoylase.

Changes in the indicated hydrogen-bonding interactions are partly responsible for switching this enzyme's active site between active (yellow) and inactive conformations. Hydrogen bonds are indicated by thin red lines. The amino acids involved in the subunit–subunit interaction are shown in red, while those that form the active site of the enzyme are shown in blue. The large drawings show the catalytic site in the interior of the enzyme; the boxed sketches show the same subunits viewed from the enzyme's external surface.

Many Changes in Proteins are driven by Phosphorylation. Enzymes are regulated by more than the binding of small molecules. A second method that is commonly used by eucaryotic cells to regulate a protein's function is the covalent addition of a phosphate group to one of its amino acid side chains. Such phosphorylation events can affect the protein in two important ways. First, because each phosphate group carries two negative charges, the enzyme catalyzed addition of a phosphate group to a protein can cause a major conformational change in the protein by, for example, attracting a cluster of positively charged amino acid side chains. This can, in turn, affect the binding of ligands elsewhere on the protein surface, dramatically changing the protein's activity through an allosteric effect. Removal of the phosphate group by a second enzyme returns the protein to its original conformation and restores its initial activity.

Second, an attached phosphate group can form part of a structure that is directly recognized by binding sites of other proteins. Certain small protein domains, called modules, appear very frequently in larger proteins. A large number of these modules provide binding sites for attaching their protein to phosphorylated peptides in other protein molecules. Several other types of modules recognize phosphorylated serine or threonine side chains in a specific context. As a result, protein phosphorylation and dephosphorylating events have a major role in driving the regulated assembly and disassembly of protein complexes.

Through such effects, reversible protein phosphorylation controls the activity structure and cellular localization of many types of proteins in eucaryotic cells. In fact, this regulation is so extensive that more than one-third of the 10,000 or so proteins in a typical mammalian cell are thought to be phosphorylated at any given time—many with more than one phosphate. As might be expected, the addition and removal of phosphate groups from specific proteins often occur in response to signals that specify some change in a cell's state. For example, the complicated series of events that takes place as a eucaryotic cell divides is largely timed in this way, and many of the signals mediating cell–cell interactions are relayed from the plasma membrane to the nucleus by a cascade of protein phosphorylation events.

A Eucaryotic Cell Contains a Large Collection of Protein Kinases and Protein Phosphatases

Protein phosphorylation involves the enzyme-catalyzed transfer of the terminal phosphate group of an ATP molecule to the hydroxyl group on a serine, threonine, or tyrosine side chain of the protein. This reaction is catalyzed by a protein kinase, and the reaction is essentially unidirectional because of the large amount of free energy released when the phosphate–phosphate bond in ATP is broken to produce ADP. The reverse reaction of phosphate removal, or dephosphorylation, is instead catalyzed by a protein phosphatase. Cells contain hundreds of different protein kinases, each responsible for phosphorylating a different protein or set of proteins. There are also many different protein phosphatases; some of these are highly specific and remove phosphate groups from only one or a few proteins, whereas others act on a broad range of proteins and are targeted to specific substrates by regulatory subunits. The state of phosphorylation of a protein at any moment, and thus its activity, depends on the relative activities of the protein kinases and phosphatases that modify it.

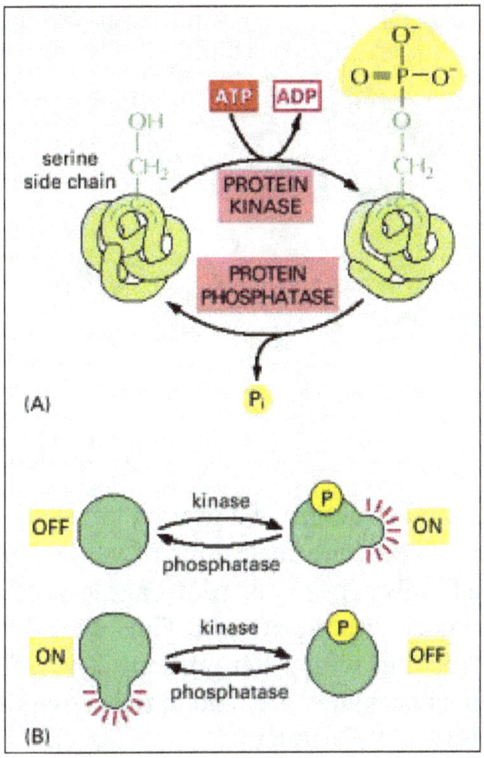

Figure shows protein phosphorylation. Many thousands of proteins in a typical eucaryotic cell are modified by the covalent addition of a phosphate group. (A) The general reaction, shown here, entails the transfer of a phosphate group from ATP to an amino acid side chain of the target protein by a protein kinase. Removal of the phosphate group is catalyzed by a second enzyme, a protein phosphatase. In this example, the phosphate is added to a serine side chain; in other cases, the phosphate is instead linked to the –OH group of a threonine or a tyrosine in the protein. (B) The phosphorylation of a protein by a protein kinase can either increase or decrease the protein's activity, depending on the site of phosphorylation and the structure of the protein.

The protein kinases that phosphorylate proteins in eucaryotic cells belong to a very large family of enzymes, which share a catalytic (kinase) sequence of 250 amino acids. The various family members contain different amino acid sequences on either side of the kinase sequence, and often have short amino acid sequences inserted into loops within it (red arrowheads in Figure). Some of these additional amino acid sequences enable each kinase to recognize the specific set of proteins it phosphorylates, or to bind to structures that localize it in specific regions of the cell. Other parts of the protein allow the activity of each enzyme to be tightly regulated, so it can be turned on and off in response to different specific signals.

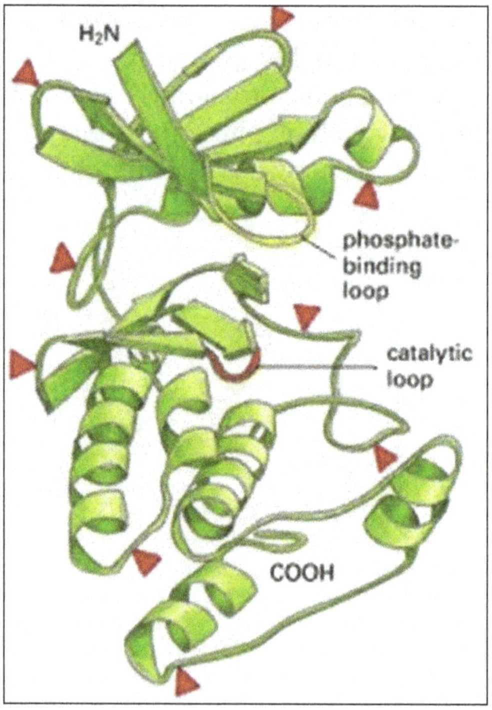

Figure shows the three-dimensional structure of a protein kinase. Superimposed on this structure are red arrowheads to indicate sites where insertions of 5–100 amino acids are found in some members of the protein kinase family. These insertions are located in loops on the surface of the enzyme where other ligands interact with the protein. Thus, they distinguish different kinases and confer on them distinctive interactions with other proteins. The ATP (which donates a phosphate group) and the peptide to be phosphorylated are held in the active site, which extends between the phosphate-binding loop (yellow) and the catalytic loop (orange).

By comparing the number of amino acid sequence differences between the various members of a protein family, one can construct an "evolutionary tree" that is thought to reflect the pattern of gene duplication and divergence that gave rise to the family. An evolutionary tree of protein kinases is shown in Figure. Not surprisingly, kinases with related functions are often located on nearby branches of the tree: the protein kinases involved in cell signaling that phosphorylate tyrosine side chains, for example, are all clustered in the top left corner of the tree. The other kinases shown phosphorylate either a serine or a threonine side chain, and many are organized into clusters that seem to reflect their function—in trans membrane signal transduction, intracellular signal amplification, cell-cycle control, and so on.

As a result of the combined activities of protein kinases and protein phosphatases, the phosphate groups on proteins are continually turning over—being added and then rapidly removed. Such phosphorylation cycles may seem wasteful, but they are important in allowing the phosphorylated proteins to switch rapidly from one state to another: the more rapid the cycle, the faster the state of phosphorylation of a population of protein molecules can change in response to a sudden stimulus that changes the phosphorylation rate. The energy required to drive the phosphorylation cycle is derived from the free energy of ATP hydrolysis, one molecule of which is consumed for each phosphorylation event.

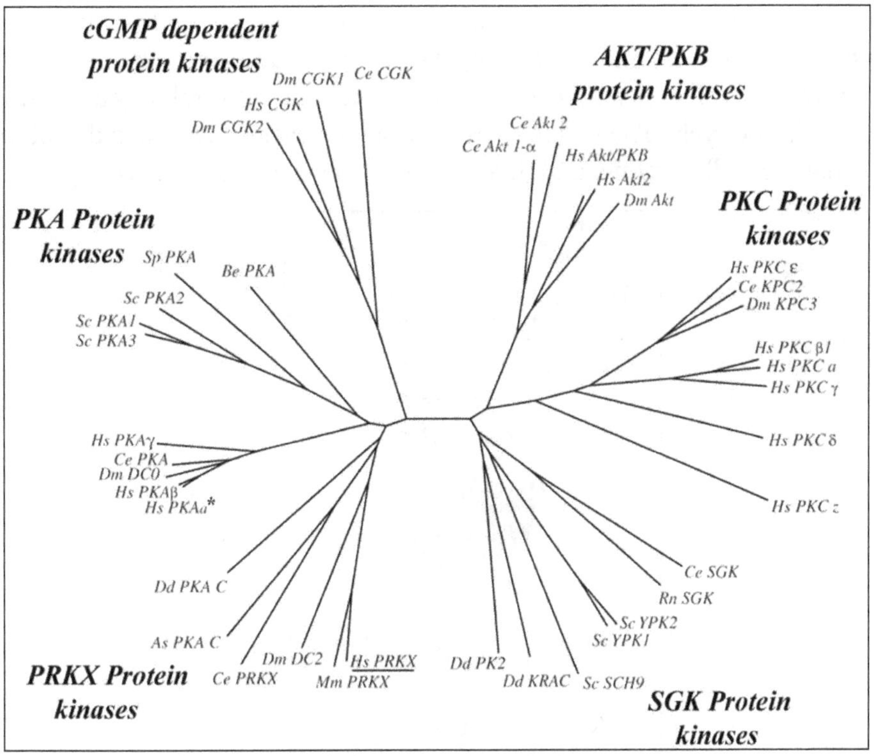

An evolutionary tree of selected protein kinases. Although a higher eucaryotic cell contains hundreds of such enzymes, and the human genome codes for more than 500.

The Regulation of CDK and SRC Protein Kinases Shows how a Protein can Function as a Microchip

The hundreds of different protein kinases in a eucaryotic cell are organized into complex networks of signaling pathways that help to coordinate the cell's activities, drive the cell cycle, and relay signals into the cell from the cell's environment. Many of the extracellular signals involved need to be both integrated and amplified by the cell. Individual protein kinases (and other signaling proteins) serve as input–output devices, or "microchips," in the integration process. An important part of the input to these proteins comes from the control that is exerted by phosphates added and removed from them by protein kinases and protein phosphatases, respectively, in the signaling network.

For a protein that is phosphorylated at multiple sites, specific sets of phosphate groups serve to activate the protein, while other sets can inactivate it. A cyclin-dependent protein kinase (Cdk) represents a good example of such a signal processing device. Kinases in this class phosphorylate serines, and they are central components of the cell-cycle control system in eucaryotic cells. In a vertebrate cell, individual Cdk enzymes turn on and off in succession, as a cell proceeds through the different phases of its division cycle. When a particular one of these kinases is on, it influences various aspects of cell behavior through effects on the proteins it phosphorylates.

A Cdk protein becomes active as a serine/threonine protein kinase only when it is bound to a second protein called a cyclin. But, as Figure shows, the binding of cyclin is only one of three distinct "inputs" required to activate the Cdk. In addition to cyclin binding, a phosphate must be added to a specific threonine side chain, and a phosphate elsewhere in the protein (covalently bound to a specific tyrosine side chain) must be removed. Cdk thus monitors a specific set of cell components—a cyclin, a protein kinase, and a protein phosphatase—and it acts as an input–output device that turns on if, and only if, each of these components has attained its appropriate activity state. Some cyclins rise and fall in concentration in step with the cell cycle, increasing gradually in amount until they are suddenly destroyed at a particular point in the cycle. The sudden destruction of a cyclin (by targeted proteolysis) immediately shuts off its partner Cdk enzyme, and this triggers a specific step in the cell cycle.

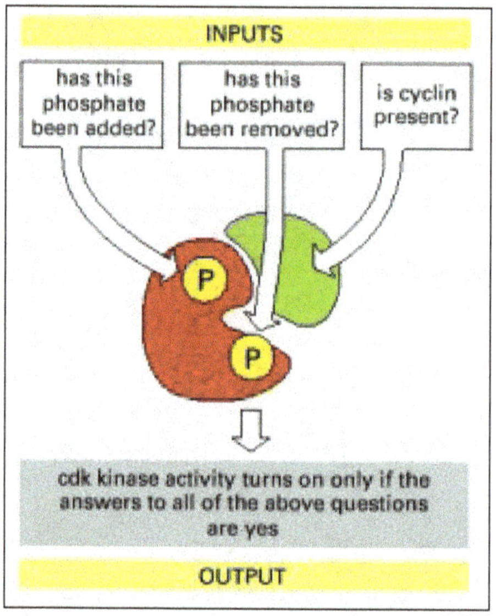

How a Cdk protein acts as an integrating device. The function of these central regulators of the cell cycle.

A similar type of microchip behavior is exhibited by the Src family of protein kinases. The Src protein (pronounced "sarc") was the first tyrosine kinase to be discovered, and it is now known to be part of a subfamily of nine very similar protein kinases, which are found only in multicellular animals. As indicated by the evolutionary tree in Figure, sequence comparisons suggest that tyrosine kinases as a group were a relatively late innovation that branched off from the serine/threonine kinases, with the Src subfamily being only one subgroup of the tyrosine kinases created in this way.

The Src protein and its homologs contain a short N-terminal region that becomes covalently linked to a strongly hydrophobic fatty acid, which holds the kinase at the cytoplasmic face of the plasma membrane. Next come two peptide-binding modules, a Src homology 3 (SH3) domains and a SH2 domain, followed by the kinase catalytic domains. These kinases normally exist in an inactive conformation, in which a phosphorylated tyrosine near the C-terminus is bound to the SH2 domain, and the SH3 domain is bound to an internal peptide in a way that distorts the active site of the enzyme and helps to render it inactive.

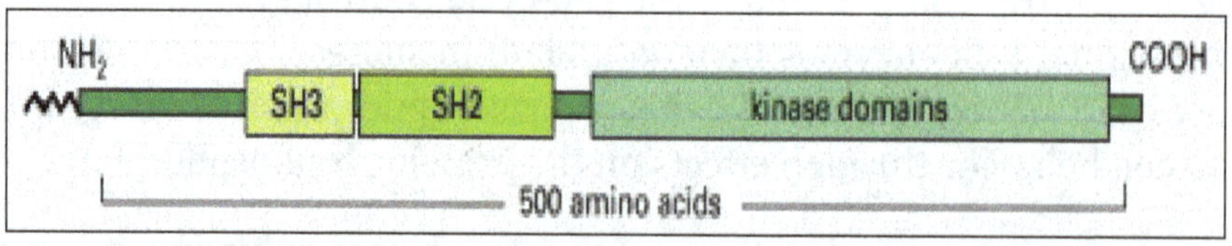

The domain structure of the Src family of protein kinases, mapped along the amino acid sequence.

Turning the kinase on involves at least two specific inputs: removal of the C-terminal phosphate and the binding of the SH3 domain by a specific activating protein. As for the Cdk protein, the activation of the Src kinase signals that a particular set of separate upstream events has been completed. Thus, both the Cdk and Src families of proteins serve as specific signal integrators, helping to generate the complex web of information-processing events that enable the cell to compute logical responses to a complex set of conditions.

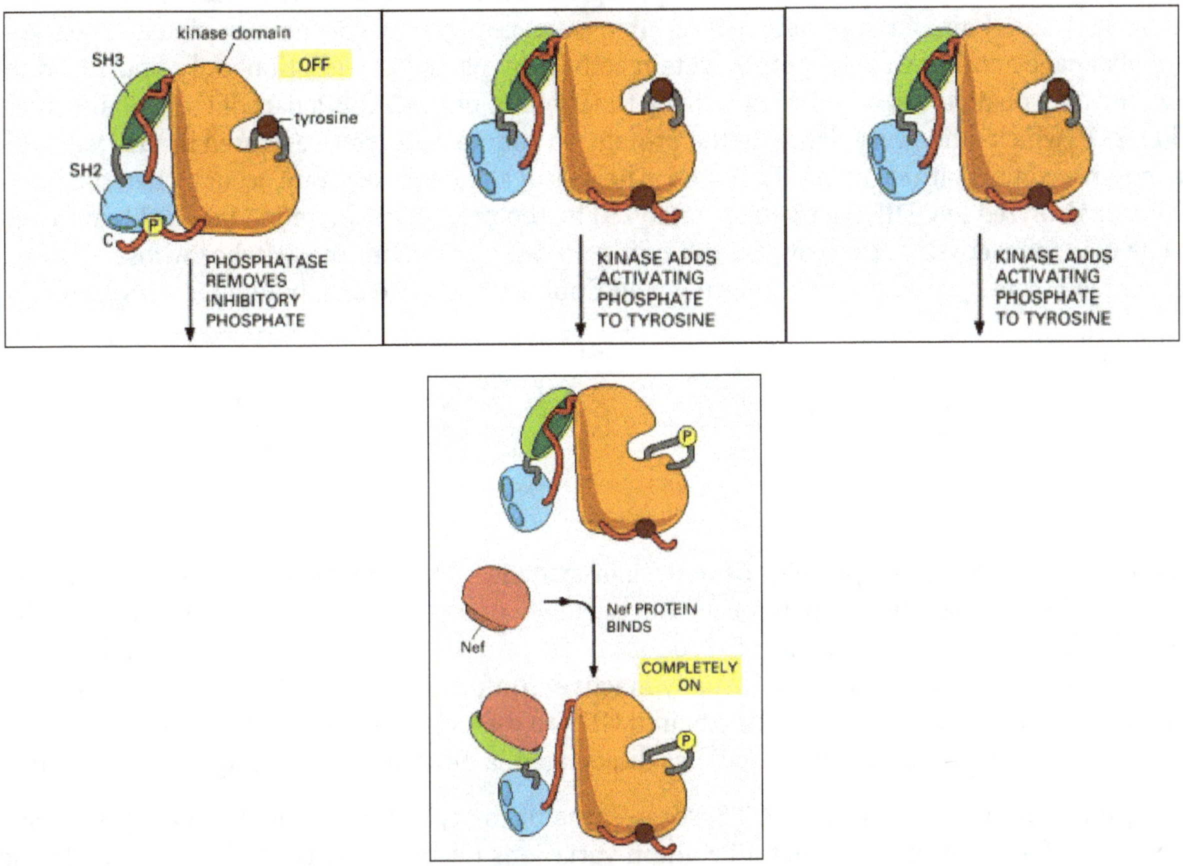

The activation of a Src-type protein kinase by two sequential events.

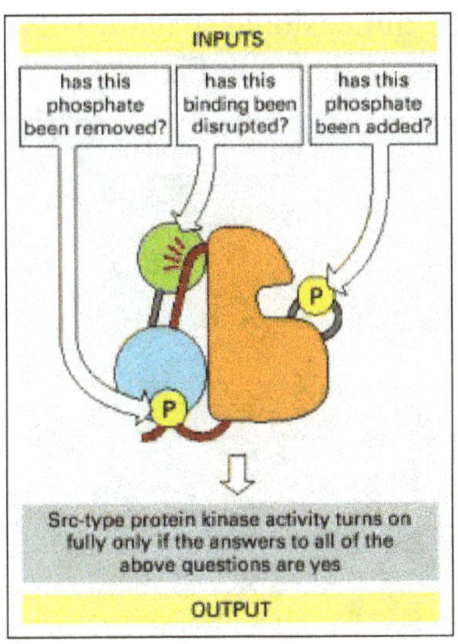

Figure shows how a Src-type protein kinase acts as an integrating device. The disruption of the SH3 domain interaction (green) can involve replacing its binding to the indicated red linker region with a tighter interaction with the Nef protein, as illustrated in Figure.

Proteins that Bind and Hydrolyze GTP are Ubiquitous Cellular Regulators

We have described how the addition or removal of phosphate groups on a protein can be used by a cell to control the protein's activity. The phosphate is transferred from an ATP molecule to an amino acid side chain of the protein in a reaction catalyzed by a specific protein kinase. Eucaryotic cells also have another way to control protein activity by phosphate addition and removal. In this case, the phosphate is not attached directly to the protein; instead, it is a part of the guanine nucleotide GTP, which binds very tightly to the protein. In general, proteins regulated in this way are in their active conformations with GTP bound. The loss of a phosphate group occurs when the bound GTP is hydrolyzed to GDP in a reaction catalyzed by the protein itself, and in its GDP bound state the protein is inactive. In this way, GTP-binding proteins act as on–off switches whose activity is determined by the presence or absence of an additional phosphate on a bound GDP molecule.

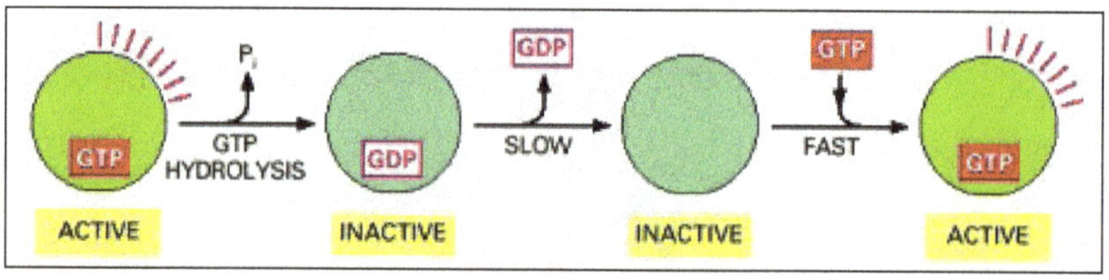

Figure shows GTP-binding proteins as molecular switches. The activity of a GTP-binding protein (also called a GTPase) generally requires the presence of a tightly bound GTP molecule (switch "on"). Hydrolysis of this GTP molecule produces GDP and inorganic phosphate (P_i), and it causes the protein to convert to a different, usually inactive, conformation (switch "off"). As shown here, resetting the switch requires the tightly bound GDP to dissociate, a slow step that is greatly accelerated by specific signals; once the GDP has dissociated, a molecule of GTP is quickly rebound.

GTP-binding proteins (also called GTPases because of the GTP hydrolysis they catalyze) constitute a large family of proteins that all contain variations on the same GTP-binding globular domain. When the tightly bound GTP is hydrolyzed to GDP, this domain undergoes a conformational change that inactivates it. The three-dimensional structure of a prototypical member of this family, the monomeric GTPase called Ras, is shown in Figure.

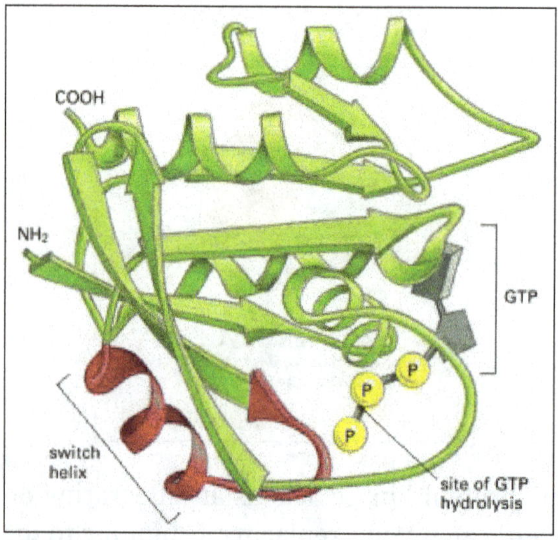

Figure shows the structure of the Ras protein in its GTP-bound form. This monomeric GTPase illustrates the structure of a GTP-binding domain, which is present in a large family of GTP-binding proteins. The red regions change their conformation when the GTP molecule is hydrolyzed to GDP and inorganic phosphate by the protein; the GDP remains bound to the protein, while the inorganic phosphate is released. The special role of the "switch helix" in proteins related to Ras is explained next.

The Ras protein has an important role in cell signaling. In its GTP-bound form, it is active and stimulates a cascade of protein phosphorylation in the cell. Most of the time, however, the protein is in its inactive, GDP-bound form. It becomes active when it exchanges its GDP for a GTP molecule in response to extracellular signals, such as growth factors, that bind to receptors in the plasma membrane.

Regulatory Proteins Control the Activity of GTP-Binding Proteins by Determining whether GTP or GDP is Bound. GTP-binding proteins are controlled by regulatory proteins that determine whether GTP or GDP is bound, just as phosphorylated proteins are turned on and off by protein kinases and protein phosphatases. Thus, Ras is inactivated by a GTPase-activating protein (GAP), which binds to the Ras protein and induces it to hydrolyze its bound GTP molecule to GDP —which remains tightly bound—and inorganic phosphate (P_i)—which is rapidly released. The Ras protein stays in its inactive, GDP-bound conformation until it encounters a guanine nucleotide exchange factor (GEF), which binds to GDP-Ras and causes it to release its GDP. Because the empty nucleotide-binding site is immediately filled by a GTP molecule (GTP is present in large excess over GDP in cells), the GEF activates Ras by indirectly adding back the phosphate removed by GTP hydrolysis. Thus, in a sense, the roles of GAP and GEF are analogous to those of a protein phosphatase and a protein kinase, respectively.

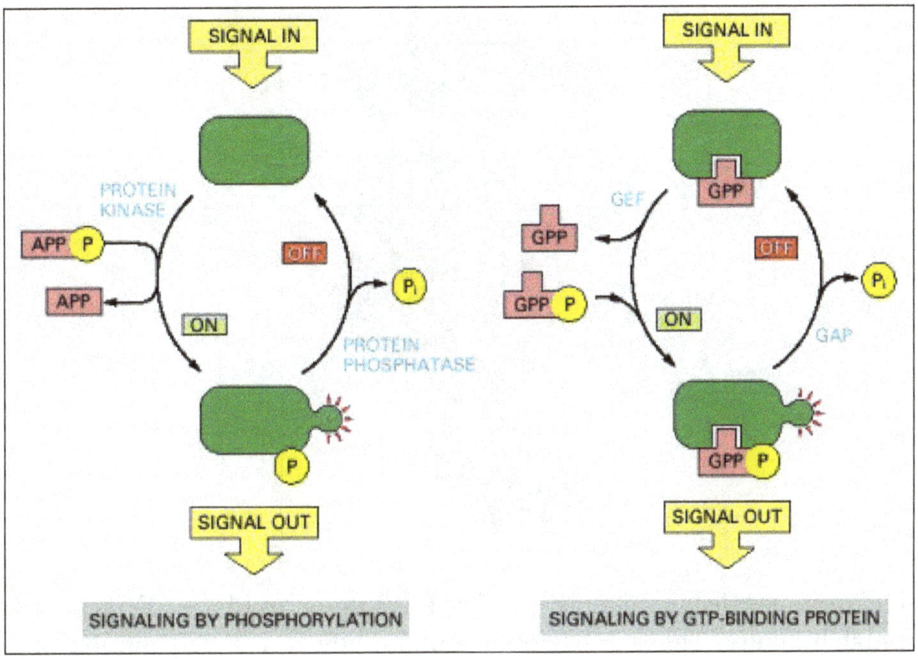

Figure shows a comparison of the two major intracellular signaling mechanisms in eucaryotic cells. In both cases, a signaling protein is activated by the addition of a phosphate group and inactivated by the removal of this phosphate. To emphasize the similarities in the two pathways,

ATP and GTP are drawn as APPP and GPPP, and ADP and GDP as APP and GPP, respectively. As shown in Figure, the addition of a phosphate to a protein can also be inhibitory.

Large Protein Movements can be Generated from Small Ones

The Ras protein belongs to a large superfamily of monomeric GTPases, each of which consists of a single GTP-binding domain of about 200 amino acids. Over the course of evolution, this domain has also become joined to larger proteins with additional domains, creating a large family of GTP-binding proteins. Family members include the receptor-associated trimeric G proteins involved in cell signaling, proteins regulating the traffic of vesicles between intracellular compartments, and proteins that bind to transfer RNA and are required as assembly factors for protein synthesis on the ribosome. In each case, an important biological activity is controlled by a change in the protein's conformation that is caused by GTP hydrolysis in a Ras-like domain.

The EF-Tu protein provides a good example of how this family of proteins works. EF-Tu is an abundant molecule that serves as an elongation factor (hence the EF) in protein synthesis, loading each amino-acyl tRNA molecule onto the ribosome. The tRNA molecule forms a tight complex with the GTP-bound form of EF-Tu. In this complex, the amino acid attached to the tRNA is masked. Unmasking is required for the tRNA to transfer its amino acid in protein synthesis, and it occurs on the ribosome when the GTP bound to EF-Tu is hydrolyzed, allowing the tRNA to dissociate. Since the GTP hydrolysis is triggered by a proper fit of the tRNA to the mRNA molecule on the ribosome, the EF-Tu serves as an assembly factor that discriminates between correct and incorrect mRNA–tRNA pairings.

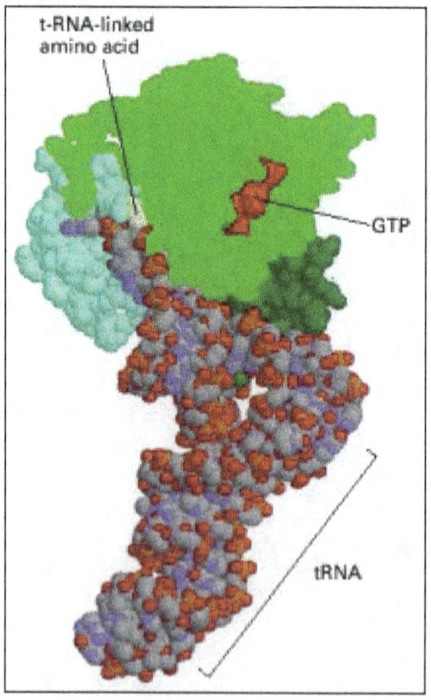

An aminoacyl tRNA molecule bound to EF-Tu: The three domains of the EF-Tu protein are colored differently, to match Figure. This is a bacterial protein; however, a very similar protein exists in eukaryotes, where it is called EF-1.

Comparison of the three-dimensional structure of EF-Tu in its GTP-bound and GDP-bound forms reveals how the unmasking of the tRNA occurs. The dissociation of the inorganic phosphate group

(P_i), which follows the reaction GTP → GDP + P_i, causes a shift of a few tenths of a nanometer at the GTP-binding site, just as it does in the Ras protein. This tiny movement, equivalent to a few times the diameter of a hydrogen atom, causes a conformational change to propagate along a crucial piece of α helix, called the switch helix, in the Ras-like domain of the protein. The switch helix seems to serve as a latch that adheres to a specific site in another domain of the molecule, holding the protein in a "shut" conformation. The conformational change triggered by GTP hydrolysis causes the switch helix to detach, allowing separate domains of the protein to swing apart, through a distance of about 4 nm. This releases the bound tRNA molecule, allowing its attached amino acid to be used.

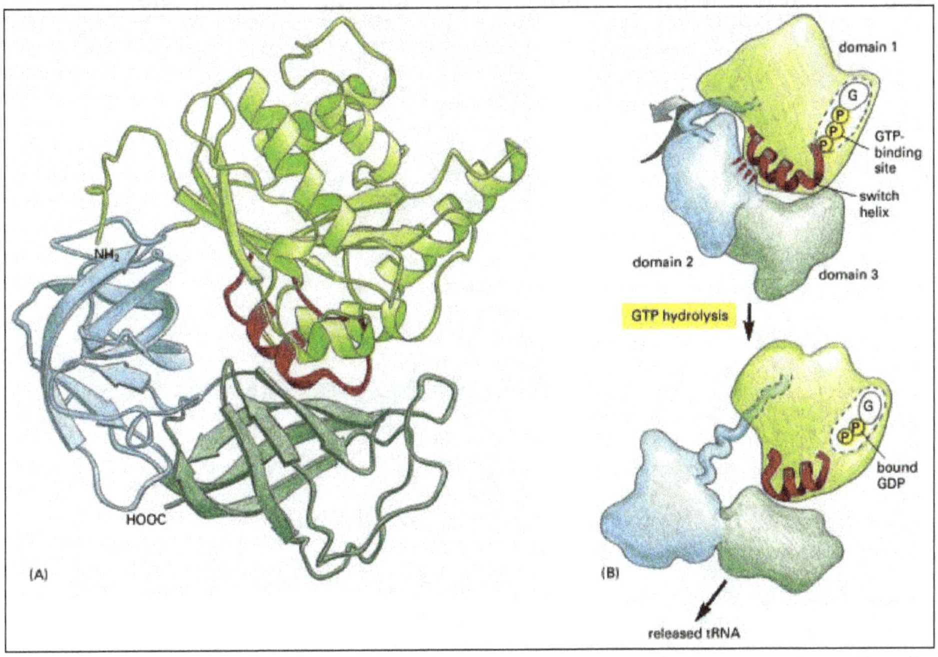

Figure shows the large conformational change in EF-Tu caused by GTP hydrolysis. (A) The three-dimensional structure of EF-Tu with GTP bound. The domain at the top is homologous to the Ras protein, and its red α helix is the switch helix, which moves after GTP hydrolysis, as shown in Figure. (B) The change in the conformation of the switch helix in domain 1 causes domains 2 and 3 to rotate as a single unit by about 90° toward the viewer, which releases the tRNA that was shown bound to this structure in Figure.

One can see from this example how cells have exploited a simple chemical change that occurs on the surface of a small protein domain to create a movement 50 times larger. Dramatic shape changes of this type also underlie the very large movements that occur in motor proteins.

Motor Proteins Produce Large Movements in Cells

These motor proteins generate the forces responsible for muscle contraction and the crawling and swimming of cells. Motor proteins also power smaller-scale intracellular movements: they help to move chromosomes to opposite ends of the cell during mitosis, to move organelles along molecular tracks within the cell, and to move enzymes along a DNA strand during the synthesis of a new DNA molecule. All these fundamental processes depend on proteins with moving parts that operate as force-generating machines.

How do these machines work? In other words, how are shape changes in proteins used to generate directed movements in cells? If, for example, a protein is required to walk along a narrow thread such as a DNA molecule, it can do this by undergoing a series of conformational changes, such as those shown in Figure. With nothing to drive these changes in an orderly sequence, however, they are perfectly reversible, and the protein can wander randomly back and forth along the thread. We can look at this situation in another way. Since the directional movement of a protein does work, the laws of thermodynamics demand that such movement utilize free energy from some other source (otherwise the protein could be used to make a perpetual motion machine). Therefore, without an input of energy, the protein molecule can only wander aimlessly.

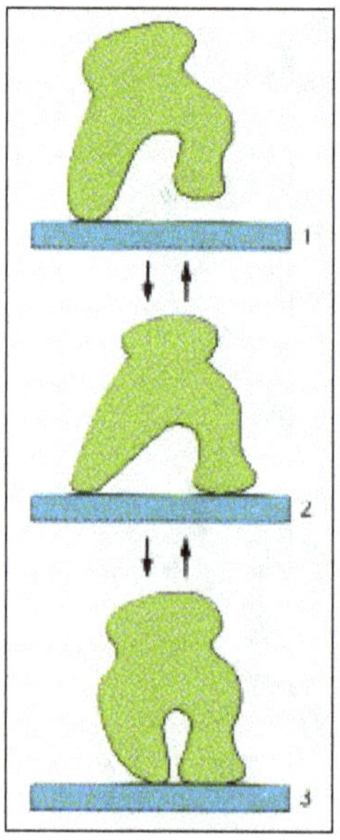

An allosteric "walking" protein. Although its three different conformations allow it to wander randomly back and forth while bound to a thread or a filament, the protein cannot move uniformly in a single direction.

How, then, can one make the series of conformational changes unidirectional? To force the entire cycle to proceed in one direction, it is enough to make any one of the changes in shape irreversible. For most proteins that are able to walk in one direction for long distances, this is achieved by coupling one of the conformational changes to the hydrolysis of an ATP molecule bound to the protein. The mechanism is similar to the one just discussed that drives allosteric protein shape changes by GTP hydrolysis. Because a great deal of free energy is released when ATP (or GTP) is hydrolyzed, it is very unlikely that the nucleotide-binding protein will undergo the reverse shape change needed for moving backward–since this would require that it also reverse the ATP hydrolysis by adding a phosphate molecule to ADP to form ATP.

In the model shown in Figure, ATP binding shifts a motor protein from conformation 1 to conformation 2. The bound ATP is then hydrolyzed to produce ADP and inorganic phosphate (Pi), causing a change from conformation 2 to conformation 3. Finally, the release of the bound ADP and P_i drives the protein back to conformation 1. Because the transition 2 $\rightarrow$ 3 is driven by the energy provided by ATP hydrolysis, this series of conformational changes is effectively irreversible. Thus, the entire cycle goes in only one direction, causing the protein molecule to walk continuously to the right in this example.

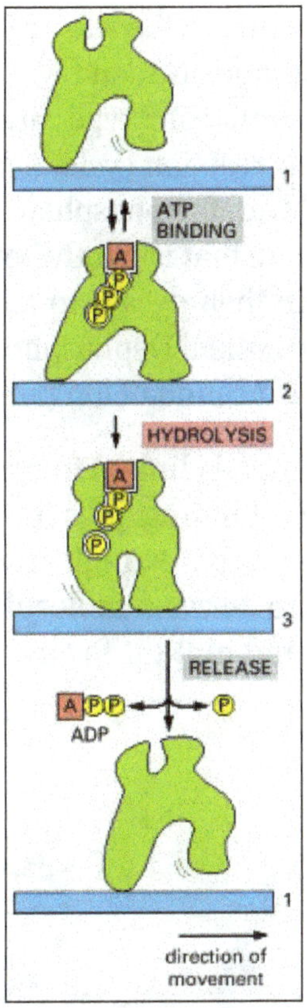

Figure shows an allosteric motor protein. The transition between three different conformations includes a step driven by the hydrolysis of a bound ATP molecule, and this makes the entire cycle essentially irreversible. By repeated cycles, the protein therefore moves continuously to the right along the thread.

Many motor proteins generate directional movement in this general way, including the muscle motor protein myosin, which walks along actin filaments to generate muscle contraction, and the kinesin proteins that walk along microtubules. These movements can be rapid: some of the motor proteins involved in DNA replication (the DNA helicases) propel themselves along a DNA strand at rates as high as 1000 nucleotides per second.

Membrane-bound Transporters Harness Energy to Pump Molecules through Membranes

We have thus far seen how allosteric proteins can act as microchips (Cdk and Src kinases), as assembly factors (EF-Tu), and as generators of mechanical force and motion (motor proteins). Allosteric proteins can also harness energy derived from ATP hydrolysis, ion gradients, or electron transport processes to pump specific ions or small molecules across a membrane.

One of the best understood pump proteins is the calcium transport protein from muscle cells. This protein, called the Ca^{2+} pump, is embedded in the membrane of a specialized organelle in muscle cells called the sarcoplasmic reticulum. The Ca^{2+} pump (also known as the Ca^{2+} ATPase) maintains the low cytoplasmic calcium concentration of resting muscle by pumping calcium out of the cytosol into the membrane-enclosed sarcoplasmic reticulum; then, in response to a nerve impulse, Ca^{2+} is rapidly released (through other channels) back into the cytosol to trigger muscle contraction. The Ca^{2+} pump is homologous to the Na^{+}-K^{+} pump that maintains Na^{+} and K^{+} concentration differences across the plasma membrane, both being members of a family of P-type cation pumps—so named because these proteins are auto phosphorylated during their reaction cycle. A large, cytoplasmic head region binds and hydrolyzes ATP, forming a covalent bond with the phosphate released. The protein thereby transiently shifts to a high-energy, phosphorylated state that is tightly bound to two Ca^{2+} ions that were picked up from the cytosol. This form of the protein then decays to a low-energy, phosphorylated state, which causes the Ca^{2+} ions to be released into the lumen. Dephosphorylating of the enzyme finally releases the phosphate and resets the protein for its next round of ion pumping.

The head region of all P-type cation pumps is linked to a series of ten trans membrane α helices, four of which form trans membrane Ca^{2+}-binding sites for the Ca^{2+} pump. The three-dimensional structure of this protein has been deciphered by high-resolution electron microscopy and x-ray diffraction. This has enabled biologists to derive a molecular model for pump action based on extensive biochemical data on its normal and mutant forms.

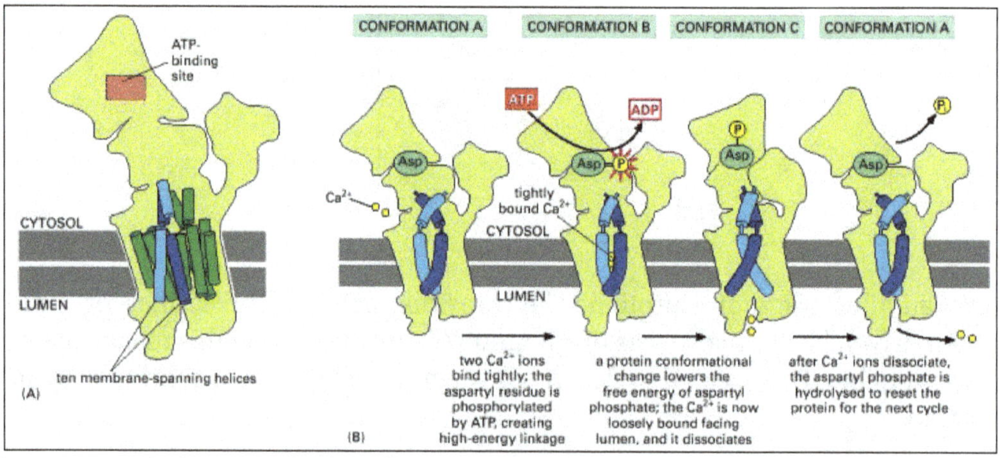

The transport of calcium ions by the Ca^{2+} pump. (A) The structure of the pump protein, formed from a single subunit of 994 amino acids. (B) A model for ion pumping. For simplicity, only two of the four transmembrane α helices that bind Ca^{2+} are shown.

As illustrated in Figure, the Trans membrane α helices that bind the Ca^{2+} wind around each other and create a cavity for Ca^{2+} between the helices. ATP hydrolysis generates a series of major conformational changes in the cytoplasmic head, which—through its stalk connection to the trans membrane helices—changes the structure and relative orientations of some of the helices in the membrane, thereby altering the cavity in a way that pushes the Ca^{2+} ions unidirectional across the membrane. As for a motor protein, unidirectional transport of an ion requires the cycle to use energy, so as to impart a preferred directionality to the protein's conformational changes.

Humans have invented many different types of mechanical pumps, and it should not be surprising that cells also contain membrane-bound pumps that function in other ways. Most notable are the rotary pumps that couple the hydrolysis of ATP to the transport of $H^{\pm}$ ions (protons). These pumps resemble miniature turbines, and they are used to acidify the interior of lysosomes and other eucaryotic organelles. Like other ion pumps that create ion gradients, they can function in reverse to catalyze the reaction ADP + Pi → ATP if there is a steep enough gradient across their membrane of the ion that they transport. One such pump, the ATP synthase, harnesses a gradient of proton concentration produced by electron transport processes to produce most of the ATP used in the living world.

Proteins often Form Large Complexes that Function as Protein Machines

As one progress from small, single-domain proteins to large proteins formed from many domains, the functions the proteins can perform become more elaborate. The most impressive tasks, however, are carried out by large protein assemblies formed from many protein molecules. Now that it is possible to reconstruct most biological processes in cell-free systems in the laboratory, it is clear that each of the central processes in a cell—such as DNA replication, protein synthesis, vesicle budding, or transmembrane signaling—is catalyzed by a highly coordinated, linked set of ten or more proteins. In most such protein machines, the hydrolysis of bound nucleoside triphosphates (ATP or GTP) drives an ordered series of conformational changes in some of the individual protein subunits, enabling the ensemble of proteins to move coordinately. In this way, each of the appropriate enzymes can be placed directly into the positions where they are needed to perform successive reactions in a series. This is what occurs, for example, in protein synthesis on a ribosome —or in DNA replication, where a large multiprotein complex moves rapidly along the DNA.

Cells have evolved protein machines for the same reason that humans have invented mechanical and electronic machines. For accomplishing almost any task, manipulations that are spatially and temporally coordinated through linked processes are much more efficient than the sequential use of individual tools.

A Complex Network of Protein Interactions Underlies Cell Function

There are many challenges facing cell biologists in this "post-genome" era when complete genome sequences are known. One is the need to dissect and reconstruct each one of the thousands of protein machines that exist in an organism such as us. To understand these remarkable protein complexes, each must be reconstituted from its purified protein parts—so that its detailed mode of operation can be studied under controlled conditions in a test tube, free from all other cell components. This alone is a massive task. But we now know that each of these subcomponents of a cell also interacts with other sets of macromolecules, creating a large network of protein–protein and protein–nucleic acid interactions throughout the cell. To understand the cell, therefore, one will need to analyze most of these other interactions as well.

Some idea of the complexity of intracellular protein networks can be gained from a particularly well-studied example: the many dozens of proteins that interact with the actin cytoskeleton in the yeast Saccharomyces cerevisiae.

The extent of such protein–protein interactions can also be estimated more generally. In particular, large-scale efforts have been undertaken to detect these interactions using the two-hybrid screening method. Thus, for example, this method is being applied to a large set of the 6000 gene products produced by S. cerevisiae. As expected, the majority of the interactions that have been observed are between proteins in the same functional group. That is, proteins involved in cell-cycle control tend to interact extensively with each other, as do proteins involved with DNA synthesis, or DNA repair, and so on. But there are also a surprisingly large number of interactions between the protein members of different functional groups. These interactions are presumably important for coordinating cell functions, but most of them are not understood.

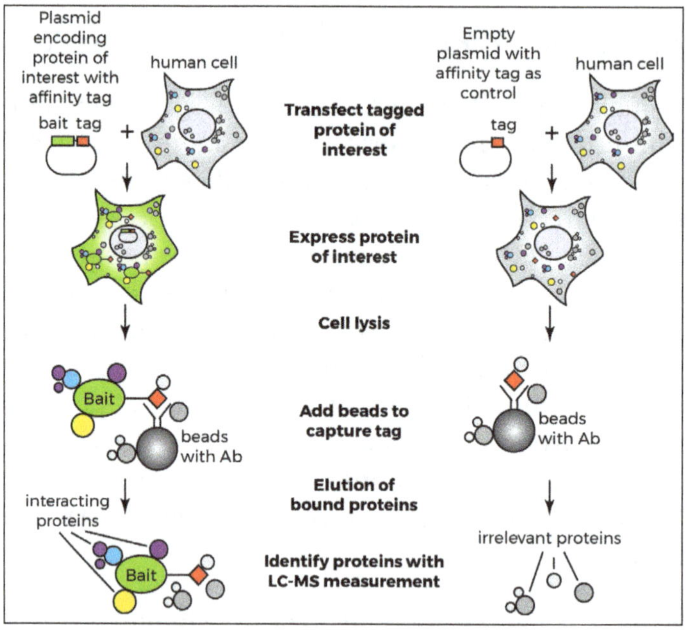

The protein–protein interactions observed between different functional groups of proteins in the yeast S. cerevisiae. To produce this map, more than 1500 proteins were assigned to the indicated 31 functional groups. About 70 percent of the known protein–protein interactions were observed to occur between proteins assigned to the same functional group. However, as indicated, a surprisingly large number of interactions crossed between these groups. Each line in this diagram designates that more than 15 protein–protein interactions were observed between proteins in the two functional groups that are connected by that line. The three functional groups highlighted in yellow display at least 15 interactions with 10 or more of the 31 functional groups defined in the study.

An examination of the range of available data suggests that an average protein in a human cell may interact with somewhere between 5 and 15 different partners. Often, a different set of partners will be bound by each of the different domains in a multi domain protein; in fact, one can speculate that the unusually extensive multi domain structures observed for human proteins may have evolved to generate these interactions. Given the enormous complexity of the interacting networks of macromolecules in cells, deciphering their full functional meaning may well keep scientists busy for centuries.

Protein Metabolism

Protein metabolism is the chemical cycle of breaking down protein (catabolism) and using the components to synthesizing (anabolism) new molecules to be used in the body. The process is also known as proteometabolism.

Proteins, fats, and carbohydrates (called macronutrients) are part of a complex metabolic cycle that is essential to life. During digestion food containing these nutrients is chemically broken down into its basic components and absorbed for use in the body. Protein molecules are split into their basic building blocks, called amino acids, which are then chemically re-arranged to synthesize new proteins that the body needs. Fats are broken down into fatty acids and cholesterol, and carbohydrates are split into simple sugars such as glucose and fructose, which provide most of the energy to drive chemical reactions in the body. These smaller, simpler molecules are absorbed in the small intestine and enter the circulatory system. They then pass through the liver where some of these "building block molecules" are synthesized into more complex compounds needed by the body.

In order for cellular metabolism to occur, two reactions happen continuously. Small molecules are build up into larger molecules, a process called anabolism or constructive metabolism, while large molecules are broken down into their component parts during a process called catabolism or destructive metabolism. This building up and breaking down is regulated by a complex set of hormones and enzymes, themselves proteins. For an individual to remain healthy, the processes of anabolism and catabolism must remain in balance.

Catabolism or the breakdown of nutrients obtained from food, releases energy that drives all metabolic activities in the body. For example, glucose is broken down to provide energy for cellular respiration that allows functions such as muscle movement. Proteins are broken into amino acids then re-synthesized into hormones and enzymes to regulate chemical reactions in the cell, and molecules used for tissue growth and repair. Carbohydrates and fats are the preferred sources of energy for cellular metabolism. When the supply of fats and carbohydrates is insufficient to meet the body' needs, proteins can be broken down to supply energy. This accounts for the loss of muscle seen in prolonged cases of starvation.

Protein is a nitrogen-containing compound found in all plants and animals. There is a continuous need for protein to make hormones, enzymes, antibodies, and to produce new tissue (growth) and repair damaged tissues (maintenance). About 75% of human body tissue is made of protein. From 1-2% of the body's total protein is broken down each day into amino acid and recycled into new proteins. About60-70% of the amino acids the body needs come from this recycling process. The rest must be supplied by diet.

A Complex Molecule

Proteins are complex nitrogen-containing molecules formed by a combination of about 20 amino acids. These twenty amino acids can be connected in thousands of different combinations to form all the different proteins in the body. During protein formation (anabolism) the amino acids are connected in long chains called polypeptides that fold into three-dimensional shapes. The combinations of amino acids produce proteins with unique shapes that perform specific functions in the body such as catalyzing metabolic reactions, repairing tissue, or stimulating glands to produce other proteins.

The body can synthesize 11 amino acids naturally. The remaining nine must be obtained from dietary protein. The nine amino acids that cannot be synthesized by the body are called essential amino acids. It is an absolute requirement for health that diet include foods that contain an adequate amount of these essential amino acids. Foods that contain all the essential amino acids are called complete proteins. Examples of complete proteins are all meats and fish, eggs, and dairy products. Foods that lack one or more essential amino acid are called incomplete proteins. Most plant proteins are incomplete proteins. Dried beans, lentils, and rice are examples of incomplete proteins. Although some excess amino acids are converted into fat and stored as a future energy source, the body cannot not store amino acids for use in making future proteins. As a result, humans need a steady supply of protein in their diets. Individuals whose diet is completely or primarily vegetarian must make sure to eat combinations of foods that provide all the essential amino acids, for example red beans and rice or corn tortillas and beans.

Digestion of Protein

The digestion of proteins begins in the stomach. When protein-rich foods enter the stomach, they are greeted by a mixture of the enzyme pepsin and hydrochloric acid (HCl; 0.5 percent). The latter produces an environmental pH of 1.5–3.5 that denatures proteins within food. Pepsin cuts proteins into smaller polypeptides and their constituent amino acids. When the food-gastric juice mixture (chyme) enters the small intestine, the pancreas releases sodium bicarbonate to neutralize the HCl. This helps to protect the lining of the intestine. The small intestine also releases digestive hormones, including secretin and CCK, which stimulate digestive processes to break down the proteins further. Secretin also stimulates the pancreas to release sodium bicarbonate. The pancreas releases most of the digestive enzymes, including the proteases trypsin, chymotrypsin, and elastase, which aid protein digestion. Together, all of these enzymes break complex proteins into smaller individual amino acids, which are then transported across the intestinal mucosa to be used to create new proteins, or to be converted into fats or acetyl CoA and used in the Krebs cycle.

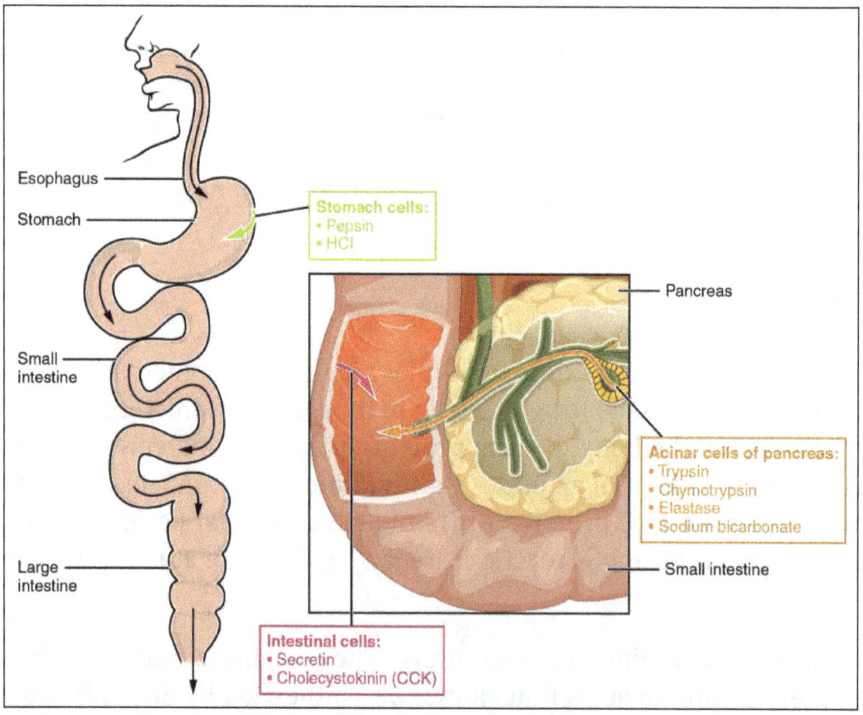

Enzymes in the stomach and small intestine break down proteins into amino acids. HCl in the stomach aids in proteolysis, and hormones secreted by intestinal cells direct the digestive processes.

In order to avoid breaking down the proteins that make up the pancreas and small intestine, pancreatic enzymes are released as inactive proenzymes that are only activated in the small intestine. In the pancreas, vesicles store trypsin and chymotrypsin as trypsinogen and chymotrypsinogen. Once released into the small intestine, an enzyme found in the wall of the small intestine, called enterokinase, binds to trypsinogen and converts it into its active form, trypsin. Trypsin then binds to chymotrypsinogen to convert it into the active chymotrypsin. Trypsin and chymotrypsin break down large proteins into smaller peptides, a process called proteolysis. These smaller peptides are catabolized into their constituent amino acids, which are transported across the apical surface of the intestinal mucosa in a process that is mediated by sodium-amino acid transporters. These transporters bind sodium and then bind the amino acid to transport it across the membrane. At the basal surface of the mucosal cells, the sodium and amino acid are released. The sodium can be reused in the transporter, whereas the amino acids are transferred into the bloodstream to be transported to the liver and cells throughout the body for protein synthesis.

Freely available amino acids are used to create proteins. If amino acids exist in excess, the body has no capacity or mechanism for their storage; thus, they are converted into glucose or ketones, or they are decomposed. Amino acid decomposition results in hydrocarbons and nitrogenous waste. However, high concentrations of nitrogen are toxic. The urea cycle processes nitrogen and facilitates its excretion from the body.

Urea Cycle

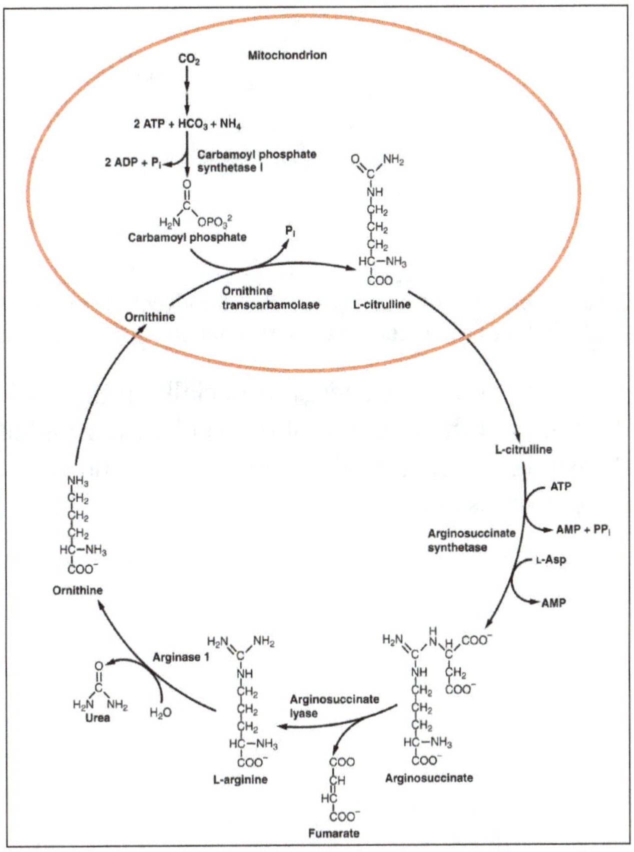

Nitrogen is transaminated, creating ammonia and intermediates of the Krebs cycle. Ammonia is processed in the urea cycle to produce urea that is eliminated through the kidneys.

The urea cycle is a set of biochemical reactions that produces urea from ammonium ions in order to prevent a toxic level of ammonium in the body. It occurs primarily in the liver and, to a lesser extent, in the kidney. Prior to the urea cycle, ammonium ions are produced from the breakdown of amino acids. In these reactions, an amine group, or ammonium ion, from the amino acid is exchanged with a keto group on another molecule. This transamination event creates a molecule that is necessary for the Krebs cycle and an ammonium ion that enters into the urea cycle to be eliminated. In the urea cycle, ammonium is combined with CO_2, resulting in urea and water. The urea is eliminated through the kidneys in the urine.

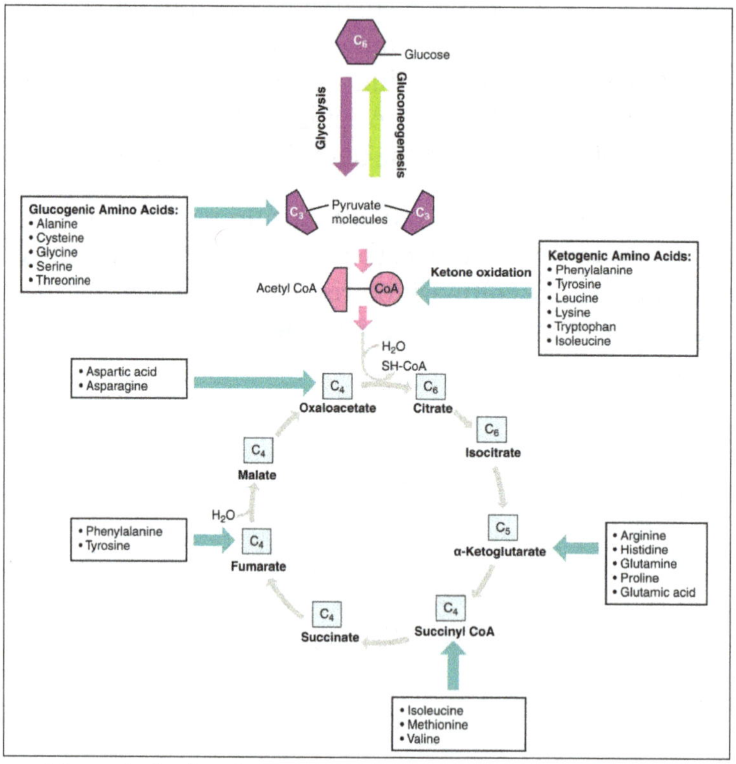

Amino acids can be broken down into precursors for glycolysis or the Krebs cycle. Amino acids (in bold) can enter the cycle through more than one pathway.

Amino acids can also be used as a source of energy, especially in times of starvation. Because the processing of amino acids results in the creation of metabolic intermediates, including pyruvate, acetyl CoA, acetoacyl CoA, oxaloacetate, and α-ketoglutarate, amino acids can serve as a source of energy production through the Krebs cycle.

4

A Comprehensive Study of Lipids

Chemically, a lipid is a macromolecule which is insoluble in water and is soluble in non-polar solvents such as chloroform, ether and alcohol. This chapter has been carefully written to provide an easy understanding of the varied facets of lipids such as their types, functions and metabolism.

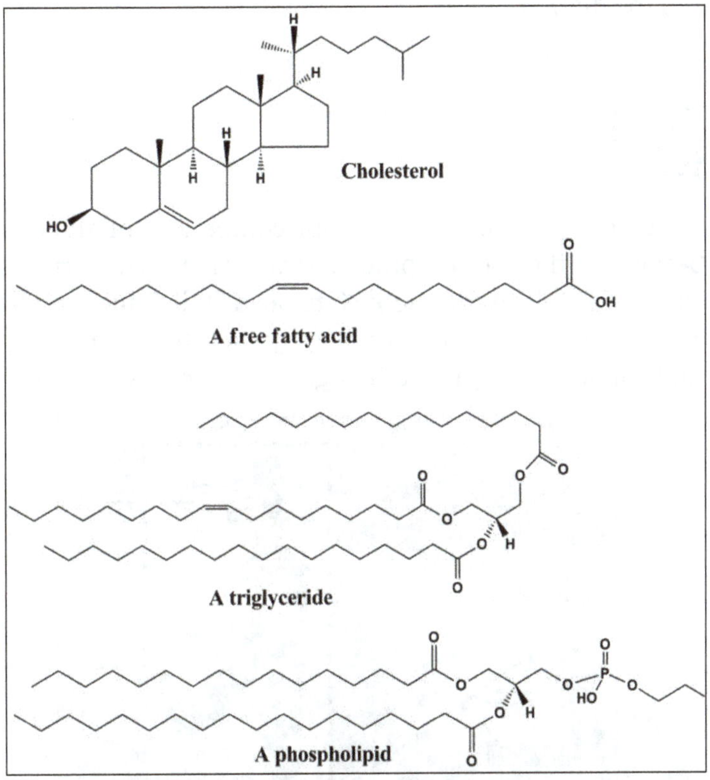

Structures of some common lipids. At the top are cholesterol and oleic acid. The middle structure is a triglyceride composed of oleoyl, stearoyl, and palmitoyl chains attached to a glycerol backbone. At the bottom is the common phospholipid phosphatidylcholine.

In biology, a lipid is a substance of biological origin that is soluble in nonpolar solvents. It comprises a group of naturally occurring molecules that include fats, waxes, sterols, fat-soluble vitamins (such as vitamins A, D, E, and K), monoglycerides, diglycerides, triglycerides, and phospholipids. The main biological functions of lipids include storing energy, signaling, and acting as structural components of cell membranes. Lipids have applications in the cosmetic and food industries as well as in nanotechnology.

Scientists sometimes broadly define lipids as hydrophobic or amphiphilic small molecules; the amphiphilic nature of some lipids allows them to form structures such as vesicles, multilamellar/unilamellar liposomes, or membranes in an aqueous environment. Biological lipids originate entirely or in part from two distinct types of biochemical subunits or "building-blocks": ketoacyl and isoprene groups. Using this approach, lipids may be divided into eight categories: fatty acids, glycerolipids, glycerophospholipids, sphingolipids, saccharolipids, and polyketides (derived from condensation of ketoacyl subunits); and sterol lipids and prenol lipids (derived from condensation of isoprene subunits).

Although the term "lipid" is sometimes used as a synonym for fats, fats are a subgroup of lipids called triglycerides. Lipids also encompass molecules such as fatty acids and their derivatives (including tri-, di-, monoglycerides, and phospholipids), as well as other sterol-containing metabolites such as cholesterol. Although humans and other mammals use various biosynthetic pathways both to break down and to synthesize lipids, some essential lipids cannot be made this way and must be obtained from the diet.

Phospholipid

Phospholipids are a class of lipids that are a major component of all cell membranes. They can form lipid bilayers because of their amphiphilic characteristic. The structure of the phospholipid molecule generally consists of two hydrophobic fatty acid "tails" and a hydrophilic "head" consisting of a phosphate group. The two components are joined together by a glycerol molecule. The phosphate groups can be modified with simple organic molecules such as choline.

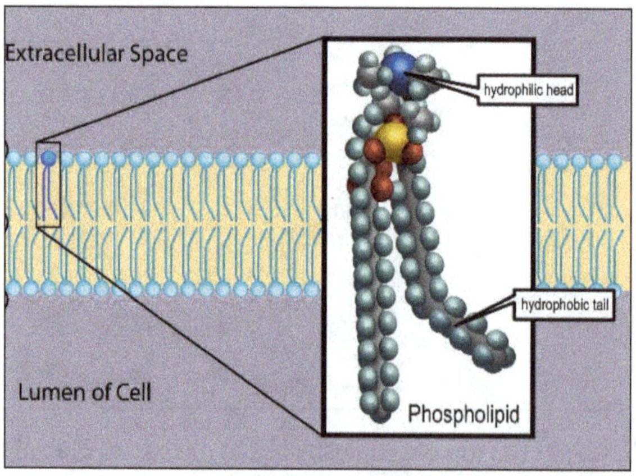

Phospholipid.

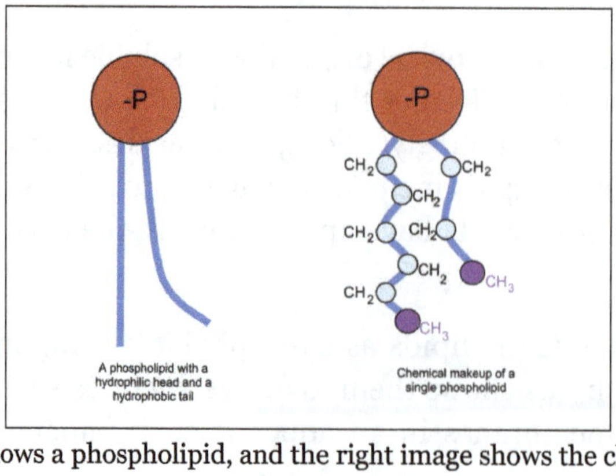

The left image shows a phospholipid, and the right image shows the chemical makeup.

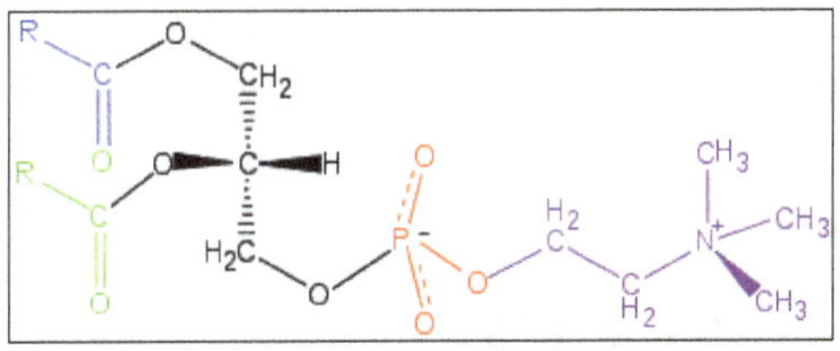

Phosphatidyl choline is the major component of lecithin. It is also a source for choline in the synthesis of acetylcholine in cholinergic neurons.

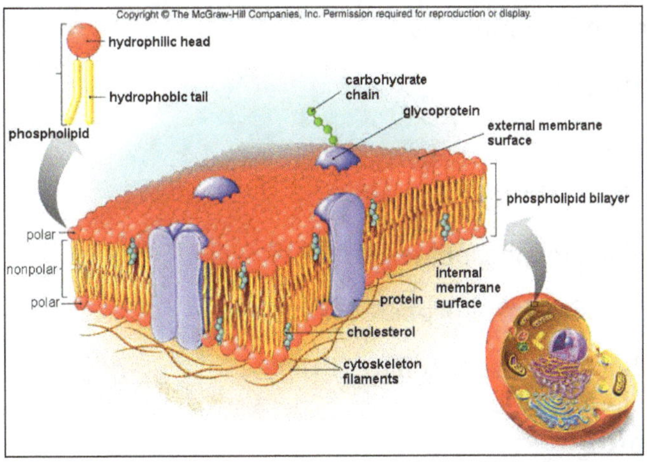

Cell membranes consist of phospholipid bilayers.

The first phospholipid identified in 1847 as such in biological tissues was lecithin, or phosphatidylcholine, in the egg yolk of chickens by the French chemist and pharmacist, Theodore Nicolas Gobley. Biological membranes in eukaryotes also contain another class of lipid, sterol, interspersed among the phospholipids and together they provide membrane fluidity and mechanical strength. Purified phospholipids are produced commercially and have found applications in nanotechnology and materials science.

Amphiphilic Character

An amphiphile is a term describing a chemical compound possessing both hydrophilic (water-loving, polar) and lipophilic (fat-loving) properties. The phospholipid head contains a negatively charged phosphate group and glycerol; it is hydrophilic (attracted to water). The phospholipid tails usually consist of 2 long fatty acid chains; they are hydrophobic and avoid interactions with water. When placed in aqueous solutions, phospholipids are driven by hydrophobic interactions that result in the fatty acid tails aggregating to minimize interactions with water molecules. These specific properties allow phospholipids to play an important role in the phospholipid bilayer. In biological systems, the phospholipids often occur with other molecules (e.g., proteins, glycolipids, sterols) in a bilayer such as a cell membrane. Lipid bilayers occur when hydrophobic tails line up against one another, forming a membrane of hydrophilic heads on both sides facing the water.

Such movement can be described by the fluid mosaic model, that describes the membrane as a mosaic of lipid molecules that act as a solvent for all the substances and proteins within it, so proteins and lipid molecules are then free to diffuse laterally through the lipid matrix and migrate over the membrane. Sterols contribute to membrane fluidity by hindering the packing together of phospholipids. However, this model has now been superseded, as through the study of lipid polymorphism it is now known that the behaviour of lipids under physiological (and other) conditions is not simple.

Diacylglyceride Structures

- Phosphatidic acid (phosphatidate) (PA)
- Phosphatidylethanolamine (cephalin) (PE)
- Phosphatidylcholine (lecithin) (PC)
- Phosphatidylserine (PS)
- Phosphoinositides:
    - Phosphatidylinositol (PI)
    - Phosphatidylinositol phosphate (PIP)
    - Phosphatidylinositol bisphosphate (PIP2) and
    - Phosphatidylinositol trisphosphate (PIP3)

Phosphosphingolipids

- Ceramide phosphorylcholine (Sphingomyelin) (SPH)
- Ceramide phosphorylethanolamine (Sphingomyelin) (Cer-PE)
- Ceramide phosphoryllipid

Applications

Phospholipids have been widely used to prepare liposomal, ethosomal and other nanoformulations of topical, oral and parenteral drugs for differing reasons like improved bio-availability, reduced toxicity and increased permeability across membranes. Liposomes are often composed of phosphatidylcholine-enriched phospholipids and may also contain mixed Phospholipid chains with surfactant properties. The ethosomal formulation of ketoconazole using phospholipids is a promising option for transdermal delivery in fungal infections.

Simulations

Computational simulations of phospholipids are often performed using molecular dynamics with force fields such as GROMOS, CHARMM, or AMBER.

Characterization

Phospholipids are optically highly birefringent, i.e. their refractive index is different along their axis as opposed to perpendicular to it. Measurement of birefringence can be achieved using cross polarisers in a microscope to obtain an image of e.g. vesicle walls or using techniques such as dual polarisation interferometry to quantify lipid order or disruption in supported bilayers.

Analysis

There are no simple methods available for analysis of phospholipids since the close range of polarity between different phospholipid species makes detection difficult. Oil chemists often use spectroscopy to determine total Phosphorus abundance and then calculate approximate mass of phospholipids based on molecular weight of expected fatty acid species. Modern lipid profiling employs more absolute methods of analysis, with nuclear magnetic resonance spectroscopy (NMR), particularly ^{31}P-NMR, while HPLC-ELSD provides relative values.

Phospholipid Synthesis

Phospholipid synthesis occurs in the cytosole adjacent to ER membrane that is studded with proteins that act in synthesis (GPAT and LPAAT acyl transferases, phosphatase and choline phosphotransferase) and allocation (flippase and floppase). Eventually a vesicle will bud off from the ER containing phospholipids destined for the cytoplasmic cellular membrane on its exterior leaflet and phospholipids destined for the exoplasmic cellular membrane on its inner leaflet.

Sources

Common sources of industrially produced phospholipids are soya, rapeseed, sunflower, chicken eggs, bovine milk, fish eggs etc. Each source has a unique profile of individual phospholipid species and consequently differing applications in food, nutrition, pharmaceuticals, cosmetics and drug delivery.

Market

The key companies operating in the global Phospholipid market include Avanti Polar Lipids, Lipoid GmbH, VAV Life Sciences Pvt. Ltd. and NOF Corporation.

In Signal Transduction

Some types of phospholipid can be split to produce products that function as second messengers in signal transduction. Examples include phosphatidylinositol (4,5)-bisphosphate (PIP_2), that can be split by the enzyme Phospholipase C into inositol triphosphate (IP_3) and diacylglycerol (DAG), which both carry out the functions of the G_q type of G protein in response to various stimuli and intervene in various processes from long term depression in neurons to leukocyte signal pathways started by chemokine receptors.

Phospholipids also intervene in prostaglandin signal pathways as the raw material used by lipase enzymes to produce the prostaglandin precursors. In plants they serve as the raw material to produce Jasmonic acid, a plant hormone similar in structure to prostaglandins that mediates defensive responses against pathogens.

Food Technology

Phospholipids can act as emulsifiers, enabling oils to form a colloid with water. Phospholipids are one of the components of lecithin which is found in egg-yolks, as well as being extracted from soy beans, and is used as a food additive in many products, and can be purchased as a dietary supplement. Lysolecithins are typically used for water-oil emulsions like margarine, due to their higher HLB ratio.

Phospholipid Derivatives

- Natural phospholipid derivates:

  egg PC (Egg lecithin), egg PG, soy PC, hydrogenated soy PC, sphingomyelin as natural phospholipids.

- Synthetic phospholipid derivates:
  - Phosphatidic acid (DMPA, DPPA, DSPA).
  - Phosphatidylcholine (DDPC, DLPC, DMPC, DPPC, DSPC, DOPC, POPC, DEPC).
  - Phosphatidylglycerol (DMPG, DPPG, DSPG, POPG).
  - Phosphatidylethanolamine (DMPE, DPPE, DSPE DOPE).
  - Phosphatidylserine (DOPS).
  - PEG phospholipid (mPEG-phospholipid, polyglycerin-phospholipid, funcitionalized-phospholipid, terminal activated-phospholipid).

Cardiolipin

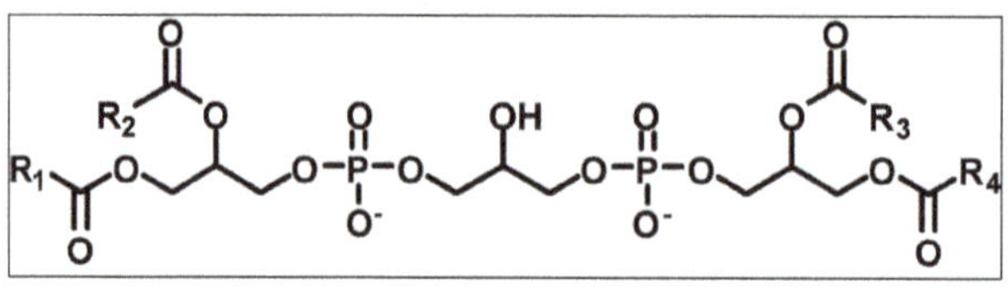

Cardiolipin.

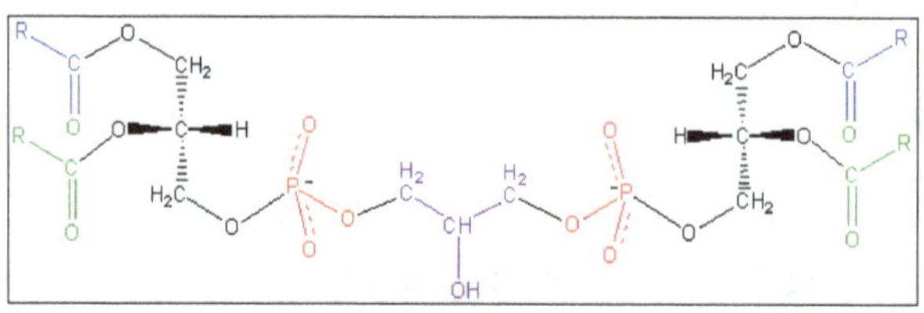

Cardiolipin (alternate image).

Cardiolipin (IUPAC name "1,3-bis(*sn*-3'-phosphatidyl)-*sn*-glycerol") is an important component of the inner mitochondrial membrane, where it constitutes about 20% of the total lipid composition. It can also be found in the membranes of most bacteria. The name 'cardiolipin' is derived from the fact that it was first found in animal hearts. It was first isolated from beef heart in the early 1940s. In mammalian cells, but also in plant cells, cardiolipin (CL) is found almost exclusively in the inner mitochondrial membrane where it is essential for the optimal function of numerous enzymes that are involved in mitochondrial energy metabolism.

Structure

Cardiolipin (CL) is a kind of diphosphatidylglycerol lipid. Two phosphatidic acid moieties connect with a glycerol backbone in the center to form a dimeric structure. So it has four alkyl groups and potentially carries two negative charges. As there are four distinct alkyl chains in cardiolipin, the potential for complexity of this molecule species is enormous. However, in most animal tissues, cardiolipin contains

18-carbon fatty alkyl chains with 2 unsaturated bonds on each of them. It has been proposed that the (18:2)4 acyl chain configuration is an important structural requirement for the high affinity of CL to inner membrane proteins in mammalian mitochondria. However, studies with isolated enzyme preparations indicate that its importance may vary depending on the protein examined.

Source	Molecular species				
	Fatty acid R1	Fatty acid R2	Fatty acid R3	Fatty acid R4	Amount (Mol%)
Bovine heart	18:2	18:2	18:2	18:2	48
	18:3	18:2	18:2	18:2	21
	18:2	18:3	18:2	18:2	
	18:2	18:2	18:2	18:3	
	18:2	18:2	18:3	18:2	
	18:2	18:1	18:2	18:2	15
	18:2	18:2	18:2	18:1	
Rat liver	18:2	18:2	18:2	18:2	57
	18:2	18:1	18:2	18:2	35
	18:2	18:2	18:2	18:1	

Cardiolipin in animal tissues.

Since there are two phosphates in the molecule, each of them can catch one proton. Although it has a symmetric structure, ionizing one phosphate happens at a very different levels of acidity than ionizing both: $pK_1 = 3$ and $pK_2 > 7.5$. So under normal physiological conditions (wherein pH is around 7), the molecule may carry only one negative charge. The hydroxyl groups (–OH and $-O^-$) on phosphate would form a stable intramolecular hydrogen bond with the centered glycerol's hydroxyl group, thus forming a bicyclic resonance structure. This structure traps one proton, which is quite helpful for oxidative phosphorylation.

As the head group forms such compact bicycle structure, the head group area is quite small relative to the big tail region consisting of 4 acyl chains. Based on this special structure, the fluorescent mitochondrial indicator, nonyl acridine orange (NAO) was introduced in 1982, and was later found to target mitochondria by binding to CL. NAO has a very large head and small tail structure which can compensate with cardiolipin's small head large tail structure, and arrange in a highly ordered way. Several studies were published utilizing NAO both as a quantitative mitochondrial indicator and an indicator of CL content in mitochondria. However, NAO is influenced by membrane potential and/or the spatial arrangement of CL, so it's not proper to use NAO for CL or mitochondria quantitative studies of intact respiring mitochondria. But NAO still represents a simple method of assessing CL content.

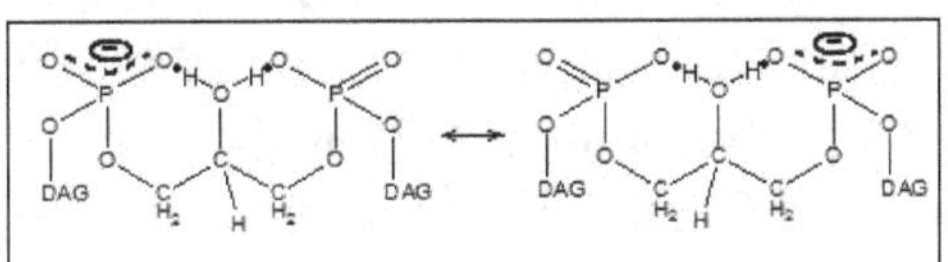

Bicyclic structure of Cardiolipin

Cardiolipin bicyclic structure.

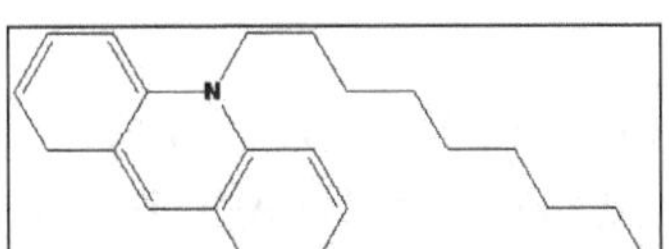

10-N-Nonyl acridine

Structure of NAO.

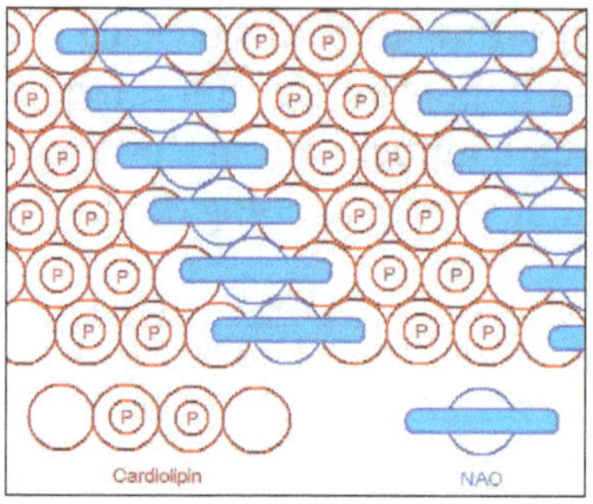

NAO & CL arranged in a highly ordered way.

Metabolism and Catabolism

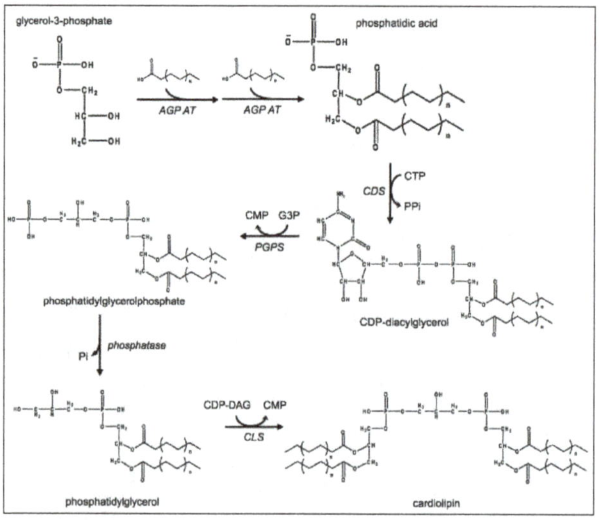

Cardiolipin synthesis in eukaryotes.

Metabolism

Eukaryotic Pathway

In eukaryotes such as yeasts, plants and animals, the synthesis processes are believed to happen in mitochondria. The first step is the acylation of glycerol-3-phosphate by a glycerol-3-phosphate acyltransferase. Then acylglycerol-3-phosphate can be once more acylated to form a phosphatidic acid (PA). With the help of the enzyme CDP-DAG synthase (CDS) (phosphatidate cytidylyltransferase), PA is converted into cytidinediphosphate-diacylglycerol (CDP-DAG). The following step is conversion of CDP-DAG to phosphatidylglycerol phosphate (PGP) by the enzyme PGP synthase, followed by dephosphorylation by PTPMT1 to form PG. Finally, a molecule of CDP-DAG is bound to PG to form one molecule of cardiolipin, catalyzed by the mitochondria-localized enzyme cardiolipin synthase (CLS).

Prokaryotic Pathway

In prokaryotes such as bacteria, diphosphatidylglycerol synthase catalyses a transfer of the phosphatidyl moiety of one phosphatidylglycerol to the free 3'-hydroxyl group of another, with the elimination of one molecule of glycerol, via the action of an enzyme related to phospholipase D. The enzyme can operate in reverse under some physiological conditions to remove cardiolipin.

Catabolism

Catabolism of cardiolipin may happen by the catalysis of phospholipase A2 (PLA) to remove fatty acyl groups. Phospholipase D (PLD) in the mitochondrion hydrolyses cardiolipin to phosphatidic acid.

Functions

Regulates Aggregate Structures

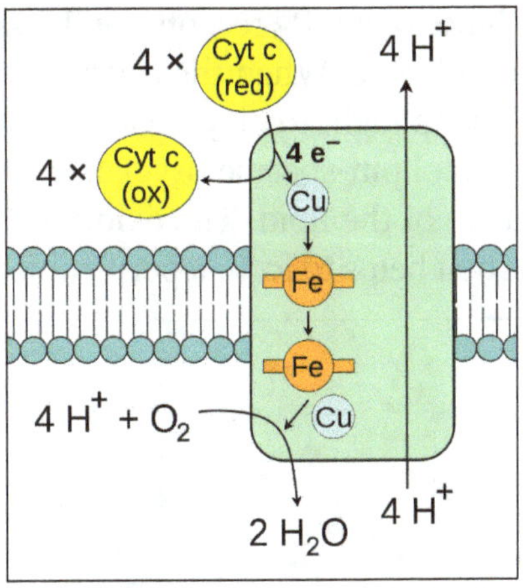

Respiratory electron transfer of Complex IV.

Because of cardiolipin's unique structure, a change in pH and the presence of divalent cations can induce a structural change. CL shows a great variety of forms of aggregates. It is found that in the presence of Ca^{2+} or other divalent cations, CL can be induced to have a lamellar-to-hexagonal (L_a-H_{II}) phase transition. And it is believed to have a close connection with membrane fusion.

Helps to Build Quaternary Structure

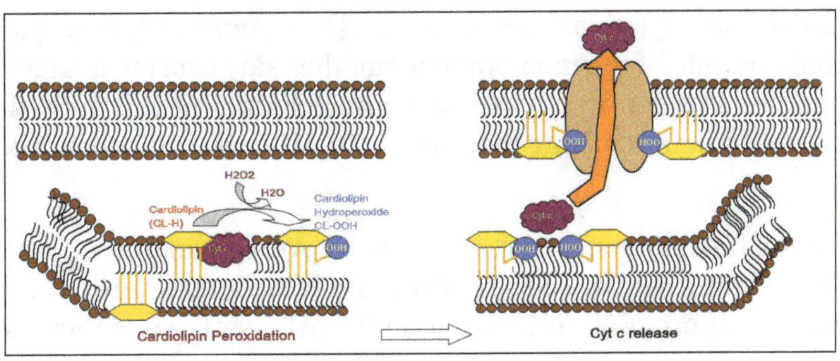

The mechanism whereby CL triggers apoptosis.

The enzyme cytochrome c oxidase or Complex IV is a large transmembrane protein complex found in bacteria and the mitochondrion. It is the last enzyme in the respiratory electron transport chain of mitochondria (or bacteria) located in the mitochondrial (or bacterial) membrane. It receives an electron from each of four cytochrome c molecules, and transfers them to one oxygen molecule, converting molecular oxygen to two molecules of water. Complex IV has been shown to require two associated CL molecules in order to maintain its full enzymatic function. Cytochrome bc1(Complex III) also needs cardiolipin to maintain its quaternary structure and functional role. Complex V of the oxidative phosphorylation machinery also displays high binding affinity for CL, binding four molecules of CL per molecule of complex V.

Triggers Apoptosis

Cardiolipin distribution to the outer mitochondrial membrane would lead to apoptosis of the cells, as evidenced by cytochrome c (cyt c) release, Caspase-8 activation, MOMP induction and NLRP3 inflammasome activation. During apoptosis, cyt c is released from the intermembrane spaces of mitochondria into the cytosol. Cyt c can then bind to the IP3 receptor on ER, stimulating calcium release, which then reacts back to cause the release of cyt c. When the calcium concentration reaches a toxic level, this causes cell death. Cytochrome c is thought to play a role in apoptosis via the release of apoptotic factors from the mitochondria. A cardiolipin-specific oxygenase produces CL hydroperoxides which can result in the conformation change of the lipid. The oxidized CL transfers from the inner membrane to the outer membrane, and then helps to form a permeable pore which releases cyt c.

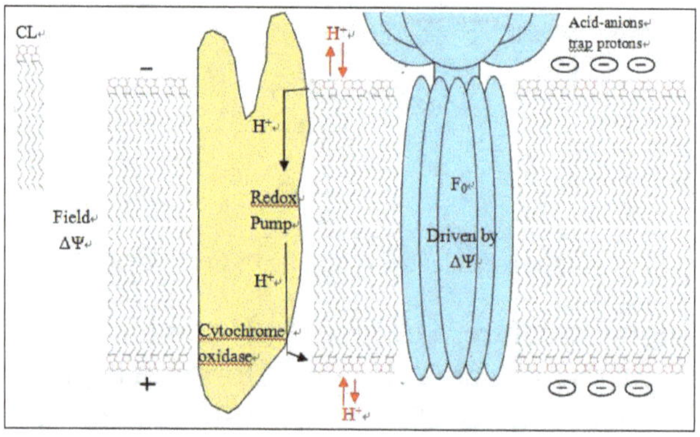

CL serves as a proton trap in oxidative phosphorylation.

Serves as Proton Trap for Oxidative Phosphorylation

During the oxidative phosphorylation process catalyzed by Complex IV, large quantities of protons are transferred from one side of the membrane to another side causing a large pH change. CL is suggested to function as a proton trap within the mitochondrial membranes, thereby strictly localizing the proton pool and minimizing the changes in pH in the mitochondrial intermembrane space.

This function is due to CL's unique structure. As stated above, CL can trap a proton within the bicyclic structure while carrying a negative charge. Thus, this bicyclic structure can serve as an electron buffer pool to release or absorb protons to maintain the pH near the membranes.

Other Functions

- Cholesterol translocation from outer to the inner membrane of mitochondrial.
- Activates mitochondrial cholesterol side-chain cleavage.
- Import protein into mitochondrial matrix.
- Anticoagulant function.

Glycolipid

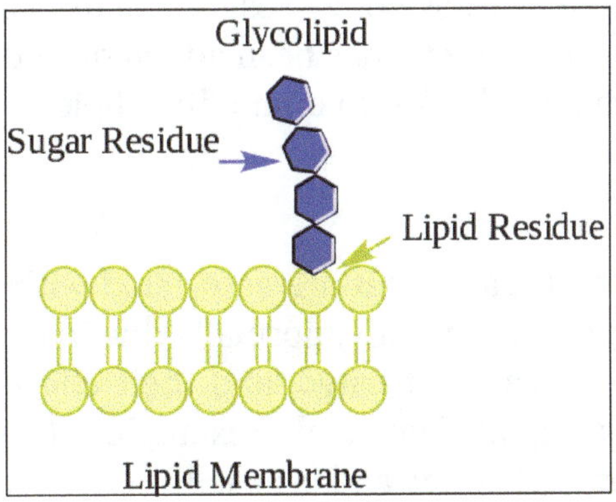

Glycolipid.

Glycolipids are lipids with a carbohydrate attached by a glycosidic bond. Their role is to maintain the stability of the cell membrane and to facilitate cellular recognition, which is crucial to the immune response and in the connections that allow cells to connect to one another to form tissues. Glycolipids are found on the surface of all eukaryotic cell membranes, where they extend from the phospholipid bilayer into the extracellular environment.

Structure

The essential feature of a glycolipid is the presence of a monosaccharide or oligosaccharide bound to a lipid moiety. The most common lipids in cellular membranes are glycerolipids and sphingolipids, which have glycerol or a sphingosine backbones, respectively. Fatty acids are connected to this backbone, so that the lipid as a whole has a polar head and a non-polar tail. The lipid bilayer of the cell membrane consists of two layers of lipids, with the inner and outer surfaces of the membrane made up of the polar head groups, and the inner part of the membrane made up of the non-polar fatty acid tails.

The saccharides that are attached to the polar head groups on the outside of the cell are the ligand components of glycolipids, and are likewise polar, allowing them to be soluble in the aqueous environment surrounding the cell. The lipid and the saccharide form a glycoconjugate through a glycosidic bond, which is a covalent bond. The anomeric carbon of the sugar binds to a free hydroxyl group on the lipid backbone. The structure of these saccharides varies depending on the structure of the molecules to which they bind.

Metabolism

Glycosyltransferases

Enzymes called glycosyltransferases link the saccharide to the lipid molecule, and also play a role in assembling the correct oligosaccharide so that the right receptor can be activated on the cell which responds to the presence of the glycolipid on the surface of the cell. The glycolipid is assembled in the Golgi apparatus and embedded in the surface of a vesicle which is then transported to the cell membrane. The vesicle merges with the cell membrane so that the glycolipid can be presented on the cell's outside surface.

Glycoside Hydrolases

Glycoside hydrolases catalyze the breakage of glycosidic bonds. They are used to modify the oligosaccharide structure of the glycan after it has been added onto the lipid. They can also remove glycans from glycolipids to turn them back into unmodified lipids.

Defects in Metabolism

Sphingolipidoses are a group of diseases that are associated with the accumulation of sphingolipids which have not been degraded correctly, normally due to a defect in a glycoside hydrolase enzyme. Sphingolipidoses are typically inherited, and their effects depend on which enzyme is affected, and the degree of impairment. One notable example is Niemann–Pick disease which can cause pain and damage to neural networks, and is usually fatal in early infancy.

Function

Cell–cell Interactions

The main function of glycolipids in the body is to serve as recognition sites for cell–cell interactions. The saccharide of the glycolipid will bind to a specific complementary carbohydrate or to a lectin (carbohydrate-binding protein), of a neighboring cell. The interaction of these cell surface markers is the basis of cell recognitions, and initiates cellular responses that contribute to activities such as regulation, growth, and apoptosis.

Immune Responses

An example of how glycolipids function within the body is the interaction between leukocytes and endothelial cells during inflammation. Selectins, a class of lectins found on the surface of leukocytes and endothelial cells bind to the carbohydrates attached to glycolipids to initiate the immune response. This binding causes leukocytes to leave circulation and congregate near the site of inflammation. This is the initial binding mechanism, which is followed by the expression of integrins which form stronger bonds and allow leukocytes to migrate toward the site of inflammation. Glycolipids are also responsible for other responses, notably the recognition of host cells by viruses.

Blood Types

Blood types are an example of how glycolipids on cell membranes mediate cell interactions with the surrounding environment. The four main human blood types (A, B, AB, O) are determined by the oligosaccharide attached to a specific glycolipid on the surface of red blood cells, which acts as an antigen. The unmodified antigen, called the H antigen, is the characteristic of type O, and is present on red blood cells of all blood types. Blood type A has an N-acetylgalactosamine added as the main determining structure, type B has a galactose, and type AB has all three of these antigens. Antigens which are not present in an individual's blood will cause antibodies to be produced, which will bind to the foreign glycolipids. For this reason, people with blood type AB can receive transfusions from all blood types (the universal acceptor), and people with blood type O can act as donors to all blood types (the universal donor).

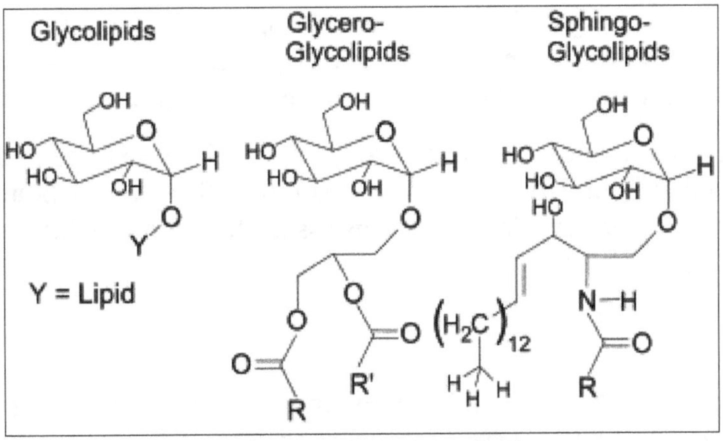

Chemical structure of glycolipids.

Types of Glycolipids

This list is incomplete; you can help by expanding it.

- Glyceroglycolipids: a sub-group of glycolipids characterized by an acetylated or non-acetylated glycerol with at least one fatty acid as the lipid complex. Glyceroglycolipids are often associated with photosynthetic membranes and their functions. The subcategories of glyceroglycolipids depend on the carbohydrate attached.
  - Galactolipids: defined by a galactose sugar attached to a glycerol lipid molecule. They are found in chloroplast membranes and are associated with photosynthetic properties.
  - Sulfolipids: have a sulfur-containing functional group in the sugar moiety attached to a lipid. An important group is the sulfoquinovosyl diacylglycerols which are associated with the sulfur cycle in plants.
- Glycosphingolipids: a sub-group of glycolipids based on sphingolipids. Glycosphingolipids are mostly located in nervous tissue and are responsible for cell signaling.
  - Cerebrosides: a group glycosphingolipids involved in nerve cell membranes.
    - Galactocerebrosides: a type of cerebroseide with galactose as the saccharide moiety.
    - Glucocerebrosides: a type of cerebroside with glucose as the saccharide moiety; often found in non-neural tissue.
    - Sulfatides: a class of glycolipids containing a sulfate group in the carbohydrate with a ceramide lipid backbone. They are involved in numerous biological functions ranging from immune response to nervous system signaling.
  - Gangliosides: the most complex animal glycolipids. They contain negatively charged oligosacchrides with one or more sialic acid residues; more than 200 different gangliosides have been identified. They are most abundant in nerve cells.
  - Globosides: glycosphingolipids with more than one sugar as part of the carbohydrate complex. They have a variety of functions; failure to degrade these molecules leads to Fabry disease.

- Glycophosphosphingolipids: complex glycophospholipids from fungi, yeasts, and plants, where they were originally called "phytoglycolipids". They may be as complicated a set of compounds as the negatively charged gangliosides in animals.
- Glycophosphatidylinositols: a sub-group of glycolipids defined by a phosphatidylinositol lipid moiety bound to a carbohydrate complex. They can be bound to the C-terminus of a protein and have various functions associated with the different proteins they can be bound to.

Sophorolipid

A sophorolipid is a surface-active glycolipid compound that can be synthesized by a selected number of non-pathogenic yeast species. The first report on sophorolipids was published in 1961, but in the last two decades, because of growing environmental awareness, they have gained increased attention as potential bio-surfactants due to their biodegradability and low eco-toxicity. Today sophorolipids are considered promising bio-surfactants.

Biosurfactants

The term "biosurfactant" can be applied to a surfactant that is obtained directly from a natural source (from plant, animal or microbial cells) by some kind of separation procedure such as extraction, precipitation or distillation. These surfactants have recently gained appeal as an alternative to surfactants organically synthesized from petrochemicals. While many bio-surfactants prove to be acceptable substitutes for their synthetic counterparts, often minor chemical processing can alter their properties to meet specific needs. There is debate regarding how much organic synthesis should be involved while still being able to call a given substance a bio-surfactant, but currently the term is not protected and often "bio-based surfactant" is used interchangeably with "bio-surfactant".

Since surface active agents are abundant in nature, the factor that limits the production of bio-surfactants is definitely not their presence, but likely the technological cost of scaled-up production, separation, and purification. In most instances this makes the cost of bio-based surfactants much higher than the cost of their synthetic equivalents. For example, most biological membranes are made up of a phospholipid bilayer which is composed of surfactant molecules having a polar head and non-polar or hydrophobic tail. However while these phospholipids are ubiquitous in nature, their intracellular location and interaction with other cellular components make their recovery and purification difficult and costly.

Other than sophorolipids, common biosurfactants include rhamnolipids which can be produced by some species of Pseudomonas, and Lecithin which can be obtained either from soybean or from egg yolk and is a common food ingredient. One of interesting surfactants producing by yeast are cellobiose lipids consist of a residue of cellobiose, the disaccharide composed of two glucose residues linked by a 1,4'-beta-glycoside bond, and fatty acid residue as an aglycone . Cellobiose lipids are effective fungicides against many species of ascomycetous and basidiomycetous yeast and mycelial fungi, including human pathogens, Candida albicans and Cryptococcus neoformans.

Microbial Biosurfactant

Usually the term bio-surfactant is used for the surfactants of microbial origin. This is because the shorter generation time of microbial biomass compared to biomass of animal or plant origin, makes the production of bio-surfactants using bacteria and fungi more industrially relevant. Furthermore, most of the bio-surfactants produced by microbes are synthesized extracellularly and many microbes are known to produce bio-surfactants in large relative quantities. Both these factors (i.e. extracellular location and high concentration) can make the process of recovery, concentration and purification of microbial bio-surfactants simple and cost effective. Recent research has meant sophorolipids can be recovered during a fermentation using a gravity separator in a loop with the bioreactor, enabling the production of >770 g/l sophorolipid at a productivity 4.24 g/l/h, some of the highest values seen in a fermentation process.

Structure and Properties

Honey like viscous, oily, acidic sophorolipids produced by solid state fermentation (SSF) of *starmerella bombicola*.

Sophorolipids are glycolipid class of microbial biosurfactants which consist of a hydrophobic fatty acid tail of 16 or 18 carbon atoms and a hydrophilic carbohydrate head, sophorose. which is a glucose di-saccharide with an unusual β-1,2 bond and can be acetylated on the 6′- and/or 6′′- positions. One terminal or sub terminal hydroxylated fatty acid is β-glycosidically linked to the sophorose molecule. The carboxylic end of this fatty acid is either free (acidic or open form) or internally esterified at the 4′′ or in some rare cases at the 6′- or 6′′-position (lactonic form). The hydroxy fatty acid itself counts in general 16 or 18 carbon atoms and can have one or more unsaturated bonds. The physicochemical and biological properties of sophorolipids are significantly influenced by the distribution of the lactone vs. acidic forms produced in the fermentative broth. In general, lactone sophorolipids are more efficient in reducing surface tension and are better antimicrobial agents, whereas acidic sophorolipids display better foaming properties. Acetyl groups can also lower the hydrophilicity of sophorolipids and enhance their antiviral and cytokine stimulating effects.

Biosurfactant Properties

Two desirable properties of biosurfactants are biodegradability and low toxicity. While molecules of biological origin are almost always bio-degradable, the rate at which they are degraded can vary significantly as does the degree to which they can be considered toxic. Though many biosurfactants are easily biodegradable and exhibit low toxicity, it is inappropriate to assume that all bio-surfactants possess these same traits under all conditions. Furthermore, even though a bio-surfactant may not be toxic itself, the producing micro-organism could be pathogenic and hence the production of such bio-surfactants may lead to environment concerns. For example, Pseudomonads,including the opportunistic pathogen *Pseudomonas aeruginosa*, can be used to produce the bio-surfactant rhamnolipid. Similar concerns have been expressed over the widespread production of citric acid by the opportunistic pathogen *Aspergillus niger* but were limited to ensuring that the final product was sufficiently pure and that the production process was contained. In the case of rhamnolipids, there are many non-pathogenic strains of Pseudomonas which can be used to produce that molecule. Sophorolipids are synthesised by various non pathogenic yeast species such as *Candida apicola, Rhodotorula bogoriensis, Wickerhamiella domercqiae*, and *Starmerella bombicola*.

Besides biodegradibility, low toxicity, and high production potential, sophorolipids also have a high surface and interfacial activity compared to many chemically synthesized surfactants. These properties help justify the higher costs of production and purification associated with large scale production. Generally a surfactant is considered as a good surfactant if it can lower surface tension (ST) of water from 72 to 30-35 mN/m and the interfacial tension (IT) water/hexadecane from 40 to 1 mN/m. Sophorolipids are reported to reduce ST up to 29 mN/m and IT to 5 mN/m, at the much lower concentrations than the critical micelle concentration(CMC) of many of the synthetic surfactants. In addition to this, sophorolipids are reported to function under wide ranges of temperatures, pressures and ionic strengths; and they also possess a number of other useful biological activities including Antimicrobial, virucidal, Anticancer, Immuno-modulatory properties.

Research

A detailed and comprehensive literature review on the various aspects of sophorolipids production (e.g. producing micro-organisms, bio-synthetic pathway, effect of medium components and other fermentation conditions and downstream process of sophorolipids is available in the published work of Van Bogaert et al. This work also discusses potential application of sophorolipids (and their derivatives) as well as the potential for genetic engineering strains to enhance sophorolipid yields. Researchers have focused on optimization of sophorolipid production in submerged fermentation, but some efforts have also investigated the possibility of sophorololipid production using solid state fermentation (SSF). The production process can be significantly impacted by the specific properties of the carbon and oil substrates used; and several inexpensive alternatives to more traditional substrates have been investigated. These potential substrates include: biodiesel by-product streams, waste frying oil, restaurant waste oil, industrial fatty acid residues, mango seed fat, and soybean dark oil. The use of most of these substrates have resulted in lower yields compared to traditional fermentation substrates.

Limitations

There no doubt that the sophorolipids are the most promising microbial biosurfactant from microbial origin. However even sophorolipids have few shortcomings. For example, sophorolipids are not readily soluble in aqueous medium at acidic pH. Their solubility increases above pH 5.0 and are freely soluble above pH 6.0. However sophorolipids shows instability at pH values higher than 7.0 to 7.5 for long term storage: beyond this point, irreversible hydrolysis of the acetyl groups and ester bonds is observed. Sophorolipids are low foaming surfactants even at high concentrations; these limitations excludes them from a number of applications (for example: preparation of soaps), but makes them suitable for non-foam requiring purposes. (for example: cleansing solution for glass and metallic surfaces, as food additives). Further, even though sophorolipids have excellent antimicrobial, anticancer and antiviral properties, its oral administration is almost impossible, since it gets readily hydrolyzed in stomach as well as intestine, thus they can either be useful for topical preparations such as antimicrobial creams, or may require development of parenteral or novel dosage forms, if to be used for systemic infections.

Chemical Modifications of Sophorolipids and Polysophorolipids

To enhance the performance of surfactant properties of natural sophorolipids, chemical modification methods have been actively pursued. Recently, researchers demonstrated the possibility of applying sophorolipids as building blocks via ring-opening metathesis polymerization for a new type of polymers, known as polysophorolipids which show promising potentials in biomaterials applications.

Sphingolipid

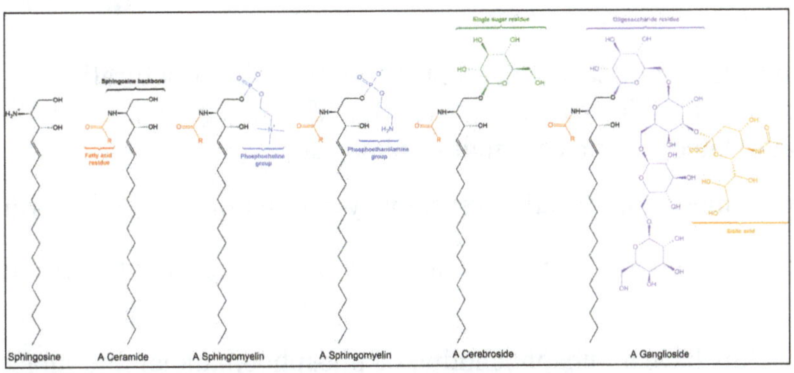

General structures of sphingolipids.

Sphingolipids, or glycosylceramides, are a class of lipids containing a backbone of sphingoid bases, a set of aliphatic amino alcohols that includes sphingosine. They were discovered in brain extracts in the 1870s and were named after the mythological Sphinx because of their enigmatic nature. These compounds play important roles in signal transmission and cell recognition. Sphingolipidoses, or disorders of sphingolipid metabolism, have particular impact on neural tissue. A sphingolipid with an R group consisting of a hydrogen atom only is a ceramide. Other common R groups include phosphocholine, yielding a sphingomyelin, and various sugar monomers or dimers, yielding cerebrosides and globosides, respectively. Cerebrosides and globosides are collectively known as glycosphingolipids.

Structure

The long-chain bases, sometimes simply known as sphingoid bases, are the first non-transient products of *de novo* sphingolipid synthesis in both yeast and mammals. These compounds, specifically known as phytosphingosine and dihydrosphingosine (also known as sphinganine, although this term is less common), are mainly C_{18} compounds, with somewhat lower levels of C_{20} bases. Ceramides and glycosphingolipids are *N*-acyl derivatives of these compounds.

The sphingosine backbone is O-linked to a (usually) charged head group such as ethanolamine, serine, or choline.

The backbone is also amide-linked to an acyl group, such as a fatty acid.

Types

Simple sphingolipids, which include the sphingoid bases and ceramides, make up the early products of the sphingolipid synthetic pathways.

- Sphingoid bases are the fundamental building blocks of all sphingolipids. The main mammalian sphingoid bases are dihydrosphingosine and sphingosine, while dihydrosphingosine and phytosphingosine are the principle sphingoid bases in yeast. Sphingosine, dihydrosphingosine, and phytosphingosine may be phosphorylated.
- Ceramides, as a general class, are *N*-acylated sphingoid bases lacking additional head groups.
    - Dihydroceramide is produced by *N*-acylation of dihydrosphingosine. Dihydroceramide is found in both yeast and mammalian systems.
    - Ceramide is produced in mammalian systems by desaturation of dihydroceramide by dihydroceramide desaturase 1 (DES1). This highly bioactive molecule may also be phosphorylated to form ceramide-1-phosphate.
    - Phytoceramide is produced in yeast by hydroxylation of dihydroceramide at C-4.

Complex sphingolipids may be formed by addition of head groups to ceramide or phytoceramide:

- Sphingomyelins have a phosphocholine or phosphoethanolamine molecule with an ester linkage to the 1-hydroxy group of a ceramide.
- Glycosphingolipids are ceramides with one or more sugar residues joined in a β-glycosidic linkage at the 1-hydroxyl position.
    - Cerebrosides have a single glucose or galactose at the 1-hydroxy position.
        - Sulfatides are sulfated cerebrosides.
    - Gangliosides have at least three sugars, one of which must be sialic acid.
- Inositol-containing ceramides, which are derived from phytoceramide, are produced in yeast. These include inositol phosphorylceramide, mannose inositol phosphorylceramide, and mannose diinositol phosphorylceramide.

Mammalian Sphingolipid Metabolism

De novo sphingolipid synthesis begins with formation of 3-keto-dihydrosphingosine by serine palmitoyltransferase. The preferred substrates for this reaction are palmitoyl-CoA and serine. However, studies have demonstrated that serine palmitoyltransferase has some activity toward other species of fatty acyl-CoA and alternative amino acids, and the diversity of sphingoid bases has recently been reviewed. Next, 3-keto-dihydrosphingosine is reduced to form dihydrosphingosine. Dihydrosphingosine is acylated by a (dihydro)-ceramide synthase, such as Lass1p or Lass2p (also termed as CerS), to form dihydroceramide. This is desaturated to form ceramide.

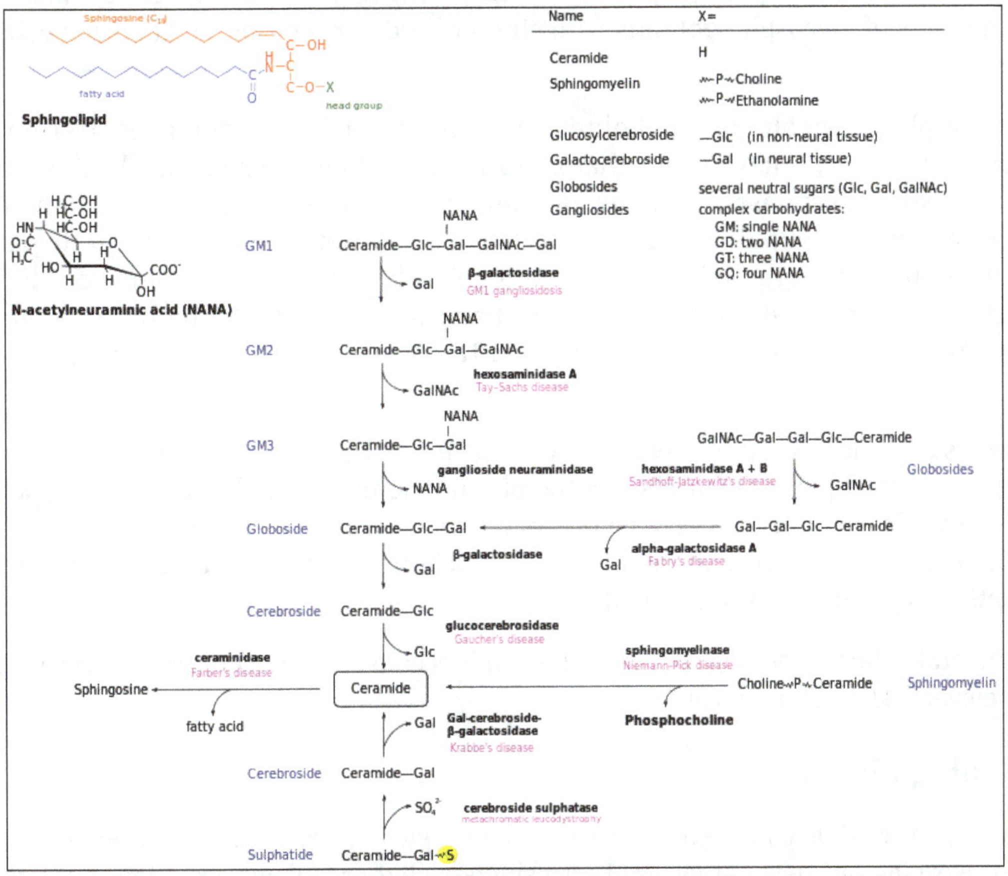

Metabolic pathways of various forms of sphingolipids. Sphingolipidoses are labeled at corresponding stages that are deficient.

Ceramide may subsequently have several fates. It may be phosphorylated by ceramide kinase to form ceramide-1-phosphate. Alternatively, it may be glycosylated by glucosylceramide synthase or galactosylceramide synthase. Additionally, it can be converted to sphingomyelin by the addition of a phosphorylcholine headgroup by sphingomyelin synthase. Diacylglycerol is generated by this process. Finally, ceramide may be broken down by a ceramidase to form sphingosine. Sphingosine may be phosphorylated to form sphingosine-1-phosphate. This may be dephosphorylated to reform sphingosine.

Breakdown pathways allow the reversion of these metabolites to ceramide. The complex glycosphingolipids are hydrolyzed to glucosylceramide and galactosylceramide. These lipids are then hydrolyzed by beta-glucosidases and beta-galactosidases to regenerate ceramide. Similarly, sphingomyelin may be broken down by sphingomyelinase to form ceramide.

The only route by which sphingolipids are converted to non-sphingolipids is through sphingosine-1-phosphate lyase. This forms ethanolamine phosphate and hexadecenal.

Functions of Mammalian Sphingolipids

Sphingolipids are commonly believed to protect the cell surface against harmful environmental factors by forming a mechanically stable and chemically resistant outer leaflet of the plasma membrane lipid bilayer. Certain complex glycosphingolipids were found to be involved in specific functions, such as cell recognition and signaling. Cell recognition depends mainly on the physical properties of the sphingolipids, whereas signaling involves specific interactions of the glycan structures of glycosphingolipids with similar lipids present on neighboring cells or with proteins.

Recently, simple sphingolipid metabolites, such as ceramide and sphingosine-1-phosphate, have been shown to be important mediators in the signaling cascades involved in apoptosis, proliferation, stress responses, necrosis, inflammation, autophagy, senescence, and differentiation. Ceramide-based lipids self-aggregate in cell membranes and form separate phases less fluid than the bulk phospholipids. These sphingolipid-based microdomains, or "lipid rafts" were originally proposed to sort membrane proteins along the cellular pathways of membrane transport. At present, most research focuses on the organizing function during signal transduction.

Sphingolipids are synthesized in a pathway that begins in the ER and is completed in the Golgi apparatus, but these lipids are enriched in the plasma membrane and in endosomes, where they perform many of their functions. Transport occurs via vesicles and monomeric transport in the cytosol. Sphingolipids are virtually absent from mitochondria and the ER, but constitute a 20-35 molar fraction of plasma membrane lipids.

In experimental animals, feeding sphingolipids inhibits colon carcinogenesis, reduces LDL cholesterol and elevates HDL cholesterol.

Yeast Sphingolipids

Because of the incredible complexity of mammalian systems, yeast are often used as a model organism for working out new pathways. These single-celled organisms are often more genetically tractable than mammalian cells, and strain libraries are available to supply strains harboring almost any non-lethal open reading frame single deletion. The two most commonly used yeasts are *Saccharomyces cerevisiae* and *Schizosaccharomyces pombe*, although research is also done in the pathogenic yeast *Candida albicans*.

In addition to the important structural functions of complex sphingolipids (inositol phosphorylceramide and its mannosylated derivatives), the sphingoid bases phytosphingosine and dihydrosphingosine (sphinganine) play vital signaling roles in *S. cerevisiae*. These effects include regulation of endocytosis, ubiquitin-dependent proteolysis (and, thus, regulation of nutrient uptake), cytoskeletal dynamics, the cell cycle, translation, posttranslational protein modification, and the heat stress response. Additionally, modulation of sphingolipid metabolism by phosphatidylinositol (4,5)-bisphosphate signaling *via* Slm1p and Slm2p and calcineurin has recently been described. Additionally, a substrate-level interaction has been shown between complex sphingolipid synthesis and cycling of phosphatidylinositol 4-phosphate by the phosphatidylinositol kinase Stt4p and the lipid phosphatase Sac1p.

Plant Sphingolipids

Higher plants contain a wider variety of sphingolipids than animals and fungi.

Disorders

There are several disorders of sphingolipid metabolism, known as sphingolipidoses. The main members of this group are Niemann-Pick disease, Fabry disease, Krabbe disease, Gaucher disease, Tay-Sachs disease and Metachromatic leukodystrophy. They are generally inherited in an autosomal recessive fashion, but notably Fabry disease is X-linked. Taken together, sphingolipidoses have an incidence of approximately 1 in 10,000, but substantially more in certain populations such as Ashkenazi Jews. Enzyme replacement therapy is available to treat mainly Fabry disease and Gaucher disease, and people with these types of sphingolipidoses may live well into adulthood. The other types are generally fatal by age 1 to 5 years for infantile forms, but progression may be mild for juvenile- or adult-onset forms.

Sphingolipids have also been implicated with the frataxin protein (Fxn), the deficiency of which is associated with Friedreich's ataxia (FRDA). Loss of Fxn in the nervous system in mice also activates an iron/sphingolipid/PDK1/Mef2 pathway, indicating that the mechanism is evolutionarily conserved. Furthermore, sphingolipid levels and PDK1 activity are also increased in hearts of FRDA patients, suggesting that a similar pathway is affected in FRDA. Other research has demonstrated that iron accumulation in the nervous systems of flies enhances the synthesis of sphingolipids, which in turn activates 3-phosphoinositide dependent protein kinase-1 (Pdk1) and myocyte enhancer factor-2 (Mef2) to trigger neurodegeneration of adult photoreceptors.

Lipid Bilayer

The lipid bilayer (or phospholipid bilayer) is a thin polar membrane made of two layers of lipid molecules. These membranes are flat sheets that form a continuous barrier around all cells. The cell membranes of almost all living organisms and many viruses are made of a lipid bilayer, as are the membranes surrounding the cell nucleus and other sub-cellular structures. The lipid bilayer is the barrier that keeps ions, proteins and other molecules where they are needed and prevents them from diffusing into areas where they should not be. Lipid bilayers are ideally suited to this role, even though they are only a few nanometers in width, they are impermeable to most water-soluble (hydrophilic) molecules. Bilayers are particularly impermeable to ions, which allows cells to regulate salt concentrations and pH by transporting ions across their membranes using proteins called ion pumps.

Biological bilayers are usually composed of amphiphilic phospholipids that have a hydrophilic phosphate head and a hydrophobic tail consisting of two fatty acid chains. Phospholipids with certain head groups can alter the surface chemistry of a bilayer and can, for example, serve as signals as well as "anchors" for other molecules in the membranes of cells. Just like the heads, the tails of lipids can also affect membrane properties, for instance by determining the phase of the bilayer. The bilayer can adopt a solid gel phase state at lower temperatures but undergo phase transition to a fluid state at higher temperatures, and the chemical properties of the lipids' tails influence at which temperature this happens. The packing of lipids within the bilayer also affects its mechanical properties, including its resistance to stretching and bending. Many of these properties have been studied with the use of artificial "model" bilayers produced in a lab. Vesicles made by model bilayers have also been used clinically to deliver drugs.

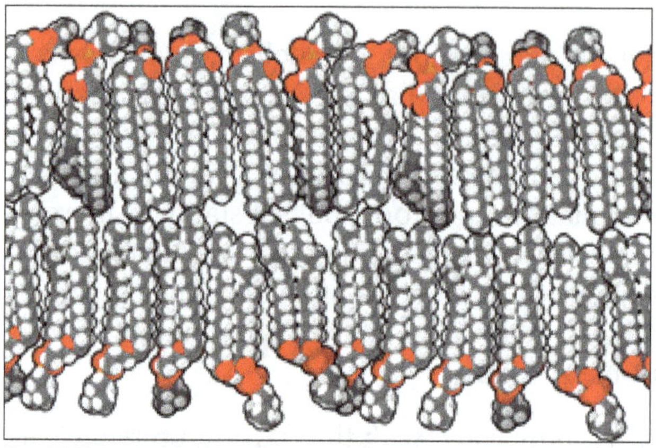

This fluid lipid bilayer cross section is made up entirely of phosphatidylcholine.

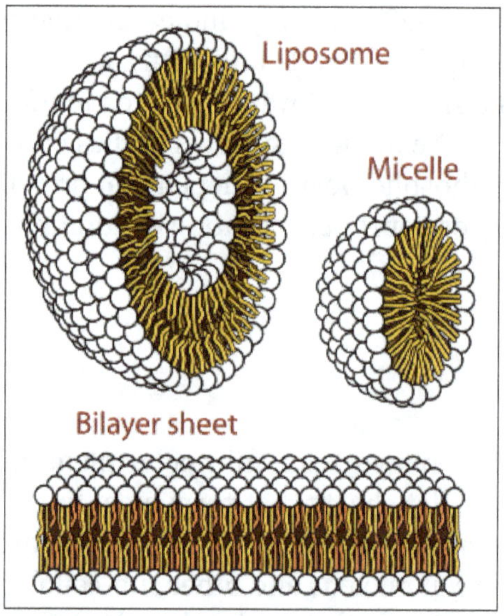

The three main structures phospholipids form in solution; the liposome (a closed bilayer), the micelle and the bilayer.

Biological membranes typically include several types of molecules other than phospholipids. A particularly important example in animal cells is cholesterol, which helps strengthen the bilayer and decrease its permeability. Cholesterol also helps regulate the activity of certain integral membrane proteins. Integral membrane proteins function when incorporated into a lipid bilayer, and they are held tightly to lipid bilayer with the help of an annular lipid shell. Because bilayers define the boundaries of the cell and its compartments, these membrane proteins are involved in many intra- and inter-cellular signaling processes. Certain kinds of membrane proteins are involved in the process of fusing two bilayers together. This fusion allows the joining of two distinct structures as in the fertilization of an egg by sperm or the entry of a virus into a cell. Because lipid bilayers are quite fragile and invisible in a traditional microscope, they are a challenge to study. Experiments on bilayers often require advanced techniques like electron microscopy and atomic force microscopy.

Structure and Organization

When phospholipids are exposed to water, they self-assemble into a two-layered sheet with the hydrophobic tails pointing toward the center of the sheet. This arrangement results in two "leaflets" that are each a single molecular layer. The center of this bilayer contains almost no water and excludes molecules like sugars or salts that dissolve in water. The assembly process is driven by interactions between hydrophobic molecules (also called the hydrophobic effect). An increase in interactions between hydrophobic molecules (causing clustering of hydrophobic regions) allows water molecules to bond more freely with each other, increasing the entropy of the system. This complex process includes non-covalent interactions such as van der Waals forces, electrostatic and hydrogen bonds.

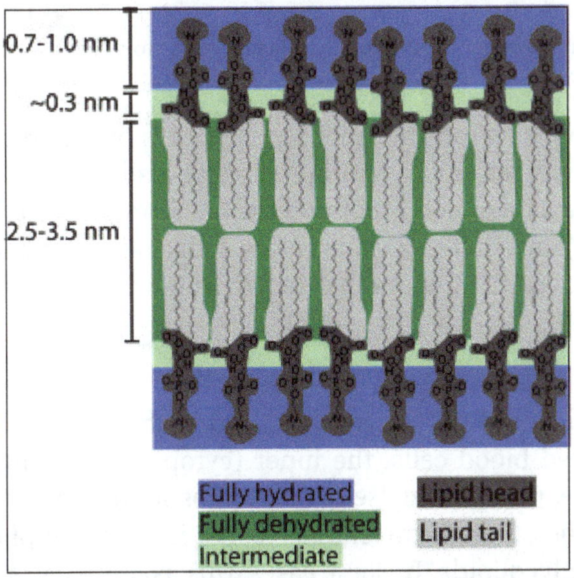

Schematic cross sectional profile of a typical lipid bilayer. There are three distinct regions: the fully hydrated head-groups, the fully dehydrated alkane core and a short intermediate region with partial hydration. Although the head groups are neutral, they have significant dipole moments that influence the molecular arrangement.

Cross Section Analysis

The lipid bilayer is very thin compared to its lateral dimensions. If a typical mammalian cell (diameter ~10 micrometers) were magnified to the size of a watermelon (~1 ft/30 cm), the lipid bilayer making up the plasma membrane would be about as thick as a piece of office paper. Despite being only a few nanometers thick, the bilayer is composed of several distinct chemical regions across its cross-section. These regions and their interactions with the surrounding water have been characterized over the past several decades with x-ray reflectometry, neutron scattering and nuclear magnetic resonance techniques.

The first region on either side of the bilayer is the hydrophilic headgroup. This portion of the membrane is completely hydrated and is typically around 0.8-0.9 nm thick. In phospholipid bilayers the phosphate group is located within this hydrated region, approximately 0.5 nm outside the hydrophobic core. In some cases, the hydrated region can extend much further, for instance in lipids with a large protein or long sugar chain grafted to the head. One common example of such a modification in nature is the lipopolysaccharide coat on a bacterial outer membrane, which helps retain a water layer around the bacterium to prevent dehydration.

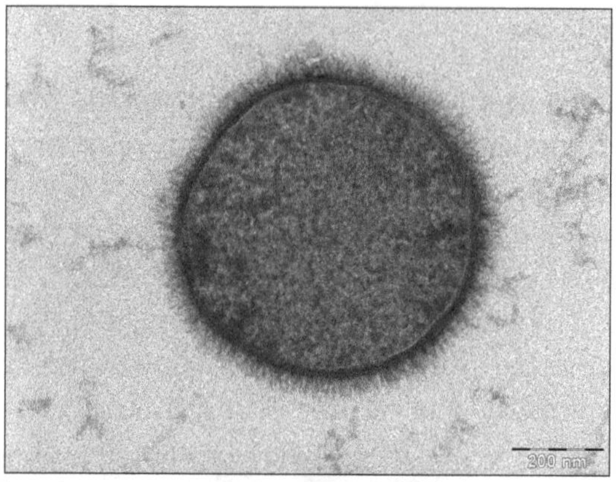

TEM image of a bacterium. The furry appearance on the outside is due to a coat of long-chain sugars attached to the cell membrane. This coating helps trap water to prevent the bacterium from becoming dehydrated.

Next to the hydrated region is an intermediate region that is only partially hydrated. This boundary layer is approximately 0.3 nm thick. Within this short distance, the water concentration drops from 2M on the headgroup side to nearly zero on the tail (core) side. The hydrophobic core of the bilayer is typically 3-4 nm thick, but this value varies with chain length and chemistry. Core thickness also varies significantly with temperature, in particular near a phase transition.

Asymmetry

In many naturally occurring bilayers, the compositions of the inner and outer membrane leaflets are different. In human red blood cells, the inner (cytoplasmic) leaflet is composed mostly of phosphatidylethanolamine, phosphatidylserine and phosphatidylinositol and its phosphorylated derivatives. By contrast, the outer (extracellular) leaflet is based on phosphatidylcholine, sphingomyelin and a variety of glycolipids, In some cases, this asymmetry is based on where the lipids are made in the cell and reflects their initial orientation. The biological functions of lipid asymmetry are imperfectly understood, although it is clear that it is used in several different situations. For example, when a cell undergoes apoptosis, the phosphatidylserine — normally localised to the cytoplasmic leaflet — is transferred to the outer surface: There, it is recognised by a macrophage that then actively scavenges the dying cell.

Lipid asymmetry arises, at least in part, from the fact that most phospholipids are synthesised and initially inserted into the inner monolayer: those that constitute the outer monolayer are then transported from the inner monolayer by a class of enzymes called flippases. Other lipids, such as sphingomyelin, appear to be synthesised at the external leaflet. Flippases are members of a larger family of lipid transport molecules that also includes floppases, which transfer lipids in the opposite direction, and scramblases, which randomize lipid distribution across lipid bilayers (as in apoptotic cells). In any case, once lipid asymmetry is established, it does not normally dissipate quickly because spontaneous flip-flop of lipids between leaflets is extremely slow.

It is possible to mimic this asymmetry in the laboratory in model bilayer systems. Certain types of very small artificial vesicle will automatically make themselves slightly asymmetric, although the mechanism by which this asymmetry is generated is very different from that in cells. By utilizing two different monolayers in Langmuir-Blodgett deposition or a combination of Langmuir-Blodgett and vesicle rupture deposition it is also possible to synthesize an asymmetric planar bilayer. This asymmetry may be lost over time as lipids in supported bilayers can be prone to flip-flop.

Phases and Phase Transitions

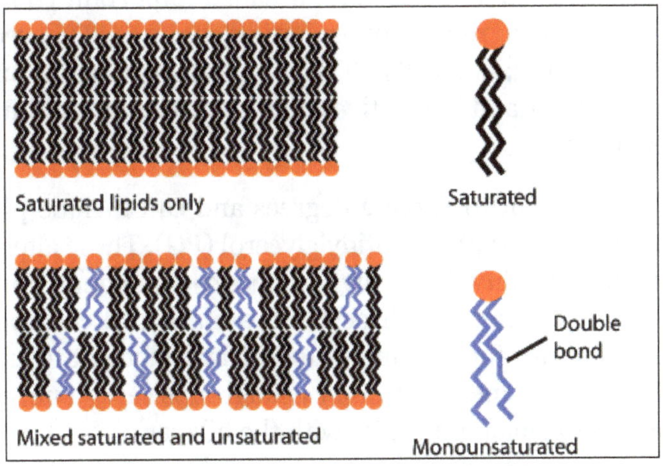

Diagram showing the effect of unsaturated lipids on a bilayer. The lipids with an unsaturated tail (blue) disrupt the packing of those with only saturated tails (black). The resulting bilayer has more free space and is, as a consequence, more permeable to water and other small molecules.

At a given temperature a lipid bilayer can exist in either a liquid or a gel (solid) phase. All lipids have a characteristic temperature at which they transition (melt) from the gel to liquid phase. In both phases the lipid molecules are prevented from flip-flopping across the bilayer, but in liquid phase bilayers a given lipid will exchange locations with its neighbor millions of times a second. This random walk exchange allows lipid to diffuse and thus wander across the surface of the membrane. Unlike liquid phase bilayers, the lipids in a gel phase bilayer have less mobility.

The phase behavior of lipid bilayers is determined largely by the strength of the attractive Van der Waals interactions between adjacent lipid molecules. Longer-tailed lipids have more area over which to interact, increasing the strength of this interaction and, as a consequence, decreasing the lipid mobility. Thus, at a given temperature, a short-tailed lipid will be more fluid than an otherwise identical long-tailed lipid. Transition temperature can also be affected by the degree of unsaturation of the lipid tails. An unsaturated double bond can produce a kink in the alkane chain, disrupting the lipid packing. This disruption creates extra free space within the bilayer that allows additional flexibility in the adjacent chains. An example of this effect can be noted in everyday life as butter, which has a large percentage saturated fats, is solid at room temperature while vegetable oil, which is mostly unsaturated, is liquid.

Most natural membranes are a complex mixture of different lipid molecules. If some of the components are liquid at a given temperature while others are in the gel phase, the two phases can coexist in spatially separated regions, rather like an iceberg floating in the ocean. This phase separation plays a critical role in biochemical phenomena because membrane components such as proteins can partition into one or the other phase and thus be locally concentrated or activated. One particularly important component of many mixed phase systems is cholesterol, which modulates bilayer permeability, mechanical strength, and biochemical interactions.

Surface Chemistry

While lipid tails primarily modulate bilayer phase behavior, it is the headgroup that determines the bilayer surface chemistry. Most natural bilayers are composed primarily of phospholipids, but sphingolipids and sterols such as cholesterol are also important components. Of the phospholipids, the most common headgroup is phosphatidylcholine (PC), accounting for about half the phospholipids in most mammalian cells. PC is a zwitterionic headgroup, as it has a negative charge on the phosphate group and a positive charge on the amine but, because these local charges balance, no net charge.

Other headgroups are also present to varying degrees and can include phosphatidylserine (PS) phosphatidylethanolamine (PE) and phosphatidylglycerol (PG). These alternate headgroups often confer specific biological functionality that is highly context-dependent. For instance, PS presence on the extracellular membrane face of erythrocytes is a marker of cell apoptosis, whereas PS in growth plate vesicles is necessary for the nucleation of hydroxyapatite crystals and subsequent bone mineralization. Unlike PC, some of the other headgroups carry a net charge, which can alter the electrostatic interactions of small molecules with the bilayer.

Biological Roles

Containment and Separation

The primary role of the lipid bilayer in biology is to separate aqueous compartments from their surroundings. Without some form of barrier delineating "self" from "non-self," it is difficult to even define the concept of an organism or of life. This barrier takes the form of a lipid bilayer in all known life forms except for a few species of archaea that utilize a specially adapted lipid monolayer. It has even been proposed that the very first form of life may have been a simple lipid vesicle with virtually its sole biosynthetic capability being the production of more phospholipids. The partitioning ability of the lipid bilayer is based on the fact that hydrophilic molecules cannot easily cross the hydrophobic bilayer core, as discussed in Transport across the bilayer below. The nucleus, mitochondria and chloroplasts have two lipid bilayers, while other sub-cellular structures are surrounded by a single lipid bilayer (such as the plasma membrane, endoplasmic reticula, Golgi apparatus and lysosomes).

Prokaryotes have only one lipid bilayer- the cell membrane (also known as the plasma membrane). Many prokaryotes also have a cell wall, but the cell wall is composed of proteins or long chain carbohydrates, not lipids. In contrast, eukaryotes have a range of organelles including the nucleus, mitochondria, lysosomes and endoplasmic reticulum. All of these sub-cellular compartments are surrounded by one or more lipid bilayers and, together, typically comprise the majority of the bilayer area present in the cell. In liver hepatocytes for example, the plasma membrane accounts for only two percent of the total bilayer area of the cell, whereas the endoplasmic reticulum contains more than fifty percent and the mitochondria a further thirty percent.

Signaling

Probably the most familiar form of cellular signaling is synaptic transmission, whereby a nerve impulse that has reached the end of one neuron is conveyed to an adjacent neuron via the release of neurotransmitters. This transmission is made possible by the action of synaptic vesicles loaded with the neurotransmitters to be released. These vesicles fuse with the cell membrane at the pre-synaptic terminal and release its contents to the exterior of the cell. The contents then diffuse across the synapse to the post-synaptic terminal.

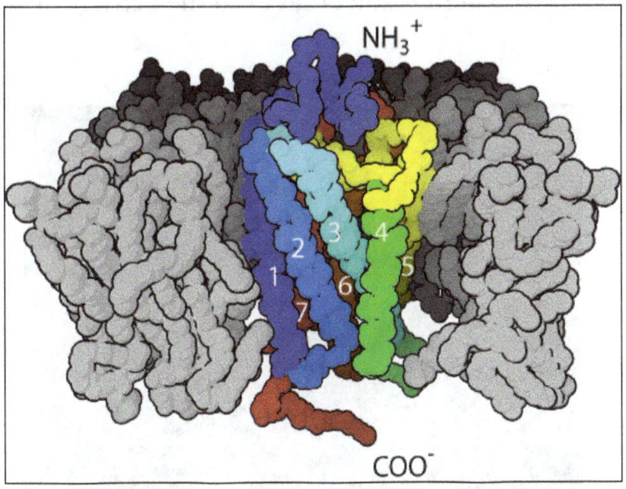

Illustration of a GPCR signaling protein. In response to a molecule such as a hormone binding to the exterior domain (blue) the GPCR changes shape and catalyzes a chemical reaction on the interior domain (red). The gray feature is the surrounding bilayer.

Lipid bilayers are also involved in signal transduction through their role as the home of integral membrane proteins. This is an extremely broad and important class of biomolecule. It is estimated that up to a third of the human proteome may be membrane proteins. Some of these proteins are linked to the exterior of the cell membrane. An example of this is the CD59 protein, which identifies cells as "self" and thus inhibits their destruction by the immune system. The HIV virus evades the immune system in part by grafting these proteins from the host membrane onto its own surface. Alternatively, some membrane proteins penetrate all the way through the bilayer and serve to relay individual signal events from the outside to the inside of the cell. The most common class of this type of protein is the G protein-coupled receptor (GPCR). GPCRs are responsible for much of the cell's ability to sense its surroundings and, because of this important role, approximately 40% of all modern drugs are targeted at GPCRs.

In addition to protein- and solution-mediated processes, it is also possible for lipid bilayers to participate directly in signaling. A classic example of this is phosphatidylserine-triggered phagocytosis. Normally, phosphatidylserine is asymmetrically distributed in the cell membrane and is present only on the interior side. During programmed cell death a protein called a scramblase equilibrates this distribution, displaying phosphatidylserine on the extracellular bilayer face. The presence of phosphatidylserine then triggers phagocytosis to remove the dead or dying cell.

Characterization Methods

The lipid bilayer is a very difficult structure to study because it is so thin and fragile. In spite of these limitations dozens of techniques have been developed over the last seventy years to allow investigations of its structure and function.

Electrical measurements are a straightforward way to characterize an important function of a bilayer: its ability to segregate and prevent the flow of ions in solution. By applying a voltage across the bilayer and measuring the resulting current, the resistance of the bilayer is determined. This resistance is typically quite high (10^8 Ohm-cm^2 or more) since the hydrophobic core is impermeable to charged species. The presence of even a few nanometer-scale holes results in a dramatic increase in current. The sensitivity of this system is such that even the activity of single ion channels can be resolved.

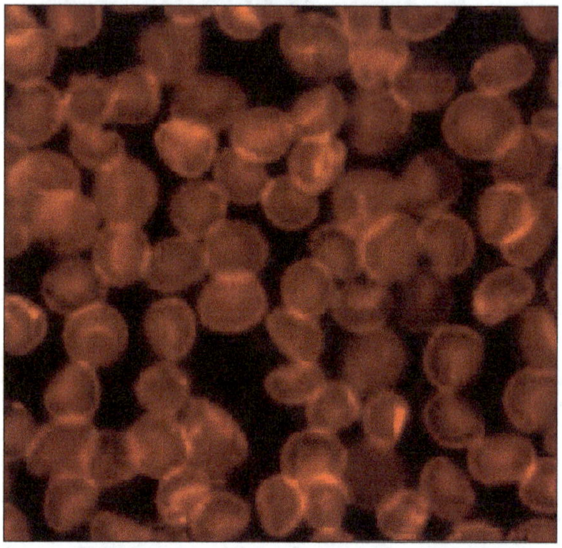

Human red blood cells viewed through a fluorescence microscope. The cell membrane has been stained with a fluorescent dye. Scale bar is 20μm.

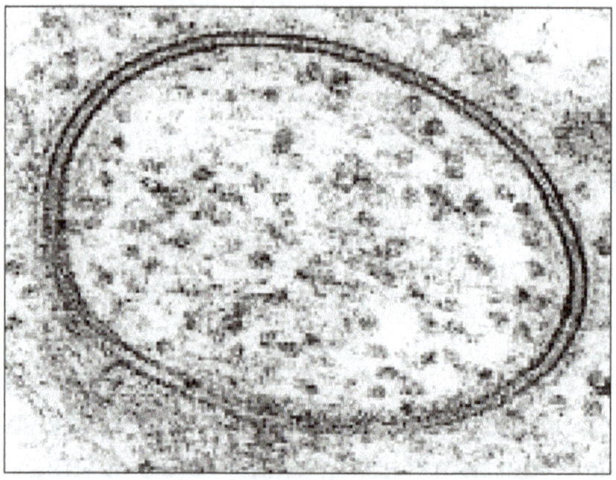

Transmission Electron Microscope (TEM) image of a lipid vesicle. The two dark bands around the edge are the two leaflets of the bilayer. Historically, similar images confirmed that the cell membrane is a bilayer.

Electrical measurements do not provide an actual picture like imaging with a microscope can. Lipid bilayers cannot be seen in a traditional microscope because they are too thin. In order to see bilayers, researchers often use fluorescence microscopy. A sample is excited with one wavelength of light and observed in a different wavelength, so that only fluorescent molecules with a matching excitation and emission profile will be seen. Natural lipid bilayers are not fluorescent, so a dye is used that attaches to the desired molecules in the bilayer. Resolution is usually limited to a few hundred nanometers, much smaller than a typical cell but much larger than the thickness of a lipid bilayer.

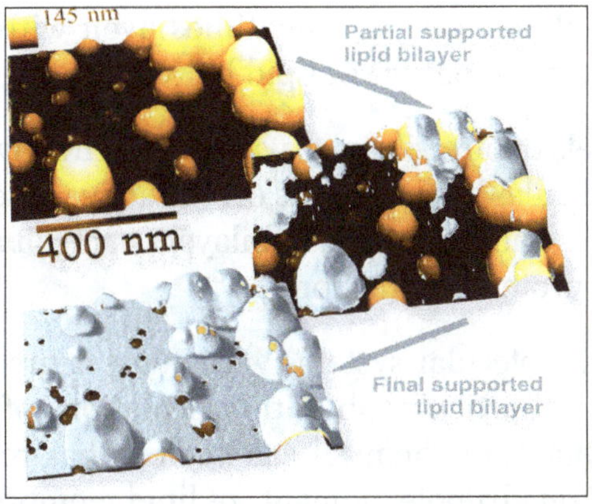

3d-Adapted AFM images showing formation of transmembrane pores (holes) in supported lipid bilayer.

Electron microscopy offers a higher resolution image. In an electron microscope, a beam of focused electrons interacts with the sample rather than a beam of light as in traditional microscopy. In conjunction with rapid freezing techniques, electron microscopy has also been used to study the mechanisms of inter- and intracellular transport, for instance in demonstrating that exocytotic vesicles are the means of chemical release at synapses.

^{31}P-NMR(nuclear magnetic resonance) spectroscopy is widely used for studies of phospholipid bilayers and biological membranes in native conditions. The analysis of ^{31}P-NMR spectra of lipids could provide a wide range of information about lipid bilayer packing, phase transitions (gel phase, physiological liquid crystal phase, ripple phases, non bilayer phases), lipid head group orientation/dynamics, and elastic properties of pure lipid bilayer and as a result of binding of proteins and other biomolecules.

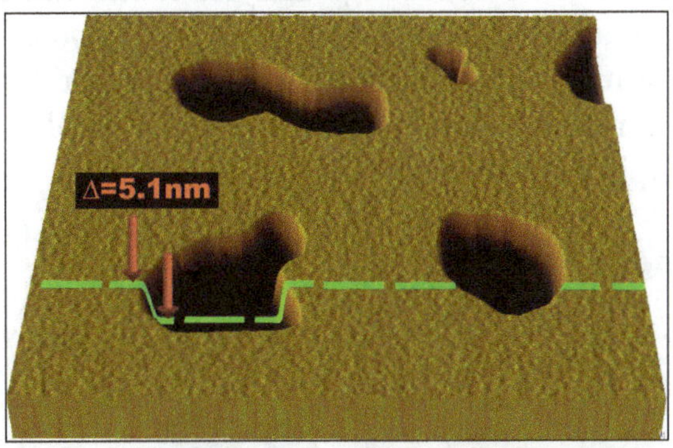

Illustration of a typical AFM scan of a supported lipid bilayer. The pits are defects in the bilayer, exposing the smooth surface of the substrate underneath.

A new method to study lipid bilayers is Atomic force microscopy (AFM). Rather than using a beam of light or particles, a very small sharpened tip scans the surface by making physical contact with the bilayer and moving across it, like a record player needle. AFM is a promising technique because it has the potential to image with nanometer resolution at room temperature and even under water or physiological buffer, conditions necessary for natural bilayer behavior. Utilizing this capability, AFM has been used to examine dynamic bilayer behavior including the formation of transmembrane pores (holes) and phase transitions in supported bilayers. Another advantage is that AFM does not require fluorescent or isotopic labeling of the lipids, since the probe tip interacts mechanically with the bilayer surface. Because of this, the same scan can image both lipids and associated proteins, sometimes even with single-molecule resolution. AFM can also probe the mechanical nature of lipid bilayers.

Lipid bilayers exhibit high levels of birefringence where the refractive index in the plane of the bilayer differs from that perpendicular by as much as 0.1 refractive index units. This has been used to characterise the degree of order and disruption in bilayers using dual polarisation interferometry to understand mechanisms of protein interaction.

Lipid bilayers are complicated molecular systems with many degrees of freedom. Thus atomistic simulation of membrane and in particular ab initio calculations of its properties is difficult and computationally expensive. Quantum chemical calculations has recently been successfully performed to estimate dipole and quadrupole moments of lipid membranes.

Transport Across the Bilayer

Passive Diffusion

Most polar molecules have low solubility in the hydrocarbon core of a lipid bilayer and, as a consequence, have low permeability coefficients across the bilayer. This effect is particularly pronounced for charged species, which have even lower permeability coefficients than neutral polar molecules. Anions typically have a higher rate of diffusion through bilayers than cations. Compared to ions, water molecules actually have a relatively large permeability through the bilayer, as evidenced by osmotic swelling. When a cell or vesicle with a high interior salt concentration is placed in a solution with a low salt concentration it will swell and eventually burst. Such a result would not be observed unless water was able to pass through the bilayer with relative ease. The anomalously large permeability of water through bilayers is still not completely understood and continues to be the subject of active debate. Small uncharged apolar molecules diffuse through lipid bilayers many orders of magnitude faster than ions or water. This applies both to fats and organic solvents like chloroform and ether. Regardless of their polar character larger molecules diffuse more slowly across lipid bilayers than small molecules.

Ion Pumps and Channels

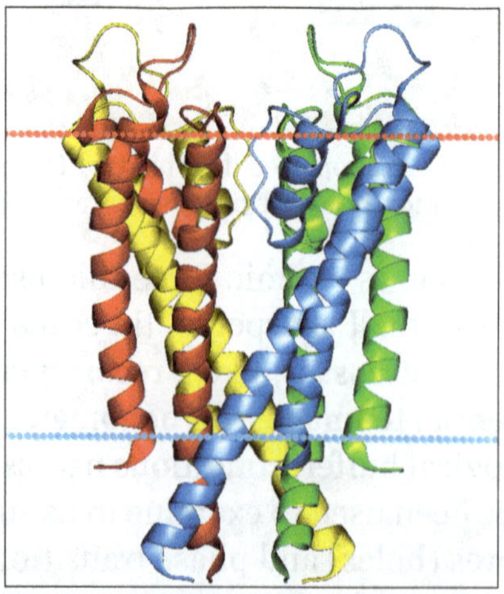

Structure of a potassium ion channel. The alpha helices penetrate the bilayer (boundaries indicated by red and blue lines), opening a hole through which potassium ions can flow.

Two special classes of protein deal with the ionic gradients found across cellular and sub-cellular membranes in nature- ion channels and ion pumps. Both pumps and channels are integral membrane proteins that pass through the bilayer, but their roles are quite different. Ion pumps are the proteins that build and maintain the chemical gradients by utilizing an external energy source to move ions against the concentration gradient to an area of higher chemical potential. The energy source can be ATP, as is the case for the Na^+-K^+ ATPase. Alternatively, the energy source can be another chemical gradient already in place, as in the Ca^{2+}/Na^+ antiporter. It is through the action of ion pumps that cells are able to regulate pH via the pumping of protons.

In contrast to ion pumps, ion channels do not build chemical gradients but rather dissipate them in order to perform work or send a signal. Probably the most familiar and best studied example is the voltage-gated Na^+ channel, which allows conduction of an action potential along neurons. All ion pumps have some sort of trigger or "gating" mechanism. In the previous example it was electrical bias, but other channels can be activated by binding a molecular agonist or through a conformational change in another nearby protein.

Endocytosis and Exocytosis

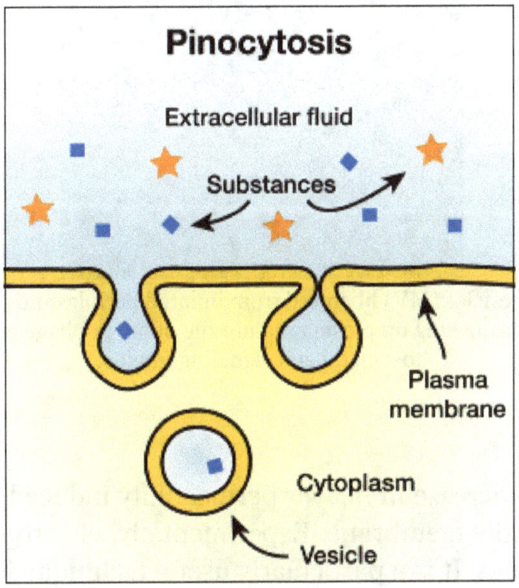

Schematic illustration of pinocytosis, a type of endocytosis.

Some molecules or particles are too large or too hydrophilic to pass through a lipid bilayer. Other molecules could pass through the bilayer but must be transported rapidly in such large numbers that channel-type transport is impractical. In both cases, these types of cargo can be moved across the cell membrane through fusion or budding of vesicles. When a vesicle is produced inside the cell and fuses with the plasma membrane to release its contents into the extracellular space, this process is known as exocytosis. In the reverse process, a region of the cell membrane will dimple inwards and eventually pinch off, enclosing a portion of the extracellular fluid to transport it into the cell. Endocytosis and exocytosis rely on very different molecular machinery to function, but the two processes are intimately linked and could not work without each other. The primary mechanism of this interdependence is the sheer volume of lipid material involved. In a typical cell, an area of bilayer equivalent to the entire plasma membrane will travel through the endocytosis/exocytosis cycle in about half an hour. If these two processes were not balancing each other, the cell would either balloon outward to an unmanageable size or completely deplete its plasma membrane within a matter of minutes.

Exocytosis in prokaryotes: Membrane vesicular exocytosis, popularly known as membrane vesicle trafficking, a Nobel prize-winning (year, 2013) process, is traditionally regarded as a prerogative of eukaryotic cells. This *myth* was however broken with the revelation that nanovesicles, popularly known as bacterial outer membrane vesicles, released by gram-negative microbes, translocate bacterial signal molecules to host or target cells to carry out multiple processes in favour of the secreting microbe e.g., in *host cell invasion* and microbe-environment interactions, in general.

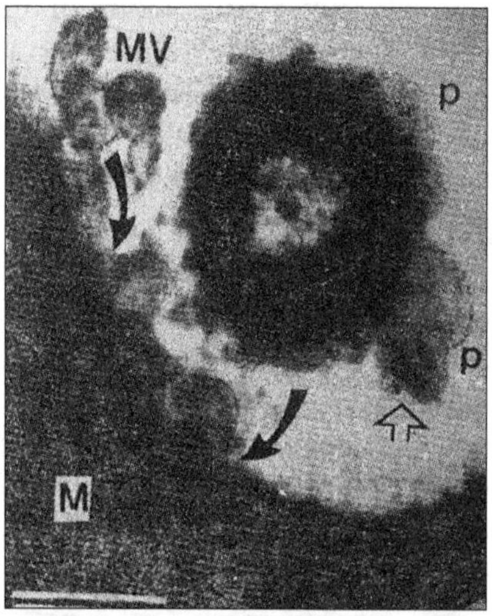

Exocytosis of outer membrane vesicles (MV) liberated from inflated periplasmic pockets (p) on surface of human *Salmonella* 3,10:r:- pathogens docking on plasma membrane of macrophage cells (M) in chicken ileum, for host-pathogen signaling *in vivo*.

Electroporation

Electroporation is the rapid increase in bilayer permeability induced by the application of a large artificial electric field across the membrane. Experimentally, electroporation is used to introduce hydrophilic molecules into cells. It is a particularly useful technique for large highly charged molecules such as DNA, which would never passively diffuse across the hydrophobic bilayer core. Because of this, electroporation is one of the key methods of transfection as well as bacterial transformation. It has even been proposed that electroporation resulting from lightning strikes could be a mechanism of natural horizontal gene transfer.

This increase in permeability primarily affects transport of ions and other hydrated species, indicating that the mechanism is the creation of nm-scale water-filled holes in the membrane. Although electroporation and dielectric breakdown both result from application of an electric field, the mechanisms involved are fundamentally different. In dielectric breakdown the barrier material is ionized, creating a conductive pathway. The material alteration is thus chemical in nature. In contrast, during electroporation the lipid molecules are not chemically altered but simply shift position, opening up a pore that acts as the conductive pathway through the bilayer as it is filled with water.

Mechanics

Lipid bilayers are large enough structures to have some of the mechanical properties of liquids or solids. The area compression modulus K_a, bending modulus K_b, and edge energy , can be used to describe them. Solid lipid bilayers also have a shear modulus, but like any liquid, the shear modulus is zero for fluid bilayers. These mechanical properties affect how the membrane functions. K_a and K_b affect the ability of proteins and small molecules to insert into the bilayer, and bilayer mechanical properties have been shown to alter the function of mechanically activated ion channels. Bilayer mechanical properties also govern what types of stress a cell can withstand without tearing. Although lipid bilayers can easily bend, most cannot stretch more than a few percent before rupturing.

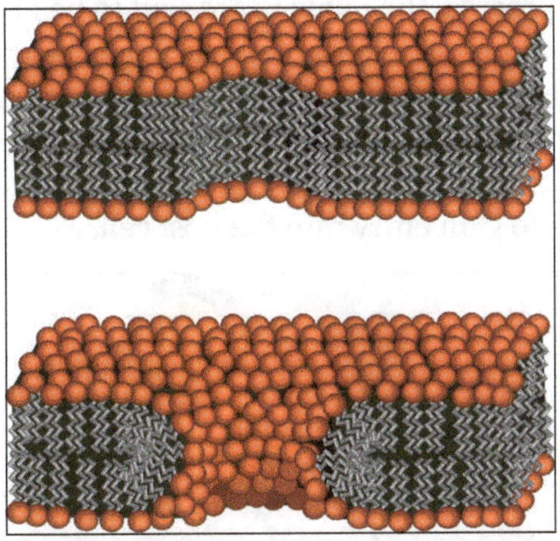

Schematic showing two possible conformations of the lipids at the edge of a pore. In the top image the lipids have not rearranged, so the pore wall is hydrophobic. In the bottom image some of the lipid heads have bent over, so the pore wall is hydrophilic.

As discussed in the Structure and organization section, the hydrophobic attraction of lipid tails in water is the primary force holding lipid bilayers together. Thus, the elastic modulus of the bilayer is primarily determined by how much extra area is exposed to water when the lipid molecules are stretched apart. It is not surprising given this understanding of the forces involved that studies have shown that K_a varies strongly with osmotic pressure but only weakly with tail length and unsaturation. Because the forces involved are so small, it is difficult to experimentally determine K_a. Most techniques require sophisticated microscopy and very sensitive measurement equipment.

In contrast to K_a, which is a measure of how much energy is needed to stretch the bilayer, K_b is a measure of how much energy is needed to bend or flex the bilayer. Formally, bending modulus is defined as the energy required to deform a membrane from its intrinsic curvature to some other curvature. Intrinsic curvature is defined by the ratio of the diameter of the head group to that of the tail group. For two-tailed PC lipids, this ratio is nearly one so the intrinsic curvature is nearly zero. If a particular lipid has too large a deviation from zero intrinsic curvature it will not form a bilayer and will instead form other phases such as micelles or inverted micelles. Addition of *small hydrophilic molecules* like *sucrose* into mixed lipid *lamellar liposomes* made from galactolipid-rich thylakoid membranes destabilises bilayers into micellar phase. Typically, K_b is not measured experimentally but rather is calculated from measurements of K_a and bilayer thickness, since the three parameters are related.

Λ is a measure of how much energy it takes to expose a bilayer edge to water by tearing the bilayer or creating a hole in it. The origin of this energy is the fact that creating such an interface exposes some of the lipid tails to water, but the exact orientation of these border lipids is unknown. There is some evidence that both hydrophobic (tails straight) and hydrophilic (heads curved around) pores can coexist.

Fusion

Fusion is the process by which two lipid bilayers merge, resulting in one connected structure. If this fusion proceeds completely through both leaflets of both bilayers, a water-filled bridge is formed and the solutions contained by the bilayers can mix. Alternatively, if only one leaflet from each bilayer is involved in the fusion process, the bilayers are said to be hemifused. Fusion is involved in many cellular processes, in particular in eukaryotes, since the eukaryotic cell is extensively sub-divided by lipid bilayer membranes. Exocytosis, fertilization of an egg by sperm and transport of waste products to the lysozome are a few of the many eukaryotic processes that rely on some form of fusion. Even the entry of pathogens can be governed by fusion, as many bilayer-coated viruses have dedicated fusion proteins to gain entry into the host cell.

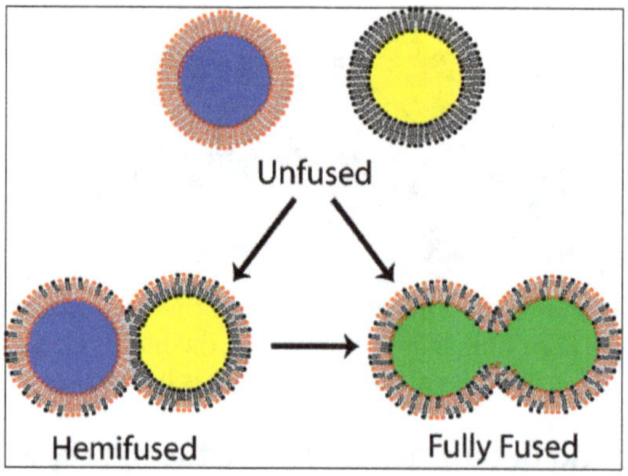

Illustration of lipid vesicles fusing showing two possible outcomes: hemifusion and full fusion. In hemifusion, only the outer bilayer leaflets mix. In full fusion both leaflets as well as the internal contents mix.

There are four fundamental steps in the fusion process. First, the involved membranes must aggregate, approaching each other to within several nanometers. Second, the two bilayers must come into very close contact (within a few angstroms). To achieve this close contact, the two surfaces must become at least partially dehydrated, as the bound surface water normally present causes bilayers to strongly repel. The presence of ions, in particular divalent cations like magnesium and calcium, strongly affects this step. One of the critical roles of calcium in the body is regulating membrane fusion. Third, a destabilization must form at one point between the two bilayers, locally distorting their structures. The exact nature of this distortion is not known. One theory is that a highly curved "stalk" must form between the two bilayers. Proponents of this theory believe that it explains why phosphatidylethanolamine, a highly curved lipid, promotes fusion. Finally, in the last step of fusion, this point defect grows and the components of the two bilayers mix and diffuse away from the site of contact.

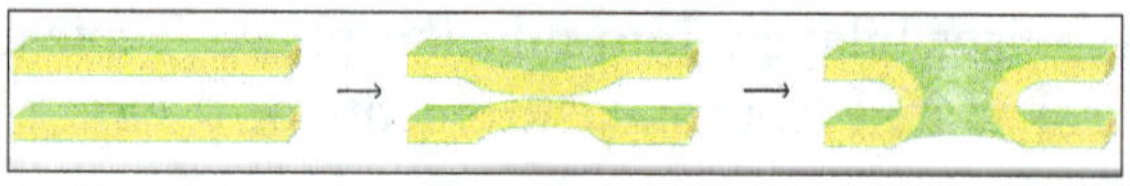

Schematic illustration of the process of fusion through stalk formation.

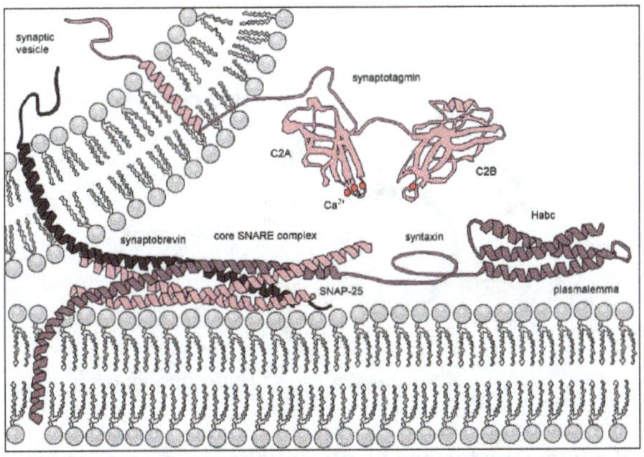

Diagram of the action of SNARE proteins docking a vesicle for exocytosis. Complementary versions of the protein on the vesicle and the target membrane bind and wrap around each other, drawing the two bilayers close together in the process.

The situation is further complicated when considering fusion *in vivo* since biological fusion is almost always regulated by the action of membrane-associated proteins. The first of these proteins to be studied were the viral fusion proteins, which allow an enveloped virus to insert its genetic material into the host cell (enveloped viruses are those surrounded by a lipid bilayer; some others have only a protein coat). Eukaryotic cells also use fusion proteins, the best-studied of which are the SNAREs. SNARE proteins are used to direct all vesicular intracellular trafficking. Despite years of study, much is still unknown about the function of this protein class. In fact, there is still an active debate regarding whether SNAREs are linked to early docking or participate later in the fusion process by facilitating hemifusion.

In studies of molecular and cellular biology it is often desirable to artificially induce fusion. The addition of polyethylene glycol (PEG) causes fusion without significant aggregation or biochemical disruption. This procedure is now used extensively, for example by fusing B-cells with myeloma cells. The resulting "hybridoma" from this combination expresses a desired antibody as determined by the B-cell involved, but is immortalized due to the melanoma component. Fusion can also be artificially induced through electroporation in a process known as electrofusion. It is believed that this phenomenon results from the energetically active edges formed during electroporation, which can act as the local defect point to nucleate stalk growth between two bilayers.

Model Systems

Lipid bilayers can be created artificially in the lab to allow researchers to perform experiments that cannot be done with natural bilayers. These synthetic systems are called model lipid bilayers. There are many different types of model bilayers, each having experimental advantages and disadvantages. They can be made with either synthetic or natural lipids. Among the most common model systems are:

- Black lipid membranes (BLM)
- Supported lipid bilayers (SLB)
- Tethered Bilayer Lipid Membranes (t-BLM)
- Vesicles

Liposome

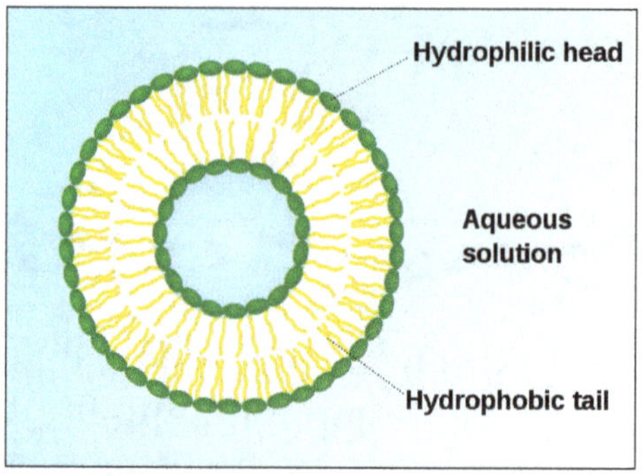

Scheme of a liposome formed by phospholipids in an aqueous solution.

A liposome is a spherical vesicle having at least one lipid bilayer. The liposome can be used as a vehicle for administration of nutrients and pharmaceutical drugs. Liposomes can be prepared by disrupting biological membranes (such as by sonication).

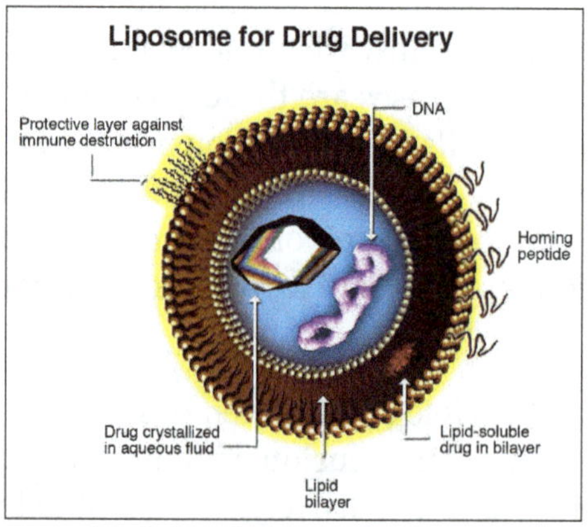

Liposomes are composite structures made of phospholipids and may contain small amounts of other molecules. Though liposomes can vary in size from low micrometer range to tens of micrometers, unilamellar liposomes, as pictured here, are typically in the lower size range with various targeting ligands attached to their surface allowing for their surface-attachment and accumulation in pathological areas for treatment of disease.

Liposomes are most often composed of phospholipids, especially phosphatidylcholine, but may also include other lipids, such as egg phosphatidylethanolamine, so long as they are compatible with lipid bilayer structure. A liposome design may employ surface ligands for attaching to unhealthy tissue.

The major types of liposomes are the multilamellar vesicle (MLV, with several lamellar phase lipid bilayers), the small unilamellar liposome vesicle (SUV, with one lipid bilayer), the large unilamellar vesicle (LUV), and the cochleate vesicle. A less desirable form are multivesicular liposomes in which one vesicle contains one or more smaller vesicles.

Liposomes should not be confused with lysosomes, or with micelles and reverse micelles composed of monolayers.

Discovery

The word *liposome* derives it is so named because its composition is primarily of phospholipid.

Liposomes were first described by British haematologist Alec D Bangham in 1961 (published 1964), at the Babraham Institute, in Cambridge. They were discovered when Bangham and R. W. Horne were testing the institute's new electron microscope by adding negative stain to dry phospholipids. The resemblance to the plasmalemma was obvious, and the microscope pictures served as the first evidence for the cell membrane being a bilayer lipid structure. Their integrity as a closed, bilayer structure, that could release its contents after detergent treatment (structure-linked latency) was established by Bangham, Standish and Weissmann in the next year. Weissmann - during a Cambridge pub discussion with Bangham - first named the structures "liposomes" after the lysosome, which his laboratory had been studying: a simple organelle the structure-linked latency of which could be disrupted by detergents and streptolysins. Liposomes can be easily distinguished from micelles and hexagonal lipid phases by negative staining transmission electron microscopy.

Alec Douglas Bangham with colleagues Jeff Watkins and Malcolm Standish wrote the 1965 paper that effectively launched the liposome "industry". Around this time he was joined at Babraham by Gerald Weissmann, an American physician with an interest in lysosomes. Now an emeritus professor at New York University School of Medicine, Weissmann recalls the two of them sitting in a Cambridge pub and reflecting on the role of lipid sheets in separating the interior of the cell from the exterior milieu. This insight, they felt, was to cell function what the discovery of the double helix had been to genetics. Bangham had called his lipid structures "multilamellar smectic mesophases" or sometimes "Banghasomes". It was Weissmann who proposed the more user-friendly term liposome.

Mechanism

A liposome has an aqueous solution core surrounded by a hydrophobic membrane, in the form of a lipid bilayer; hydrophilic solutes dissolved in the core cannot readily pass through the bilayer. Hydrophobic chemicals associate with the bilayer. A liposome can be hence loaded with hydrophobic and/or hydrophilic molecules. To deliver the molecules to a site of action, the lipid bilayer can fuse with other bilayers such as the cell membrane, thus delivering the liposome contents; this is a complex and non-spontaneous event, however. By preparing liposomes in a solution of DNA or drugs (which would normally be unable to diffuse through the membrane) they can be (indiscriminately) delivered past the lipid bilayer, but are then typically distributed non-homogeneously.

Liposomes are used as models for artificial cells. Liposomes can also be designed to deliver drugs in other ways. Liposomes that contain low (or high) pH can be constructed such that dissolved aqueous drugs will be charged in solution (i.e., the pH is outside the drug's pI range). As the pH naturally neutralizes within the liposome (protons can pass through some membranes), the drug will also be neutralized, allowing it to freely pass through a membrane. These liposomes work to deliver drug by diffusion rather than by direct cell fusion.

A similar approach can be exploited in the biodetoxification of drugs by injecting empty liposomes with a transmembrane pH gradient. In this case the vesicles act as sinks to scavenge the drug in the blood circulation and prevent its toxic effect. Another strategy for liposome drug delivery is to target endocytosis events. Liposomes can be made in a particular size range that makes them viable targets for natural macrophage phagocytosis. These liposomes may be digested while in the macrophage's phagosome, thus releasing its drug. Liposomes can also be decorated with opsonins and ligands to activate endocytosis in other cell types.

The use of liposomes for transformation or transfection of DNA into a host cell is known as lipofection.

In addition to gene and drug delivery applications, liposomes can be used as carriers for the delivery of dyes to textiles, pesticides to plants, enzymes and nutritional supplements to foods, and cosmetics to the skin.

Liposomes are also used as outer shells of some microbubble contrast agents used in contrast-enhanced ultrasound.

Lipid Signaling

Lipid signaling, broadly defined, refers to any biological signaling event involving a lipid messenger that binds a protein target, such as a receptor, kinase or phosphatase, which in turn mediate the effects of these lipids on specific cellular responses. Lipid signaling is thought to be qualitatively different from other classical signaling paradigms (such as monoamine neurotransmission) because lipids can freely diffuse through membranes. One consequence of this is that lipid messengers cannot be stored in vesicles prior to release and so are often biosynthesized "on demand" at their intended site of action. As such, many lipid signaling molecules cannot circulate freely in solution but, rather, exist bound to special carrier proteins in serum.

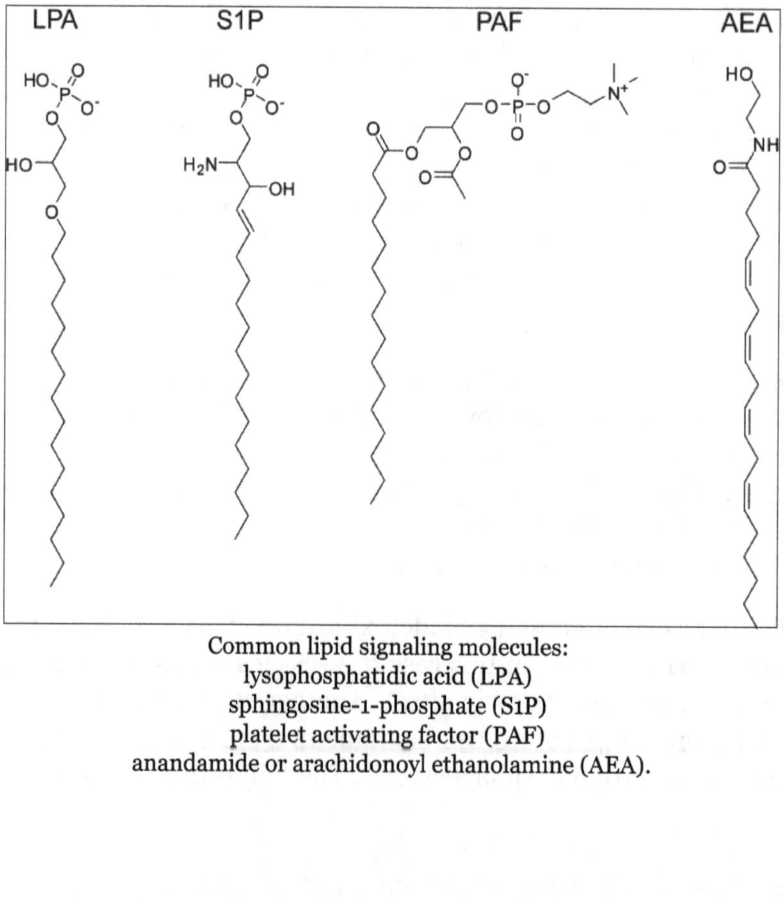

Common lipid signaling molecules:
lysophosphatidic acid (LPA)
sphingosine-1-phosphate (S1P)
platelet activating factor (PAF)
anandamide or arachidonoyl ethanolamine (AEA).

Sphingolipid Second Messengers

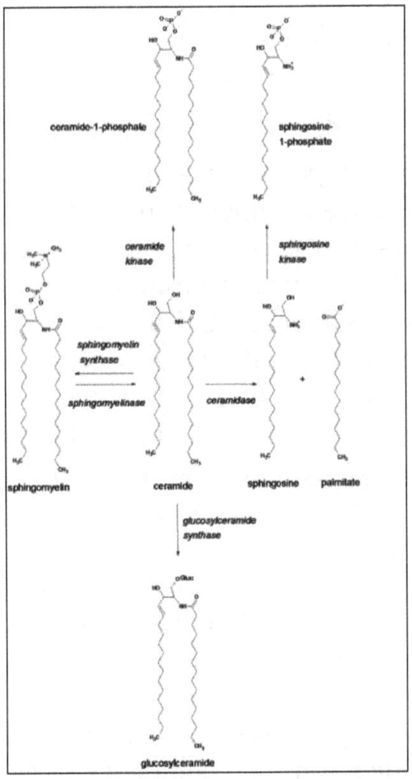

Sphingolipid second messengers. Ceramide is at the metabolic hub, leading to the formation of other sphingolipids.

Ceramide

Ceramide (Cer) can be generated by the breakdown of sphingomyelin (SM) by sphingomyelinases (SMases), which are enzymes that hydrolyze the phosphocholine group from the sphingosine backbone. Alternatively, this sphingosine-derived lipid (sphingolipid) can be synthesized from scratch (*de novo*) by the enzymes serine palmitoyl transferase (SPT) and ceramide synthase in organelles such as the endoplasmic reticulum (ER) and possibly, in the mitochondria-associated membranes (MAMs) and the perinuclear membranes. Being located in the metabolic hub, ceramide leads to the formation of other sphingolipids, with the C1 hydroxyl (-OH) group as the major site of modification. A sugar can be attached to ceramide (glycosylation) through the action of the enzymes, glucosyl or galactosyl ceramide synthases. Ceramide can also be broken down by enzymes called ceramidases, leading to the formation of sphingosine, Moreover, a phosphate group can be attached to ceramide (phosphorylation) by the enzyme, ceramide kinase. It is also possible to regenerate sphingomyelin from ceramide by accepting a phosphocholine headgroup from phosphatidylcholine (PC) by the action of an enzyme called sphingomyelin synthase. The latter process results in the formation of diacylglycerol (DAG) from PC.

Ceramide contains two hydrophobic ("water-fearing") chains and a neutral headgroup. Consequently, it has limited solubility in water and is restricted within the organelle where it was formed. Also, because of its hydrophobic nature, ceramide readily flip-flops across membranes as supported by studies in membrane models and membranes from red blood cells (erythrocytes). However, ceramide can possibly interact with other lipids to form bigger regions called microdomains which restrict its flip-flopping abilities. This could have immense effects on the signaling functions of ceramide because it is known that ceramide generated by acidic SMase enzymes in the outer leaflet of an organelle membrane may have different roles compared to ceramide that is formed in the inner leaflet by the action of neutral SMase enzymes.

Ceramide mediates many cell-stress responses, including the regulation of programmed cell death (apoptosis) and cell aging (senescence). Numerous research works have focused interest on defining the direct protein targets of action of ceramide. These include enzymes called ceramide-activated Ser-Thr phosphatases (CAPPs), such as protein phosphatase 1 and 2A (PP1 and PP2A), which were found to interact with ceramide in studies done in a controlled environment outside of a living organism (*in vitro*). On the other hand, studies in cells have shown that ceramide-inducing agents such as tumor necrosis factor-alpha α (TNFα) and palmitate induce the ceramide-dependent removal of a phosphate group (dephosphorylation) of the retinoblastoma gene product RB and the enzymes, protein kinases B (AKT protein family) and C α (PKB and PKCα). Moreover, there is also sufficient evidence which implicates ceramide to the activation of the kinase suppressor of Ras (KSR), PKCζ, and cathepsin D. Interestingly, cathepsin D has been proposed as the main target for ceramide formed in organelles called lysosomes, making lysosomal acidic SMase enzymes one of the key players in the mitochondrial pathway of apoptosis. Ceramide was also shown to activate PKCζ, implicating it to the inhibition of AKT, regulation of the voltage difference between the interior and exterior of the cell (membrane potential) and signaling functions that favor apoptosis. Chemotherapeutic agents such as daunorubicin and etoposide enhance the *de novo* synthesis of ceramide in studies done on mammalian cells. The same results were found for certain inducers of apoptosis particularly stimulators of receptors in a class of lymphocytes (a type of white blood cell) called B-cells. Regulation of the *de novo* synthesis of ceramide by palmitate may have a key role in diabetes and the metabolic syndrome. Experimental evidence shows that there is substantial increase of ceramide levels upon adding palmitate. Ceramide accumulation activates PP2A and the subsequent dephosphorylation and inactivation of AKT, a crucial mediator in metabolic control and insulin signaling. This results in a substantial decrease in insulin responsiveness (i.e. to glucose) and in the death of insulin-producing cells in the pancreas called islets of Langerhans. Inhibition of ceramide synthesis in mice via drug treatments or gene-knockout techniques prevented insulin resistance induced by fatty acids, glucocorticoids or obesity.

An increase in *in vitro* activity of acid SMase has been observed after applying multiple stress stimuli such as ultraviolet (UV) and ionizing radiation, binding of death receptors and chemotherapeutic agents such as platinum, histone deacetylase inhibitors and paclitaxel. In some studies, SMase activation results to its transport to the plasma membrane and the simultaneous formation of ceramide.

Ceramide transfer protein (CERT) transports ceramide from ER to the Golgi for the synthesis of SM. CERT is known to bind phosphatidylinositol phosphates, hinting its potential regulation via phosphorylation, a step of the ceramide metabolism that can be enzymatically regulated by protein kinases and phosphatases, and by inositol lipid metabolic pathways. Up to date, there are at least 26 distinct enzymes with varied subcellular localizations, that act on ceramide as either a substrate or product. Regulation of ceramide levels can therefore be performed by one of these enzymes in distinct organelles by particular mechanisms at various times.

Sphingosine

Sphingosine (Sph) is formed by the action of ceramidase (CDase) enzymes on ceramide in the lysosome. Sph can also be formed in the extracellular (outer leaflet) side of the plasma membrane by the action of neutral CDase enzyme. Sph then is either recycled back to ceramide or phosphorylated by one of the sphingosine kinase enzymes, SK1 and SK2. The product sphingosine-1-phosphate (S1P) can be dephosphorylated in the ER to regenerate sphingosine by certain S1P phosphatase enzymes within cells, where the salvaged Sph is recycled to ceramide. Sphingosine is a single-chain lipid (usually 18 carbons in length), rendering it to have sufficient solubility in water. This explains its ability to move between membranes and to flip-flop across a membrane. Estimates conducted at physiological pH show that approximately 70% of sphingosine remains in membranes while the remaining 30% is water-soluble. Sph that is formed has sufficient solubility in the liquid found inside cells (cytosol). Thus, Sph may come out of the lysosome and move to the ER without the need for transport via proteins or membrane-enclosed sacs called vesicles. However, its positive charge favors partitioning in lysosomes. It is proposed that the role of SK1 located near or in the lysosome is to 'trap' Sph via phosphorylation.

It is important to note that since sphingosine exerts surfactant activity, it is one of the sphingolipids found at lowest cellular levels. The low levels of Sph and their increase in response to stimulation of cells, primarily by activation of ceramidase by growth-inducing proteins such as platelet-derived growth factor and insulin-like growth factor, is consistent with its function as a second messenger. It was found that immediate hydrolysis of only 3 to 10% of newly generated ceramide may double the levels of Sph. Treatment of HL60 cells (a type of leukemia cell line) by a plant-derived organic compound called phorbol ester increased Sph levels threefold, whereby the cells differentiated into white blood cells called macrophages. Treatment of the same cells by exogenous Sph caused apoptosis. A specific protein kinase phosphorylates 14-3-3, otherwise known as sphingosine-dependent protein kinase 1 (SDK1), only in the presence of Sph.

Sph is also known to interact with protein targets such as the protein kinase H homologue (PKH) and the yeast protein kinase (YPK). These targets in turn mediate the effects of Sph and its related sphingoid bases, with known roles in regulating the actin cytoskeleton, endocytosis, the cell cycle and apoptosis. It is important to note however that the second messenger function of Sph is not yet established unambiguously.

Sphingosine-1-Phosphate

Sphingosine-1-phosphate (S1P), like Sph, is composed of a single hydrophobic chain and has sufficient solubility to move between membranes. S1P is formed by phosphorylation of sphingosine by sphingosine kinase (SK). The phosphate group of the product can be detached (dephosphorylated) to regenerate sphingosine via S1P phosphatase enzymes or S1P can be broken down by S1P lyase enzymes to ethanolamine phosphate and hexadecenal. Similar to Sph, its second messenger function is not yet clear. However, there is substantial evidence that implicates S1P to cell survival, cell migration, and inflammation. Certain growth-inducing proteins such as platelet-derived growth factor (PDGF), insulin-like growth factor (IGF) and vascular endothelial growth factor (VEGF) promote the formation of SK enzymes, leading to increased levels of S1P. Other factors that induce SK include cellular communication molecules called cytokines, such as tumor necrosis factor α (TNFα) and interleukin-1 (IL-1), hypoxia or lack of oxygen supply in cells, oxidized low-density lipoproteins (oxLDL) and several immune complexes.

S1P is probably formed at the inner leaflet of the plasma membrane in response to TNFα and other receptor activity-altering compounds called agonists. S1P, being present in low nanomolar concentrations in the cell, has to interact with high-affinity receptors that are capable of sensing their low levels. So far, the only identified receptors for S1P are the high-affinity G protein-coupled receptors (GPCRs), also known as S1P receptors (S1PRs). S1P is required to reach the extracellular side (outer leaflet) of the plasma membrane to interact with S1PRs and launch typical GPCR signaling pathways. However, the zwitterionic headgroup of S1P makes it unlikely to flip-flop spontaneously. To overcome this difficulty, the ATP-binding cassette (ABC) transporter C1 (ABCC1) serves as the "exit door" for S1P. On the other hand, the cystic fibrosis transmembrane regulator (CFTR) serves as the means of entry for S1P into the cell. In contrast to its low intracellular concentration, S1P is found in high nanomolar concentrations in serum where it is bound to albumin and lipoproteins. Inside the cell, S1P can induce calcium release independent of the S1PRs—the mechanism of which remains unknown. To date, the intracellular molecular targets for S1P are still unidentified.

The SK1-S1P pathway has been extensively studied in relation to cytokine action, with multiple functions connected to effects of TNFα and IL-1 favoring inflammation. Studies show that knockdown of key enzymes such as S1P lyase and S1P phosphatase increased prostaglandin production, parallel to increase of S1P levels. This strongly suggests that S1P is the mediator of SK1 action and not subsequent compounds. Research done on endothelial and smooth muscle cells is consistent to the hypothesis that S1P has a crucial role in regulating endothelial cell growth, and movement. Recent work on a sphingosine analogue, FTY270, demonstrates its ability to act as a potent compound that alters the activity of S1P receptors (agonist). FTY270 was further verified in clinical tests to have roles in immune modulation, such as that on multiple sclerosis. This highlights the importance of S1P in the regulation of lymphocyte function and immunity. Most of the studies on S1P are used to further understand diseases such as cancer, arthritis and inflammation, diabetes, immune function and neurodegenerative disorders.

Glucosylceramide

Glucosylceramides (GluCer) are the most widely distributed glycosphingolipids in cells serving as precursors for the formation of over 200 known glycosphingolipids. GluCer is formed by the glycosylation of ceramide in an organelle called Golgi via enzymes called glucosylceramide synthase (GCS) or by the breakdown of complex glycosphingolipids (GSLs) through the action of specific hydrolase enzymes. In turn, certain β-glucosidases hydrolyze these lipids to regenerate ceramide. GluCer appears to be synthesized in the inner leaflet of the Golgi. Studies show that GluCer has to flip to the inside of the Golgi or transfer to the site of GSL synthesis to initiate the synthesis of complex GSLs. Transferring to the GSL synthesis site is done with the help of a transport protein known as four phosphate adaptor protein 2 (FAPP2) while the flipping to the inside of the Golgi is made possible by the ABC transporter P-glycoprotein, also known as the multi-drug resistance 1 transporter (MDR1). GluCer is implicated in post-Golgi trafficking and drug resistance particularly to chemotherapeutic agents. For instance, a study demonstrated a correlation between cellular drug resistance and modifications in GluCer metabolism.

In addition to their role as building blocks of biological membranes, glycosphingolipids have long attracted attention because of their supposed involvement in cell growth, differentiation, and formation of tumors. The production of GluCer from Cer was found to be important in the growth of neurons or brain cells. On the other hand, pharmacological inhibition of GluCer synthase is being considered a technique to avoid insulin resistance.

Ceramide-1-Phosphate

Ceramide-1-phosphate (C1P) is formed by the action of ceramide kinase (CK) enzymes on Cer. C1P carry ionic charge at neutral pH and contain two hydrophobic chains making it relatively insoluble in aqueous environment. Thus, C1P reside in the organelle where it was formed and is unlikely to spontaneously flip-flop across membrane bilayers.

C1P activate phospholipase A2 and is found, along with CK, to be a mediator of arachidonic acid released in cells in response to a protein called interleukin-1β (IL-1β) and a lipid-soluble molecule that transports calcium ions (Ca^{2+}) across the bilayer, also known as calcium ionophore. C1P was also previously reported to encourage cell division (mitogenic) in fibroblasts, block apoptosis by inhibiting acid SMase in white blood cells within tissues (macrophages) and increase intracellular free calcium concentrations in thyroid cells. C1P also has known roles in vesicular trafficking, cell survival, phagocytosis ("cell eating") and macrophage degranulation.

Phosphatidylinositol bisphosphate (PIP2) Lipid Agonist

PIP_2 binds directly to ion channels and modulates their activity. PIP_2 was shown to directly agonizes Inward rectifying potassium channels(K_{ir}). In this regard intact PIP_2 signals as a bona fide neurotransmitter-like ligand. PIP_2's interaction with many ion channels suggest that the intact form of PIP_2 has an important signaling role independent of second messenger signaling.

Second Messengers from Phosphatidylinositol

Phosphatidylinositol Bisphosphate (PIP2) Second Messenger Systems

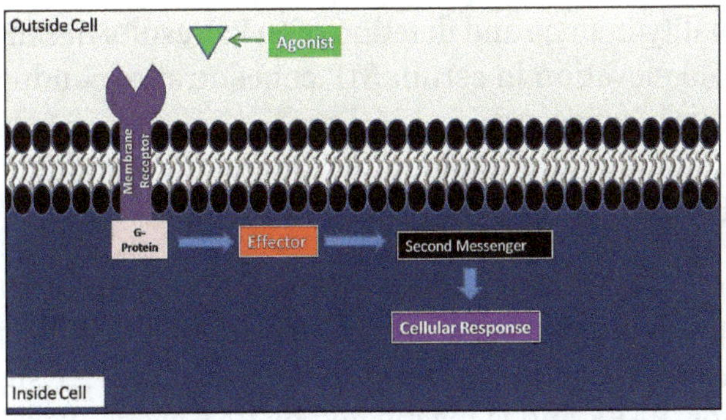

Cartoon of second messenger systems. Figure adapted From Barbraham Institute Mike Berridge.

A general second messenger system mechanism can be broken down into four steps. First, the agonist activates a membrane-bound receptor. Second, the activated G-protein produces a primary effector. Third, the primary effect stimulates the second messenger synthesis. Fourth, the second messenger activates a certain cellular process.

The G-protein coupled receptors for the PIP_2 messenger system produces two effectors, phospholipase C (PLC) and phosphoinositide 3-kinase (PI3K). PLC as an effector produces two different second messengers, inositol triphosphate (IP_3) and Diacylglycerol (DAG).

IP_3 is soluble and diffuses freely into the cytoplasm. As a second messenger, it is recognized by the inositol triphosphate receptor (IP3R), a Ca^{2+} channel in the endoplasmic reticulum (ER) membrane, which stores intracellular Ca^{2+}. The binding of IP_3 to IP3R releases Ca^{2+} from the ER into the normally Ca^{2+}-poor cytoplasm, which then triggers various events of Ca^{2+} signaling. Specifically in blood vessels, the increase in Ca^{2+} concentration from IP_3 releases nitric oxide, which then diffuses into the smooth muscle tissue and causes relaxation.

DAG remains bound to the membrane by its fatty acid "tails" where it recruits and activates both conventional and novel members of the protein kinase C family. Thus, both IP_3 and DAG contribute to activation of PKCs.

Phosphoinositide 3-kinase (PI3K) as an effector phosphorylates phosphatidylinositol bisphosphate (PIP_2) to produce phosphatidylinositol (3,4,5)-trisphosphate (PIP_3). PIP_3 has been shown to activate protein kinase B, increase binding to extracellular proteins and ultimately enhance cell survival.

Activators of G-protein Coupled Receptors

Lysophosphatidic Acid (LPA)

LPA is the result of phospholipase A2 action on phosphatidic acid. The SN-1 position can contain either an ester bond or an ether bond, with ether LPA being found at elevated levels in certain cancers. LPA binds the high-affinity G-protein coupled receptors LPA1, LPA2, and LPA3 (also known as EDG2, EDG4, and EDG7, respectively).

Sphingosine-1-phosphate (S1P)

S1P is present at high concentrations in plasma and secreted locally at elevated concentrations at sites of inflammation. It is formed by the regulated phosphorylation of sphingosine. It acts through five dedicated high-affinity G-protein coupled receptors, S1P1 - S1P5. Interestingly, targeted deletion of S1P1 results in lethality in mice and deletion of S1P2 results in seizures and deafness. Additionally, a mere 3- to 5-fold elevation in serum S1P concentrations induces sudden cardiac death by an S1P3-receptor specific mechanism.

Platelet Activating Factor (PAF)

PAF is a potent activator of platelet aggregation, inflammation, and anaphylaxis. It is similar to the ubiquitous membrane phospholipid phosphatidylcholine except that it contains an acetyl-group in the SN-2 position and the SN-1 position contains an ether-linkage. PAF signals through a dedicated G-protein coupled receptor, PAFR and is inactivated by PAF acetylhydrolase.

Endocannabinoids

The endogenous cannabinoids, or endocannabinoids, are endogenous lipids that activate cannabinoid receptors. The first such lipid to be isolated was anandamide which is the arachidonoyl amide of ethanolamine. Anandamide is formed via enzymatic release from N-arachidonoyl phosphatidylethanolamine by the N-acyl phosphatidylethanolamine phospholipase D (NAPE-PLD). Anandamide activates both the CB1 receptor, found primarily in the central nervous system, and the CB2 receptor which is found primarily in lymphocytes and the periphery. It is found at very low levels (nM) in most tissues and is inactivated by the fatty acid amide hydrolase. Subsequently, another endocannabinoid was isolated, 2-arachidonoylglycerol, which is produced when phospholipase C releases diacylglycerol which is then converted to 2-AG by diacylglycerol lipase. 2-AG can also activate both cannabinoid receptors and is inactivated by monoacylglycerol lipase. It is present at approximately 100-times the concentration of anandamide in most tissues. Elevations in either of these lipids causes analgesia and anti-inflammation and tissue protection during states of ischemia, but the precise roles played by these various endocannabinoids are still not totally known and intensive research into their function, metabolism, and regulation is ongoing. One saturated lipid from this class, often called an endocannabinoid, but with no relevant affinity for the CB1 and CB 2 receptor is palmitoylethanolamide. This signaling lipid has great affinity for the GRP55 receptor and the PPAR alpha receptor. It has been identified as an anti-inflammatory compound already in 1957, and as an analgesic compound in 1975. It was Rita Levi-Montalcini. who first identified one of its biological mechanisms of action, the inhibition of activated mast cells. Palmitoylethanolamide is the only endocannabinoid available on the market for treatment, as a food supplement.

Prostaglandins

Prostaglandins are formed through oxidation of arachidonic acid by cyclooxygenases and other prostaglandin synthases. There are currently nine known G-protein coupled receptors (eicosanoid receptors) that largely mediate prostaglandin physiology.

FAHFA

FAHFAs (fatty acid esters of hydroxy fatty acids) are formed in adipose tissue, improve glucose tolerance and also reduce adipose tissue inflammation. Palmitic acid esters of hydroxy-stearic acids (PAHSAs) are among the most bioactive members able to activate G-protein coupled receptors 120. Docosahexaenoic acid ester of hydroxy-linoleic acid (DHAHLA) exert anti-inflammatory and pro-resolving properties.

Retinol Derivatives

Retinaldehyde is a retinol (vitamin A) derivative responsible for vision. It binds rhodopsin, a well-characterized GPCR that binds all-cis retinal in its inactive state. Upon photoisomerization by a photon the cis-retinal is converted to trans-retinal causing activation of rhodopsin which ultimately leads to depolarization of the neuron thereby enabling visual perception.

Activators of Nuclear Receptors

Steroid Hormones

This large and diverse class of steroids are biosynthesized from isoprenoids and structurally resemble cholesterol. Mammalian steroid hormones can be grouped into five groups by the receptors to which they bind: glucocorticoids, mineralocorticoids, androgens, estrogens, and progestogens.

Retinoic Acid

Retinol (vitamin A) can be metabolized to retinoic acid which activates nuclear receptors such as the RAR to control differentiation and proliferation of many types of cells during development.

Prostaglandins

The majority of prostaglandin signaling occurs via GPCRs although certain prostaglandins activate nuclear receptors in the PPAR family.

Lipid Metabolism

Lipid metabolism is the synthesis and degradation of lipids in cells. Lipid metabolism is the break down or storage of fats for energy; these fats are obtained from consuming food and absorbing them or they are synthesized by an animal's liver. Lipogenesis is the process of synthesizing these fats. The majority of lipids found in the human body from ingesting food are Triglycerides. Since lipids are fats, lipid metabolism is often considered the digestion and absorption process of dietary fats. Lipid metabolism often begins with hydrolysis, which occurs when a chemical breaks down as a reaction to coming in contact with water. Lipid metabolism does exist in plants, though the processes differ in some ways when compared to animals.

Lipid Digestion

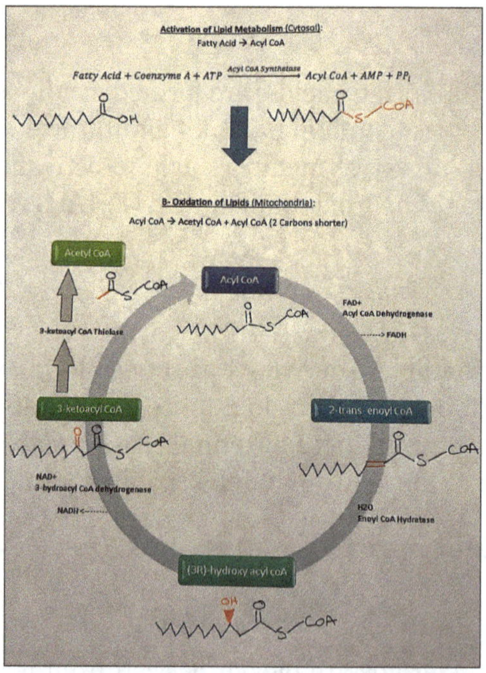

Digestion is the first step to lipid metabolism, and is the process of breaking the triglycerides down into smaller monoglyceride units with the help of lipase enzymes. Digestion of fats begin in the mouth through chemical digestion by lingual lipase. Lipids then continue to the stomach where chemical digestion continues by gastric lipase and mechanical digestion begins (Peristalsis). The majority of lipid digestion and absorption, however, occurs once the fats reach the small intestines. Chemicals from the pancreas (pancreatic lipase family and Bile salt-dependent lipase) are transported to the small intestines to help breakdown the triglycerols, along with further mechanical digestion, until they are individual fatty acid units able to be absorbed into the small intestines. It is the pancreatic lipase that is responsible for signaling for the hydrolysis of the triglycerides into separate free fatty acids and glycerol units.

Lipid Absorption

The second step in lipid metabolism is the absorption of fats. Absorption of fats occurs only in the small intestines, and only after the fats are broken down into fatty acids. The diagram to the left shows how fatty acids are absorbed into the lining of the small intestines and further broken down into smaller components to be used to create energy to fuel the body. Fatty acid metabolism begins in the cytoplasm of epithelial cells as Acyl-CoA synthetase and hydrolysis of ATP cleaves a phosphate group and catalyzes the addition of Coenzyme A to the fatty acid. The resulting Acyl-CoA activates the process of beta oxidation to further break down into Acetyl-CoA (which is used in the Citric acid cycle to produce energy). The process of beta oxidation occurs in the mitochondria of the small intestine epithelium, and require the following enzymes: Acyl CoA dehydrogenase, Enoyl-CoA hydratase, 3-hydroxyacyl-CoA dehydrogenase, and 3-ketoacyl-CoA thiolase.

Transporting Lipids

Since lipids (fats) are hydrophobic, hydrolysis in lipid metabolism occurs in the cytoplasm which ends up creating glycerol and fatty acids. Due to the hydrophobic nature of lipids they require special transport proteins known as lipoproteins, which are hydrophilic. Lipoproteins are categorized by their density levels. The varying densities between the types of lipoproteins are characteristic to what type of fats they transport. A number of these lipoproteins are synthesized in the liver, but not all of them originate from this organ.

Lipid Catabolism

Once the chilomicrons (or other lipoproteins) travel through the tissues, these particles will be broken down by lipoprotein lipase in the luminal surface of endothelial cells in capillaries to release tryglycerides.Tryglycerides will get broken down into fatty acids and glycerol before entering cells and remaining cholesterol will again travel through the blood to the liver.

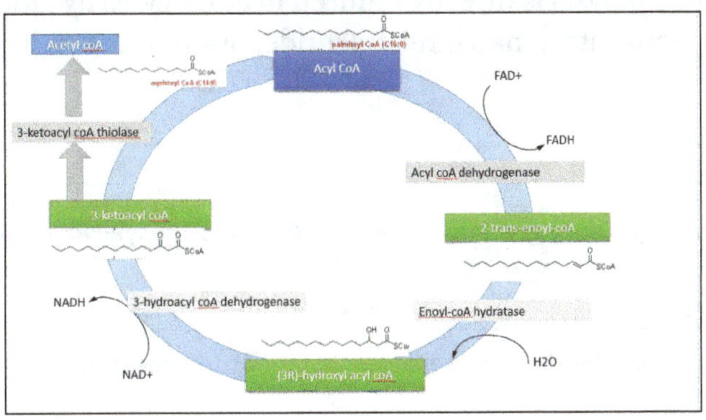

In the cytosol of the cell (for example a muscle cell), the glycerol will be converted to glyceraldehyde 3-phosphate, which is an intermediate in the glycolysis, to get further oxidized and produce energy. However, the main steps of fatty acids catabolism occur in the mitochondria. Long chain fatty acids (more than 14 carbon) need to be converted to Fatty acyl-CoA in order to pass across the mitochondria membrane. Fatty acid catabolism begins in the cytoplasm of cells as Acyl-CoA synthetase uses the energy from cleavage of an ATP to catalyze the addition of Coenzyme A to the fatty acid. The resulting Acyl-CoA cross the mitochondria membrane and enter the process of beta oxidation. The main products of the beta oxidation pathway are Acetyl-CoA (which is used in the Citric acid cycle to produce energy), NADH and FADH. The process of beta oxidation requires the following enzymes: Acyl CoA dehydrogenase, Enoyl-CoA hydratase, 3-hydroxyacyl-CoA dehydrogenase, and 3-ketoacyl-CoA thiolase. The diagram to the left shows how fatty acids are converted into Acetyl-coA. The overall net reaction, using palmitoyl CoA (16:0) as a model substrate is:

$$7\,FAD + 7\,NAD^+ + 7\,CoASH + 7\,H_2O + H(CH_2CH_2)_7CH_2CO-SCoA \rightarrow 8\,CH_3CO-SCoA + 7\,FADH_2 + 7\,NADH + 7\,H^+$$

Lipid Biosynthesis

In addition to dietary fats, storage lipids stored in the adipose tissues are one of the main sources of energy for living organisms. Triacylglycerols, lipd membrane and cholesterol can be synthesized by the organisms through various pathways.

Lipid Metabolism Disorders

Lipid Metabolism Disorders are illnesses where trouble occurs in breaking down or synthesizing fats (or fat-like substances)· A good deal of the time these disorders are hereditary, meaning it's a condition that is passed along from parent to child through their genes. Gaucher's Disease (Type I, Type II, and Type III), Neimann-Pick Disease, Tay-Sachs Disease, and Fabry's Disease are all diseases where those afflicted can have a disorder of their body's lipid metabolism. Rarer diseases concerning a disorder of the lipid metabolism are Sitosterolemia, Wolman's Disease, Refsum's Disease, and Cerebrotendinous Xanthomatosis.

Lipolysis

Lipolysis is the breakdown of lipids and involves hydrolysis of triglycerides into glycerol and free fatty acids. Predominantly occurring in adipose tissue, lipolysis is used to mobilize stored energy during fasting or exercise. Lipolysis is directly induced in adipocytes by glucagon, epinephrine, norepinephrine, growth hormone, atrial natriuretic peptide, brain natriuretic peptide, and cortisol.

Mechanisms

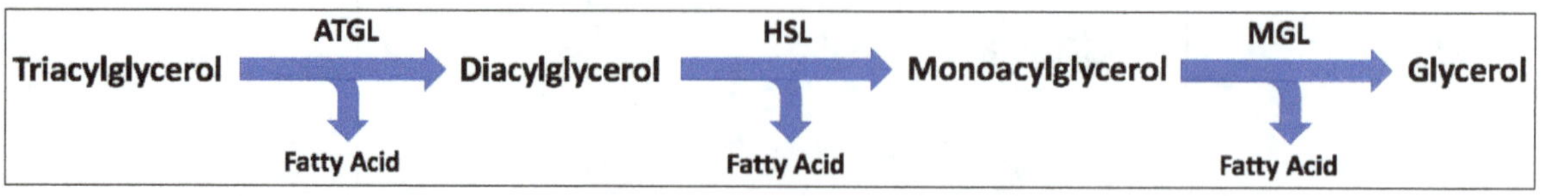

This image illustrates the three separate steps of hydrolysis involved in lipolysis.

In the first step, triacylglycerol is hydrolyzed to make diacylglycerol and this is catalyzed by adipose triglyceride lipase (ATGL). In the second step, diacylglycerol is hydrolyzed to make monoacylglycerol and this is catalyzed by hormone-sensitive lipase (HSL). In the last step, monoacylglycerol is hydrolyzed to make glycerol and this is catalyzed by monoacylglycerol lipase (MGL).

In adipose tissue, intracellular triglycerides are stored in cytoplasmic lipid droplets. When lipases are phosphorylated, they access lipid droplets and through multiple steps of hydrolysis, breakdown triglycerides into fatty acids and glycerol. Each step of hydrolysis leads to the removal of one fatty acid. The first step and the rate-limiting step of lipolysis is carried out by adipose triglyceride lipase (ATGL). This enzyme catalyzes the hydrolysis of triacylglycerol to diacylglycerol. Subsequently, hormone-sensitive lipase (HSL) catalyzes the hydrolysis of diacylglycerol to monoacylglycerol and monoacylglycerol lipase (MGL) catalyzes the hydrolysis of monoacylglycerol to glycerol. Perilipin 1A is a key protein regulator of lipolysis in adipose tissue. This lipid droplet-associated protein, when deactivated, will prevent the interaction of lipases with triglycerides in the lipid droplet and grasp the ATGL co-activator, comparative gene identification 58 (CGI-58) (a.k.a. ABHD5). When perilipin 1A is phosphorylated by PKA, it releases CGI-58 and it expedites the docking of phosphorylated lipases to the lipid droplet. CGI-58 can be further phosphorylated by PKA to assist in its dispersal to the cytoplasm. In the cytoplasm, CGI-58 can co-activate ATGL. ATGL activity is also impacted by the negative regulator of lipolysis, G0/G1 switch gene 2 (G0S2). When expressed, G0S2 acts as a competitive inhibitor in the binding of CGI-58. Fat-specific protein 27 (FSP-27) (a.k.a CIDEC) is also a negative regulator of lipolysis. FSP-27 expression is negatively correlated with ATGL mRNA levels.

Regulation

Lipolysis can be regulated through cAMP's binding and activation of protein kinase A (PKA). PKA can phosphorylate lipases, perilipin 1A, and CGI-58 to increase the rate of lipolysis. Catecholamines bind to 7TM receptors (G protein-coupled receptors) on the adipocyte cell membrane, which activate adenylate cyclase. This results in increased production of cAMP, which activates PKA and leads to an increased rate of lipolysis. Insulin counter-regulates this increase in lipolysis when it binds to insulin receptors on the adipocyte cell membrane. Insulin receptors activate insulin-like receptor substrates. These substrates activate phosphoinositide 3-kinases (PI-3K) which then phosphorylate protein kinase B (PKB) (a.k.a. Akt). PKB subsequently phosphorylates phosphodiesterase 3B (PD3B) which converts cAMP, produced by adenylate cyclase, into 5'AMP. Due to the reduced levels of cAMP, insulin decreases the rate of lipolysis. Insulin has additional actions in the mediobasal hypothalamus. It has been shown to suppress lipolysis due to lower sympathetic nervous outflow to white adipose tissue. The regulation of this process involves interactions between insulin receptors and gangliosides present in the neuronal cell membrane.

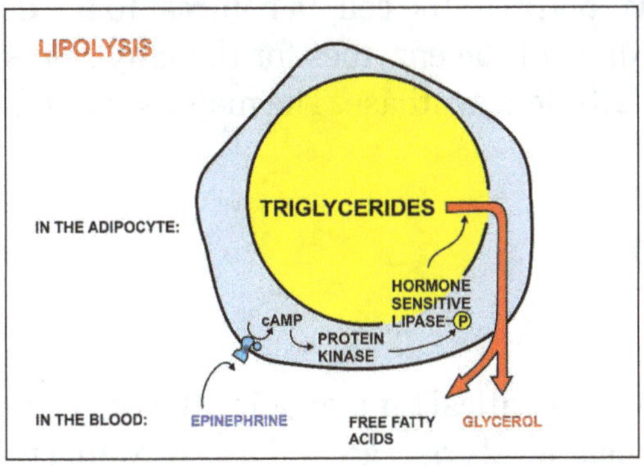

Illustration of the activation of lipolysis in an adipocyte.

Induced by high epinephrine and low insulin levels in the blood. Epinephrine binds to beta-adrenergic receptors on the cell membrane of the adipocyte, which causes cAMP to be generated inside the cell. The cAMP activate protein kinases, which phosphorylate and thus, activate hormone-sensitive lipases in the adipocyte. These lipases cleave free fatty acids from their attachment to glycerol in the lipid droplet of the adipocyte. The free fatty acids and glycerol are then released into the blood. The activity of hormone sensitive lipase is regulated by the circulating hormones, insulin, glucagon, norepinephrine, and epinephrine.

In Blood

Triglycerides are transported through the blood to appropriate tissues (adipose, muscle, etc.) by lipoproteins such as Very-Low-Density-Lipoproteins (VLDL). Triglycerides present on the VLDL undergo lipolysis by the cellular lipases of target tissues, which yields glycerol and free fatty acids. Free fatty acids released into the blood are then available for cellular uptake. Free fatty acids not immediately taken up by cells may bind to albumin for transport to surrounding tissues that require energy. Serum albumin is the major carrier of free fatty acids in the blood. The glycerol also enters the bloodstream and is absorbed by the liver or kidney where it is converted to glycerol 3-phosphate by the enzyme glycerol kinase. Hepatic glycerol 3-phosphate is converted mostly into dihydroxyacetonephosphate (DHAP) and then glyceraldehyde 3-phosphate (GA3P) to rejoin the glycolysis and gluconeogenesis pathway.

Lipogenesis

Lipogenesis is the process by which acetyl-CoA is converted to fatty acids. The former is an intermediate stage in metabolism of simple sugars, such as glucose, a source of energy of living organisms. Through lipogenesis and subsequent triglyceride synthesis, the energy can be efficiently stored in the form of fats.

Lipogenesis encompasses both the process of fatty acid synthesis and triglyceride synthesis (where fatty acids are esterified to glycerol). The products are secreted from the liver in the form of very-low-density lipoproteins (VLDL). VLDL particles are secreted directly into blood, where they mature and function to deliver the endogenously derived lipids to peripheral tissues.

Fatty Acid Synthesis

Fatty acids synthesis starts with acetyl-CoA and builds up by the addition of two-carbon units. The synthesis occurs in the cytoplasm of the cell, in contrast to the degradation (oxidation), which occurs in the mitochondria. Many of the enzymes for the fatty acid synthesis are organized into a multienzyme complex called fatty acid synthase. The major sites of fatty acid synthesis are adipose tissue and the liver.

Control and Regulation

Hormonal Regulation

Insulin is a peptide hormone that is critical for managing the body's metabolism. Insulin is released by the pancreas when blood sugar levels rise, and it has many effects that broadly promote the absorption and storage of sugars, including lipogenesis.

Insulin stimulates lipogenesis primarily by activating two enzymatic pathways. Pyruvate dehydrogenase (PDH), converts pyruvate into acetyl-CoA. Acetyl-CoA carboxylase (ACC), converts acetyl-CoA produced by PDH into malonyl-CoA. Malonyl-CoA provides the two-carbon building blocks that are used to create larger fatty acids.

Insulin stimulation of lipogenesis also occurs through the promotion of glucose uptake by adipose tissue. The increase in the uptake of glucose can occur through the use of glucose transporters directed to the plasma membrane or through the activation of lipogenic and glycolytic enzymes via covalent modification. The hormone has also been found to have long term effects on lipogenic gene expression. It is hypothesized that this effect occurs through the transcription factor SREBP-1, where the association of insulin and SREBP-1 lead to the gene expression of glucokinase. The interaction of glucose and lipogenic gene expression is assumed to be managed by the increasing concentration of an unknown glucose metabolite through the activity of glucokinase.

Another hormone that may affects lipogenesis through the SREBP-1 pathway is leptin. It is involved in the process by limiting fat storage through inhibition of glucose intake and interfering with other adipose metabolic pathways. The inhibition of lipogenesis occurs through the down regulation of fatty acid and triglyceride gene expression. Through the promotion of fatty acid oxidation and lipogenesis inhibition, leptin was found to control the release of stored glucose from adipose tissues.

Other hormones that prevent the stimulation of lipogenesis in adipose cells are growth hormones (GH). Growth hormones result in loss of fat but stimulates muscle gain. One proposed mechanism for how the hormone works is that growth hormones affects insulin signaling thereby decreasing insulin sensitivity and in turn down regulating fatty acid synthase expression. Another proposed mechanism suggests that growth hormones may phosphorylate with STAT5A and STAT5B, transcription factors that are a part of the Signal Transducer And Activator Of Transcription (STAT) family.

There is also evidence suggesting that acylation stimulating protein (ASP) promotes the aggregation of triglycerides in adipose cells. This aggregation of triglycerides occurs through the increase in the synthesis of triglyceride production.

Transcriptional Regulation

SREBPs have been found to play a role with the nutritional or hormonal effects on the lipogenic gene expression. SREBP-2 has a well defined mode of action of the different members of this transcriptional family. At high levels of free cholesterol in the cell SREBP-2 is found bound to the endoplasmic reticulum as an immature precursor. when the level of cholesterol drops then the SREBP-2 is proteolytically cleaved releasing the mature fragment so it can move to the nucleus and bind to the sterol response element in the promoter region of target genes. These genes are then activated for transcription.

It has been indicated that SREBP-2 promote the expression of genes involved in cholesterol metabolism in liver cells. It has been indicated that SREBP-1 plays a role in the activation of genes connected with lipogenesis in liver. Studies have found that an over expression of SREBP-1a or SREBP-1c in mouse liver cells results in the build-up of hepatic triglycerides and higher expression levels of lipogenic genes.

Lipogenic gene expression in the liver via glucose and insulin is moderated by SREBP-1. The effect of glucose and insulin on the transcriptional factor can occur through various pathways. There is evidence suggesting that insulin promotes SREBP-1 mRNA expression in adipocytes and hepatocytes, it has also been suggested that the hormone increases transcriptional activation by SREBP-1 through MAP-kinase-dependent phosphorylation regardless of changes in the mRNA levels. Along with insulin glucose also have been shown to promote SREBP-1 activity and mRNA expression.

PDH Dephosphorylation

Insulin stimulates the activity of pyruvate dehydrogenase phosphatase. The phosphatase removes the phosphate from pyruvate dehydrogenase activating it and allowing for conversion of pyruvate to acetyl-CoA. This mechanism leads to the increased rate of catalysis of this enzyme, so increases the levels of acetyl-CoA. Increased levels of acetyl-CoA will increase the flux through not only the fat synthesis pathway but also the citric acid cycle.

Acetyl-CoA Carboxylase

Insulin affects ACC in a similar way to PDH. It leads to its dephosphorylation via activation of PP2A phosphatase whose activity results in the activation of the enzyme. Glucagon has an antagonistic effect and increases phosphorylation, deactivation, thereby inhibiting ACC and slowing fat synthesis.

Affecting ACC affects the rate of acetyl-CoA conversion to malonyl-CoA. Increased malonyl-CoA level pushes the equilibrium over to increase production of fatty acids through biosynthesis. Long chain fatty acids are negative allosteric regulators of ACC and so when the cell has sufficient long chain fatty acids, they will eventually inhibit ACC activity and stop fatty acid synthesis.

AMP and ATP concentrations of the cell act as a measure of the ATP needs of a cell. When ATP is depleted, there is a rise in 5'AMP. This rise activates AMP-activated protein kinase, which phosphorylates ACC and thereby inhibits fat synthesis. This is a useful way to ensure that glucose is not diverted down a storage pathway in times when energy levels are low.

ACC is also activated by citrate. When there is abundant acetyl-CoA in the cell cytoplasm for fat synthesis, it proceeds at an appropriate rate.

Note: Research now shows that glucose metabolism (exact metabolite to be determined), aside from insulin's influence on lipogenic enzyme genes, can induce the gene products for liver's pyruvate kinase, acetyl-CoA carboxylase, and fatty acid synthase. These genes are induced by the transcription factors ChREBP/Mlx via high blood glucose levels and presently unknown signaling events. Insulin induction of SREBP-1c is also involved in cholesterol metabolism.

Fatty Acid Esterification

Experiments were conducted to study in vivo the over-all fatty acid specificity of the mechanisms involved in chylomicron cholesterol ester and triglyceride formation during fat absorption in the rat. Mixtures containing similar amounts of two, three, or four C14-labeled fatty acids (palmitic, stearic, oleic, and linoleic acids), but with varying ratios of unlabeled fatty acids, were given by gastric intubation to rats with cannulated thoracic ducts. The chyle or chylomicron lipid so obtained was chromatographed on silicic acid columns to separate cholesterol esters and glycerides (the latter being 98.2% triglycerides). After assaying each lipid class for total radioactivity, gas-liquid chromatography was employed to measure the total mass and the distribution of mass and of radioactivity in the individual fatty acid components of each lipid fraction. The specific radioactivity of each fatty acid in each fraction could then be calculated. The data provided quantitative information on the relative specificity of incorporation of each fatty acid into each chylomicron lipid class and on the relative extent to which each fatty acid in each lipid fraction was diluted with endogenous fatty acid. With the exception of a slight discrimination against stearic acid, the processes of fatty acid absorption and chylomicron triglyceride formation displayed no specificity for one fatty acid relative to another. In contrast, chylomicron cholesterol ester formation showed marked specificity for oleic acid, relative to the other three fatty acids. This specificity was not significantly altered by varying the composition of the test meal, by including cholesterol in the test meal, or by feeding the animal a high-cholesterol diet for several weeks preceding the study. Considerable dilution of the dietary fatty acids with endogenous fatty acids was observed. In one experiment, 43% of the chylomicron triglyceride fatty acids was of endogenous origin. Relatively more (54%) of the cholesterol ester fatty acids was of endogenous origin.

References

- Vance JE, Vance DE (2002). Biochemistry of Lipids, Lipoproteins and Membranes. Amsterdam: Elsevier. ISBN 978-0-444-51139-3
- Subramaniam S, Fahy E, Gupta S, Sud M, Byrnes RW, Cotter D, Dinasarapu AR, Maurya MR (October 2011). "Bioinformatics and systems biology of the lipidome". Chemical Reviews. 111 (10): 6452–90. doi:10.1021/cr200295k. PMC 3383319. PMID 21939287
- Geoff Watts (2010-06-12). "Alec Douglas Bangham". The Lancet. 375(9731): 2070. doi:10.1016/S0140-6736(10)60950-6. Retrieved 2014-10-01
- Barani, H; Montazer, M (2008). "A review on applications of liposomes in textile processing". Journal of liposome research. 18 (3): 249–62. doi:10.1080/08982100802354665. PMID 18770074
- Michelle A, Hopkins J, McLaughlin CW, Johnson S, Warner MQ, LaHart D, Wright JD (1993). Human Biology and Health. Englewood Cliffs, New Jersey, USA: Prentice Hall. ISBN 978-0-13-981176-0
- Mashaghi S, Jadidi T, Koenderink G, Mashaghi A (February 2013). "Lipid nanotechnology". International Journal of Molecular Sciences. 14 (2): 4242–82. doi:10.3390/ijms14024242. PMC 3588097. PMID 23429269
- "Hydrolysis - Chemistry Encyclopedia - structure, reaction, water, proteins, examples, salt, molecule". www.chemistryexplained.com. Retrieved 2016-11-01
- Murray, I.; Sniderman, A. D.; Cianflone, K. (1999-09-01). "Enhanced triglyceride clearance with intraperitoneal human acylation stimulating protein in C57BL/6 mice". The American Journal of Physiology. 277 (3 Pt 1): E474–480. ISSN 0002-9513. PMID 10484359
- Meure, LA; Knott, R; Foster, NR; Dehghani, F (2009). "The depressurization of an expanded solution into aqueous media for the bulk production of liposomes". Langmuir : the ACS journal of surfaces and colloids. 25 (1): 326–37. doi:10.1021/la802511a. PMID 19072018
- Hunter JE (November 2006). "Dietary trans fatty acids: review of recent human studies and food industry responses". Lipids. 41 (11): 967–92. doi:10.1007/s11745-006-5049-y. PMID 17263298
- IUPAC Compendium of Chemical Terminology (2nd ed.). International Union of Pure and Applied Chemistry. 1997. doi:10.1351/pac199567081307. ISBN 0-521-51150-X. Retrieved 2007-10-31
- Altar, Ted. "More Than You Wanted To Know About Fats/Oils". Sundance Natural Foods. Retrieved 2006-08-31

- Fezza F, De Simone C, Amadio D, Maccarrone M (2008). "Fatty acid amide hydrolase: a gate-keeper of the endocannabinoid system". Sub-Cellular Biochemistry. Subcellular Biochemistry. 49: 101–32. doi:10.1007/978-1-4020-8831-5_4. ISBN 978-1-4020-8830-8. PMID 18751909
- Coleman RA, Lee DP (March 2004). "Enzymes of triacylglycerol synthesis and their regulation". Progress in Lipid Research. 43 (2): 134–76. doi:10.1016/S0163-7827(03)00051-1. PMID 14654091
- Etherton, T. D. (2000-11-01). "The biology of somatotropin in adipose tissue growth and nutrient partitioning". The Journal of Nutrition. 130 (11): 2623–2625. doi:10.1093/jn/130.11.2623. ISSN 0022-3166. PMID 11053496

5

Nucleic Acids

Nucleic acids are the chemical compounds which are made up of complex nucleotides and carry primary information related to various activities of the cell. Some of the major focus areas related to nucleic acids are nucleic acid design, nucleic acid structure, nucleic acid structure determination and nucleic acid structure prediction, etc. This chapter discusses in detail the design, structure and metabolism of nucleic acids.

Nucleic acids are biopolymers, or small biomolecules, essential to all known forms of life. They are composed of nucleotides, which are monomers made of three components: a 5-carbon sugar, a phosphate group and a nitrogenous base. If the sugar is a simple ribose, the polymer is RNA (ribonucleic acid); if the sugar is derived from ribose as deoxyribose, the polymer is DNA (deoxyribonucleic acid).

Nucleic acids are the most important of all biomolecules. They are found in abundance in all living things, where they function to create and encode and then store information in the nucleus of every living cell of every life-form organism on Earth. In turn, they function to transmit and express that information inside and outside the cell nucleus—to the interior operations of the cell and ultimately to the next generation of each living organism. The encoded information is contained and conveyed via the nucleic acid sequence, which provides the 'ladder-step' ordering of nucleotides within the molecules of RNA and DNA.

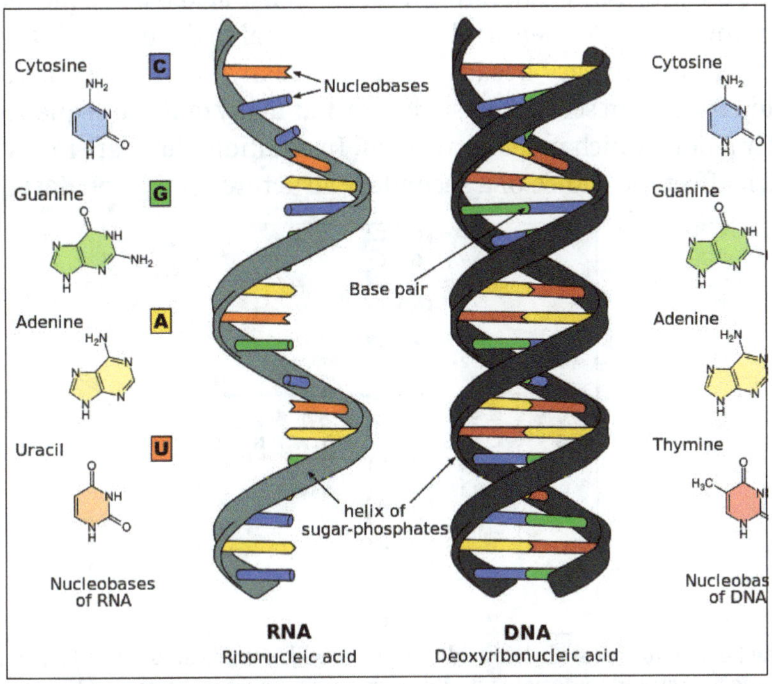

A comparison of the two principal nucleic acids: RNA (*left*) and DNA (*right*), showing the helices and nucleobases each employs.

The Swiss scientist Friedrich Miescher discovered nucleic acids (DNA). Later, he raised the idea that they could be involved in heredity.

Strings of nucleotides are bonded to form helical backbones—typically, one for RNA, two for DNA—and assembled into chains of base-pairs selected from the five primary, or canonical, nucleobases, which are: adenine, cytosine, guanine, thymine, and uracil; note, thymine occurs only in DNA and uracil only in RNA. Using amino acids and the process known as protein synthesis, the specific sequencing in DNA of these nucleobase-pairs enables storing and transmitting coded instructions as genes. In RNA, base-pair sequencing provides for manufacturing new proteins that determine the frames and parts and most chemical processes of all life forms.

Nucleic Acid Design

Nucleic acid design is the process of generating a set of nucleic acid base sequences that will associate into a desired conformation. Nucleic acid design is central to the fields of DNA nanotechnology and DNA computing. It is necessary because there are many possible sequences of nucleic acid strands that will fold into a given secondary structure, but many of these sequences will have undesired additional interactions which must be avoided. In addition, there are many tertiary structure considerations which affect the choice of a secondary structure for a given design.

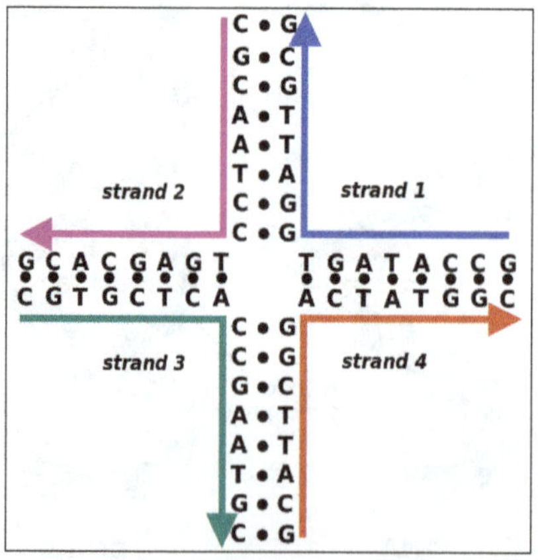

Nucleic acid design can be used to create nucleic acid complexes with complicated secondary structures such as this four-arm junction. These four strands associate into this structure because it maximizes the number of correct base pairs, with A's matched to T's and C's matched to G's. Image from Mao, 2004.

Nucleic acid design has similar goals to protein design: in both, the sequence of monomers is rationally designed to favor the desired folded or associated structure and to disfavor alternate structures. However, nucleic acid design has the advantage of being a much computationally simpler problem, since the simplicity of Watson-Crick base pairing rules leads to simple heuristic methods which yield experimentally robust designs. Computational models for protein folding require tertiary structure information whereas nucleic acid design can operate largely on the level of secondary structure. However, nucleic acid structures are less versatile than proteins in their functionality.

Nucleic acid design can be considered the inverse of nucleic acid structure prediction. In structure prediction, the structure is determined from a known sequence, while in nucleic acid design, a sequence is generated which will form a desired structure.

Fundamental Concepts

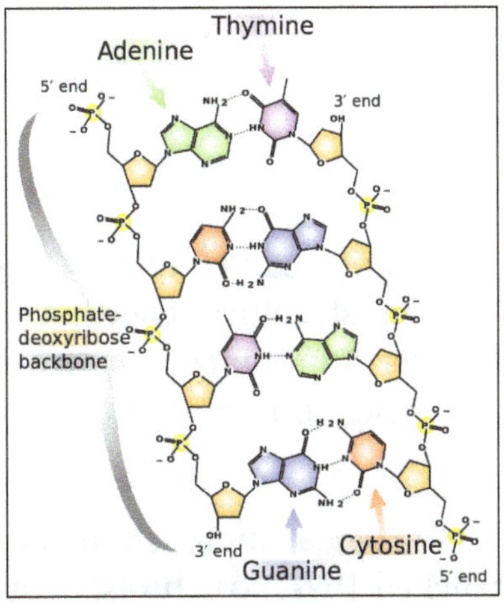

Chemical structure of DNA. Nucleic acid double helices will only form between two strands of complementary sequences, where the bases are matched into only A-T or G-C pairs.

The structure of nucleic acids consists of a sequence of nucleotides. There are four types of nucleotides distinguished by which of the four nucleobases they contain: in DNA these are adenine (A), cytosine (C), guanine (G), and thymine (T). Nucleic acids have the property that two molecules will bind to each other to form a double helix only if the two sequences are complementary, that is, they can form matching sequences of base pairs. Thus, in nucleic acids the sequence determines the pattern of binding and thus the overall structure.

Nucleic acid design is the process by which, given a desired target structure or functionality, sequences are generated for nucleic acid strands which will self-assemble into that target structure. Nucleic acid design encompasses all levels of nucleic acid structure:

- Primary structure—the raw sequence of nucleobases of each of the component nucleic acid strands.
- Secondary structure—the set of interactions between bases, i.e., which parts of which strands are bound to each other.
- Tertiary structure—the locations of the atoms in three-dimensional space, taking into consideration geometrical and steric constraints.

One of the greatest concerns in nucleic acid design is ensuring that the target structure has the lowest free energy (i.e. is the most thermodynamically favorable) whereas misformed structures have higher values of free energy and are thus unfavored. These goals can be achieved through the use of a number of approaches, including heuristic, thermodynamic, and geometrical ones. Almost all nucleic acid design tasks are aided by computers, and a number of software packages are available for many of these tasks.

Two considerations in nucleic acid design are that desired hybridizations should have melting temperatures in a narrow range, and any spurious interactions should have very low melting temperatures (i.e. they should be very weak). There is also a contrast between affinity-optimizing "positive design", seeks to minimize the energy of the desired structure in an absolute sense, and specificity-optimizing "negative design", which considers the energy of the target structure relative to those of undesired structures. Algorithms which implement both kinds of design tend to perform better than those that consider only one type.

Approaches

Heuristic Methods

Heuristic methods use simple criteria which can be quickly evaluated to judge the suitability of different sequences for a given secondary structure. They have the advantage of being much less computationally expensive than the energy minimization algorithms needed for thermodynamic or geometrical modeling, and being easier to implement, but at the cost of being less rigorous than these models.

Sequence symmetry minimization is the oldest approach to nucleic acid design and was first used to design immobile versions of branched DNA structures. Sequence symmetry minimization divides the nucleic acid sequence into overlapping subsequences of a fixed length, called the criterion length. Each of the 4^N possible subsequences of length N is allowed to appear only once in the sequence. This ensures that no undesired hybridizations can occur which have a length greater than or equal to the criterion length.

A related heuristic approach is to consider the "mismatch distance", meaning the number of positions in a certain frame where the bases are not complementary. A greater mismatch distance lessens the chance that a strong spurious interaction can happen. This is related to the concept of Hamming distance in information theory. Another related but more involved approach is to use methods from coding theory to construct nucleic acid sequences with desired properties.

Thermodynamic Models

Information about the secondary structure of a nucleic acid complex along with its sequence can be used to predict the thermodynamic properties of the complex.

When thermodynamic models are used in nucleic acid design, there are usually two considerations: desired hybridizations should have melting temperatures in a narrow range, and any spurious interactions should have very low melting temperatures (i.e. they should be very weak). The Gibbs free energy of a perfectly matched nucleic acid duplex can be predicted using a nearest neighbor model. This model considers only the interactions between a nucleotide and its nearest neighbors on the nucleic acid strand, by summing the free energy of each of the overlapping two-nucleotide subwords of the duplex. This is then corrected for self-complementary monomers and for GC-content. Once the free energy is known, the melting temperature of the duplex can be determined. GC-content alone can also be used to estimate the free energy and melting temperature of a nucleic acid duplex. This is less accurate but also much less computationally costly.

Software for thermodynamic modeling of nucleic acids includes Nupack, mfold/UNAFold, and Vienna.

A related approach, inverse secondary structure prediction, uses stochastic local search which improves a nucleic acid sequence by running a structure prediction algorithm and the modifying the sequence to eliminate unwanted features.

Geometrical Models

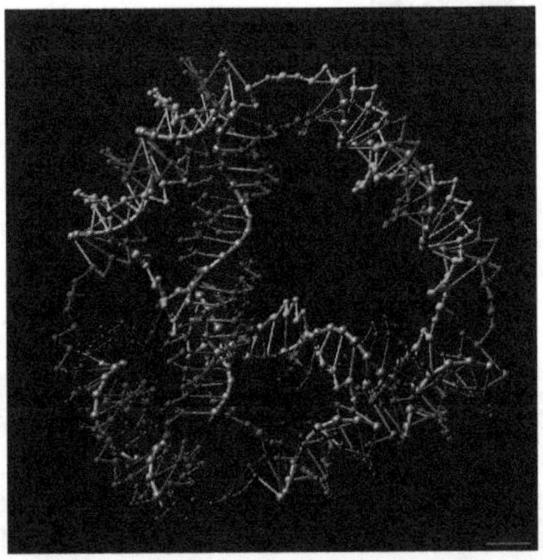

A geometrical model of a DNA tetrahedron described in Goodman. Models of this type are useful for ensuring that tertiary structure constraints do not cause excessive strain to the molecule.

Geometrical models of nucleic acids are used to predict tertiary structure. This is important because designed nucleic acid complexes usually contain multiple junction points, which introduces geometric constraints to the system. These constraints stem from the basic structure of nucleic acids, mainly that the double helix formed by nucleic acid duplexes has a fixed helicity of about 10.4 base pairs per turn, and is relatively stiff. Because of these constraints, the nucleic acid complexes are sensitive to the relative orientation of the major and minor grooves at junction points. Geometrical modeling can detect strain stemming from misalignments in the structure, which can then be corrected by the designer.

Geometric models of nucleic acids for DNA nanotechnology generally use reduced representations of the nucleic acid, because simulating every atom would be very computationally expensive for such large systems. Models with three pseudo-atoms per base pair, representing the two backbone sugars and the helix axis, have been reported to have a sufficient level of detail to predict experimental results. However, models with five pseudo-atoms per base pair, explicitly including the backbone phosphates, are also used.

Software for geometrical modeling of nucleic acids includes GIDEON, Tiamat, Nanoengineer-1, and UNIQUIMER 3D. Geometrical concerns are especially of interest in the design of DNA origami, because the sequence is predetermined by the choice of scaffold strand. Software specifically for DNA origami design has been made, including caDNAno and SARSE.

Applications

Nucleic acid design is used in DNA nanotechnology to design strands which will self-assemble into a desired target structure. These include examples such as DNA machines, periodic two- and three-dimensional lattices, polyhedra, and DNA origami. It can also be used to create sets of nucleic acid strands which are "orthogonal", or non-interacting with each other, so as to minimize or eliminate spurious interactions. This is useful in DNA computing, as well as for molecular barcoding applications in chemical biology and biotechnology.

Nucleic Acid Structure

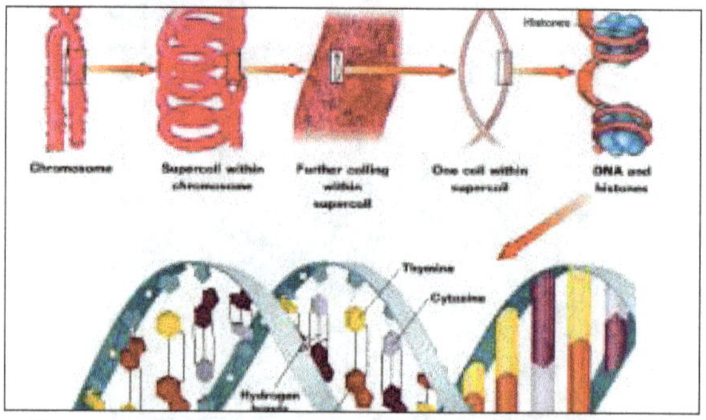

Interactive image of nucleic acid structure (primary, secondary, tertiary, and quaternary) using DNA helices and examples from the VS ribozyme and telomerase and nucleosome. (PDB: ADNA, 1BNA, 4OCB, 4R4V, 1YMO, 1EQZ).

Nucleic acid structure refers to the structure of nucleic acids such as DNA and RNA. Chemically speaking, DNA and RNA are very similar. Nucleic acid structure is often divided into four different levels: primary, secondary, tertiary and quaternary.

Primary Structure

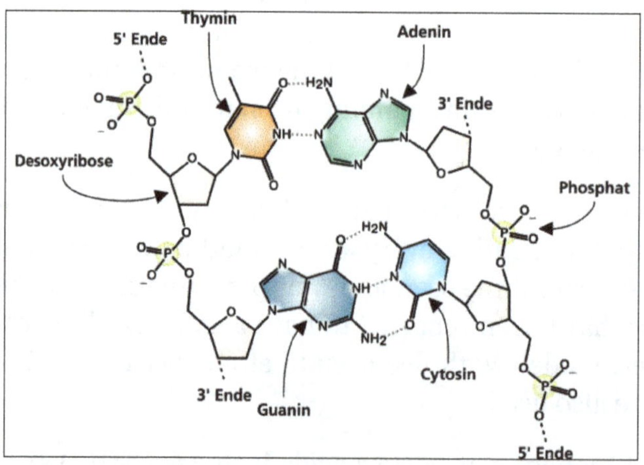

Chemical structure of DNA.

Primary structure consists of a linear sequence of nucleotides that are linked together by phosphodiester bonds. It is this linear sequence of nucleotides that make up the Primary structure of DNA or RNA. Nucleotides consist of 3 components:

1. Nitrogenous base:
   - Adenine
   - Guanine
   - Cytosine
   - Thymine (present in DNA only)
   - Uracil (present in RNA only)
2. 5-carbon sugar which is called deoxyribose (found in DNA) and ribose (found in RNA).
3. One or more phosphate groups.

The nitrogen bases adenine and guanine are purine in structure and form a glycosidic bond between their 9 nitrogen and the 1' -OH group of the deoxyribose. Cytosine, thymine and uracil are pyrimidines, hence the glycosidic bonds forms between their 1 nitrogen and the 1' -OH of the deoxyribose. For both the purine and pyrimidine bases, the phosphate group forms a bond with the deoxyribose sugar through an ester bond between one of its negatively charged oxygen groups and the 5' -OH of the sugar. The polarity in DNA and RNA is derived from the oxygen and nitrogen atoms in the backbone. Nucleic acids are formed when nucleotides come together through phosphodiester linkages between the 5' and 3' carbon atoms. A Nucleic acid sequence is the order of nucleotides within a DNA (GACT) or RNA (GACU) molecule that is determined by a series of letters. Sequences are presented from the 5' to 3' end and determine the covalent structure of the entire molecule. Sequences can be complementary to another sequence in that the base on each position is complementary as well as in the reverse order. An example of a complementary sequence to AGCT is TCGA. DNA is double-stranded containing both a sense strand and an antisense strand. Therefore, the complementary sequence will be to the sense strand.

Complexes with Alkali Metal Ions

There are three potential metal binding groups on nucleic acids: phosphate, sugar and base moieties. Solid-state structure of complexes with alkali metal ions have been reviewed.

Secondary Structure

Secondary structure is the set of interactions between bases, i.e., which parts of strands are bound to each other. In DNA double helix, the two strands of DNA are held together by hydrogen bonds. The nucleotides on one strand base pairs with the nucleotide on the other strand. The secondary structure is responsible for the shape that the nucleic acid assumes. The bases in the DNA are classified as purines and pyrimidines. The purines are adenine and guanine. Purines consist of a double ring structure, a six membered and a five membered ring containing nitrogen. The pyrimidines are cytosine and thymine. It has a single ringed structure, a six membered ring containing nitrogen. A purine base always pairs with a pyrimidine base (guanine (G) pairs with cytosine (C) and adenine (A) pairs with thymine (T) or uracil (U)). DNA's secondary structure is predominantly determined by base-pairing of the two polynucleotide strands wrapped around each other to form a double helix. Although the two strands are aligned by hydrogen bonds in base pairs, the stronger forces holding the two strands together are stacking interactions between the bases. These stacking interactions are stabilized by Van der Waals forces and hydrophobic interactions, and show a large amount of local structural variability. There are also two grooves in the double helix, which are called major groove and minor groove based on their relative size.

The secondary structure of RNA consists of a single polynucleotide. Base pairing in RNA occurs when RNA folds between complementarity regions. Both single- and double-stranded regions are often found in RNA molecules. The antiparallel strands form a helical shape. The four basic elements in the secondary structure of RNA are helices, loops, bulges, and junctions. Stem-loop or hairpin loop is the most common element of RNA secondary structure. Stem-loop is formed when the RNA chains fold back on themselves to form a double helical tract called the stem, the unpaired nucleotides forms single stranded region called the loop. Secondary structure of RNA can be predicted by experimental data on the secondary structure elements, helices, loops and bulges. Bulges and internal loops are formed by separation of the double helical tract on either one strand (bulge) or on both strands (internal loops) by unpaired nucleotides. A tetraloop is a four-base pairs hairpin RNA structure. There are three common families of tetraloop in ribosomal RNA: UNCG, GNRA, and CUUG (N is one of the four nucleotides and R is a purine).UNCG is the most stable tetraloop. Pseudoknot is a RNA secondary structure first identified in turnip yellow mosaic virus. Pseudoknots are formed when nucleotides from the hairpin loop pairs with a single stranded region outside of the hairpin to form a helical segment. H-type fold pseudoknots are best characterized. In H-type fold, nucleotides in the hairpin loop pairs with the bases outside the hairpin stem forming second stem and loop. This causes formation of pseudoknots with two stems and two loops. Pseudoknots are functional elements in RNA structure having diverse function and found in most classes of RNA. DotKnot-PW method is used for comparative pseudoknots prediction. The main points in the DotKnot-PW method is scoring the similarities found in stems, secondary elements and H-type pseudoknots.

Tertiary Structure

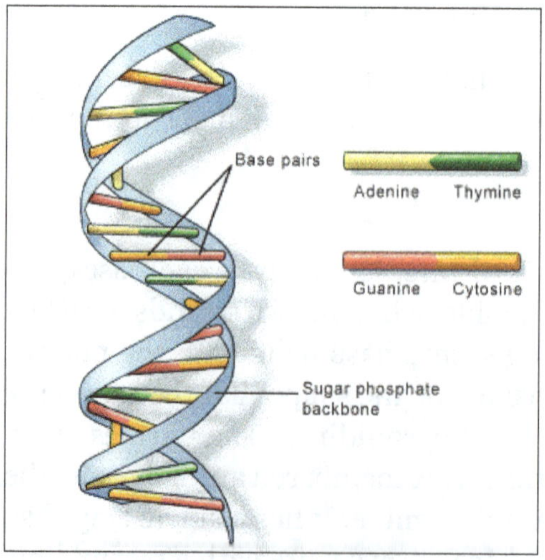

DNA structure and bases.

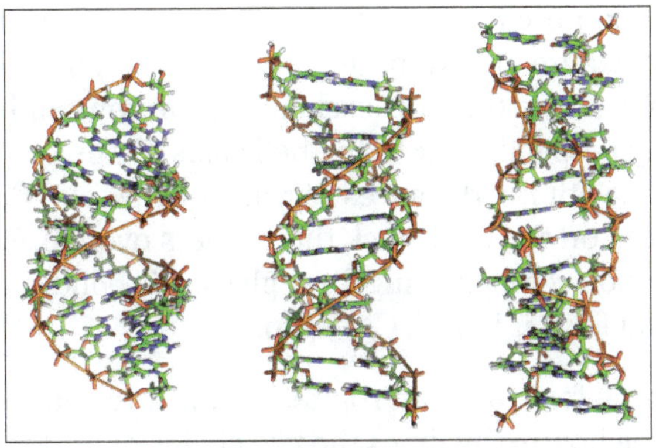

A-B-Z-DNA Side View.

Tertiary structure refers to the locations of the atoms in three-dimensional space, taking into consideration geometrical and steric constraints. It is a higher order than the secondary structure, in which large-scale folding in a linear polymer occurs and the entire chain is folded into a specific 3-dimensional shape. There are 4 areas in which the structural forms of DNA can differ.

1. Handedness – right or left.
2. Length of the helix turn.
3. Number of base pairs per turn.
4. Difference in size between the major and minor grooves.

The tertiary arrangement of DNA's double helix in space includes B-DNA, A-DNA and Z-DNA.

B-DNA is the most common form of DNA in vivo and is a more narrow, elongated helix than A-DNA. Its wide major groove makes it more accessible to proteins. On the other hand, it has a narrow minor groove. B-DNA's favored conformations occur at high water concentrations; the hydration of the minor groove appears to favor B-DNA. B-DNA base pairs are nearly perpendicular to the helix axis. The sugar pucker which determines the shape of the a-helix, whether the helix will exist in the A-form or in the B-form, occurs at the C2'-endo.

A-DNA, is a form of the DNA duplex observed under dehydrating conditions. It is shorter and wider than B-DNA. RNA adopts this double helical form, and RNA-DNA duplexes are mostly A-form, but B-form RNA-DNA duplexes have been observed. In localized single strand dinucleotide contexts, RNA can also adopt the B-form without pairing to DNA. A-DNA has a deep, narrow major groove which does not make it easily accessible to proteins. On the other hand, its wide, shallow minor groove makes it accessible to proteins but with lower information content than the major groove. Its favored conformation is at low water concentrations. A-DNAs base pairs are tilted relative to the helix axis, and are displaced from the axis. The sugar pucker occurs at the C3'-endo and in RNA 2'-OH inhibits C2'-endo conformation. Long considered little more than a laboratory artifice, A-DNA is now known to have several biological functions.

Z-DNA is a relatively rare left-handed double-helix. Given the proper sequence and superhelical tension, it can be formed in vivo but its function is unclear. It has a more narrow, more elongated helix than A or B. Z-DNA's major groove is not really a groove, and it has a narrow minor groove.

The most favored conformation occurs when there are high salt concentrations. There are some base substitutions but they require an alternating purine-pyrimidine sequence. The N2-amino of G H-bonds to 5' PO, which explains the slow exchange of protons and the need for the G purine. Z-DNA base pairs are nearly perpendicular to the helix axis. Z-DNA does not contain single base-pairs but rather a GpC repeat with P-P distances varying for GpC and CpG. On the GpC stack there is good base overlap, whereas on the CpG stack there is less overlap. Z-DNA's zigzag backbone is due to the C sugar conformation compensating for G glycosidic bond conformation. The conformation of G is syn, C2'-endo and for C it is anti, C3'-endo.

A linear DNA molecule having free ends can rotate, to adjust to changes of various dynamic processes in the cell, by changing how many times the two chains of its double helix twist around each other. Some DNA molecules are circular and are topologically constrained. More recently circular RNA was described as well to be a natural pervasive class of nucleic acids, expressed in many organisms.

A covalently closed, circular DNA (also known as cccDNA) is topologically constrained as the number of times the chains coiled around one other cannot change. This cccDNA can be supercoiled, which is the tertiary structure of DNA. Supercoiling is characterized by the linking number, twist and writhe. The linking number (Lk) for circular DNA is defined as the number of times one strand would have to pass through the other strand to completely separate the two strands. The linking number for circular DNA can only be changed by breaking of a covalent bond in one of the two strands. Always an integer, the linking number of a cccDNA is the sum of two components: twists (Tw) and writhes (Wr).

$$Lk = Tw + Wr$$

Twists are the number of times the two strands of DNA are twisted around each other. Writhes are number of times the DNA helix crosses over itself. DNA in cells is negatively supercoiled and has the tendency to unwind. Hence the separation of strands is easier in negatively supercoiled DNA than in relaxed DNA. The two components of supercoiled DNA are solenoid and plectonemic. The plectonemic supercoil is found in prokaryotes, while the solenoidal supercoiling is mostly seen in eukaryotes.

Quaternary Structure

The quaternary structure of nucleic acids is similar to that of protein quaternary structure. Although some of the concepts are not exactly the same, the quaternary structure refers to a higher-level of organization of nucleic acids. Moreover, it refers to interactions of the nucleic acids with other molecules. The most commonly seen form of higher-level organization of nucleic acids is seen in the form of chromatin which leads to its interactions with the small proteins histones. Also, the quaternary structure refers to the interactions between separate RNA units in the ribosome or spliceosome.

Nucleic Acid Sequence

A nucleic acid sequence is a succession of letters that indicate the order of nucleotides within a DNA (using GACT) or RNA (GACU) molecule. By convention, sequences are usually presented from the 5' end to the 3' end. For DNA, the sense strand is used. Because nucleic acids are normally linear (unbranched) polymers, specifying the sequence is equivalent to defining the covalent structure of the entire molecule. For this reason, the nucleic acid sequence is also termed the primary structure.

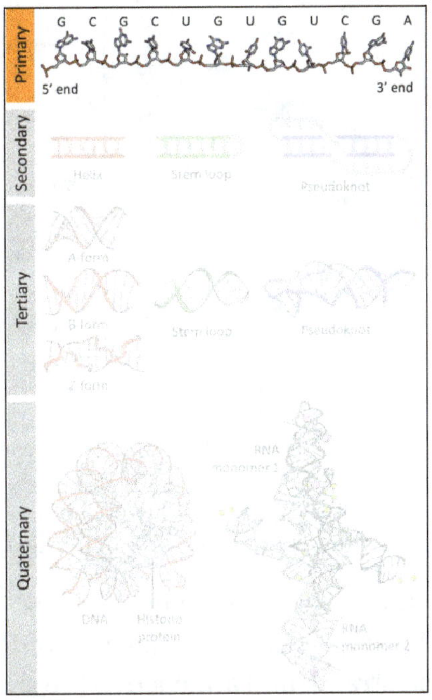

Interactive image of nucleic acid structure (primary, secondary, tertiary, and quaternary) using DNA helices and examples from the VS ribozyme and telomerase and nucleosome. (PDB: ADNA, 1BNA, 4OCB, 4R4V, 1YMO, 1EQZ).

The sequence has capacity to represent information. Biological deoxyribonucleic acid represents the information which directs the functions of a living thing.

Nucleic acids also have a secondary structure and tertiary structure. Primary structure is sometimes mistakenly referred to as *primary sequence*. Conversely, there is no parallel concept of secondary or tertiary sequence.

Nucleotides

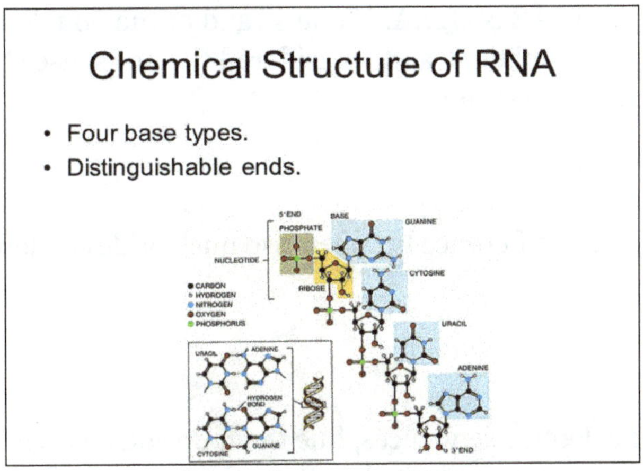

Chemical structure of RNA.

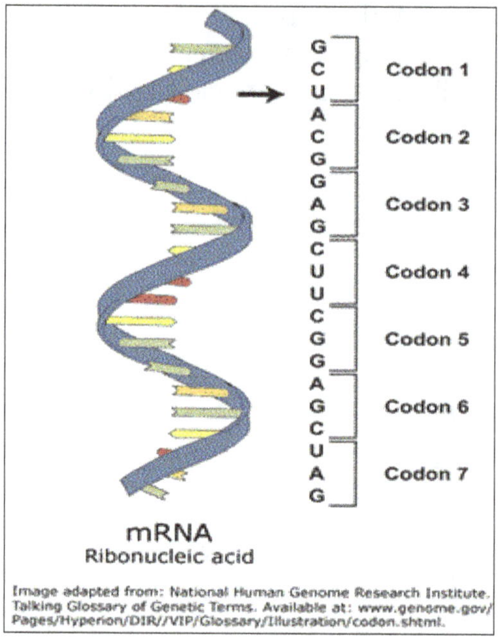

A series of codons in part of a mRNA molecule. Each codon consists of three nucleotides, usually representing a single amino acid.

Nucleic acids consist of a chain of linked units called nucleotides. Each nucleotide consists of three subunits: a phosphate group and a sugar (ribose in the case of RNA, deoxyribose in DNA) make up the backbone of the nucleic acid strand, and attached to the sugar is one of a set of nucleobases. The nucleobases are important in base pairing of strands to form higher-level secondary and tertiary structure such as the famed double helix.

The possible letters are *A*, *C*, *G*, and *T*, representing the four nucleotide bases of a DNA strand — adenine, cytosine, guanine, thymine — covalently linked to a phosphodiester backbone. In the typical case, the sequences are printed abutting one another without gaps, as in the sequence AAAGTCTGAC, read left to right in the 5' to 3' direction. With regards to transcription, a sequence is on the coding strand if it has the same order as the transcribed RNA.

One sequence can be complementary to another sequence, meaning that they have the base on each position in the complementary (i.e. A to T, C to G) and in the reverse order. For example, the complementary sequence to TTAC is GTAA. If one strand of the double-stranded DNA is considered the sense strand, then the other strand, considered the antisense strand, will have the complementary sequence to the sense strand.

Notation

Comparing and determining % difference between two nucleotide sequences.

- AATCCGCTAG
- AAACCCTTAG
- Given the two 10-nucleotide sequences, line them up and compare the differences between them. Calculate the percent similarity by taking the number of different DNA bases divided

by the total number of nucleotides. In the above case, there are three differences in the 10 nucleotide sequence. Therefore, divide 7/10 to get the 70% similarity and subtract that from 100% to get a 30% difference.

While A, T, C, and G represent a particular nucleotide at a position, there are also letters that represent ambiguity which are used when more than one kind of nucleotide could occur at that position. The rules of the International Union of Pure and Applied Chemistry (IUPAC) are as follows:

- A = adenine
- C = cytosine
- G = guanine
- T = thymine
- R = G A (purine)
- Y = T C (pyrimidine)
- K = G T (keto)
- M = A C (amino)
- S = G C (strong bonds)
- W = A T (weak bonds)
- B = G T C (all but A)
- D = G A T (all but C)
- H = A C T (all but G)
- V = G C A (all but T)
- N = A G C T (any)

These symbols are also valid for RNA, except with U (uracil) replacing T (thymine).

Apart from adenine (A), cytosine (C), guanine (G), thymine (T) and uracil (U), DNA and RNA also contain bases that have been modified after the nucleic acid chain has been formed. In DNA, the most common modified base is 5-methylcytidine (m5C). In RNA, there are many modified bases, including pseudouridine (Ψ), dihydrouridine (D), inosine (I), ribothymidine (rT) and 7-methylguanosine (m7G). Hypoxanthine and xanthine are two of the many bases created through mutagen presence, both of them through deamination (replacement of the amine-group with a carbonyl-group). Hypoxanthine is produced from adenine, xanthine from guanine. Similarly, deamination of cytosine results in uracil.

Biological Significance

In biological systems, nucleic acids contain information which is used by a living cell to construct specific proteins. The sequence of nucleobases on a nucleic acid strand is translated by cell machinery into a sequence of amino acids making up a protein strand. Each group of three bases, called a codon, corresponds to a single amino acid, and there is a specific genetic code by which each possible combination of three bases corresponds to a specific amino acid.

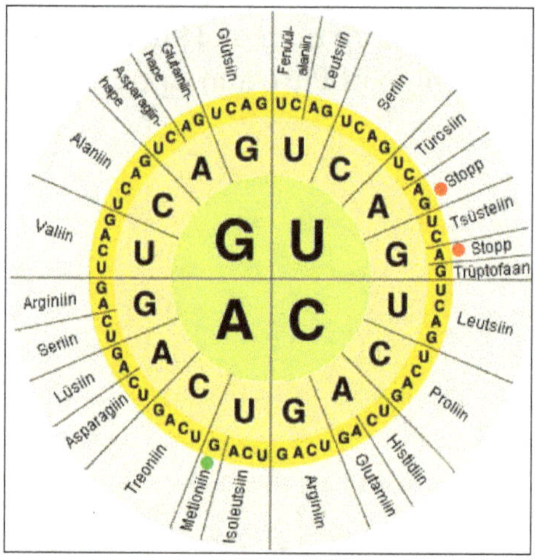

A depiction of the genetic code, by which the information contained in nucleic acids are translated into amino acid sequences in proteins.

The central dogma of molecular biology outlines the mechanism by which proteins are constructed using information contained in nucleic acids. DNA is transcribed into mRNA molecules, which travels to the ribosome where the mRNA is used as a template for the construction of the protein strand. Since nucleic acids can bind to molecules with complementary sequences, there is a distinction between "sense" sequences which code for proteins, and the complementary "antisense" sequence which is by itself nonfunctional, but can bind to the sense strand.

Sequence Determination

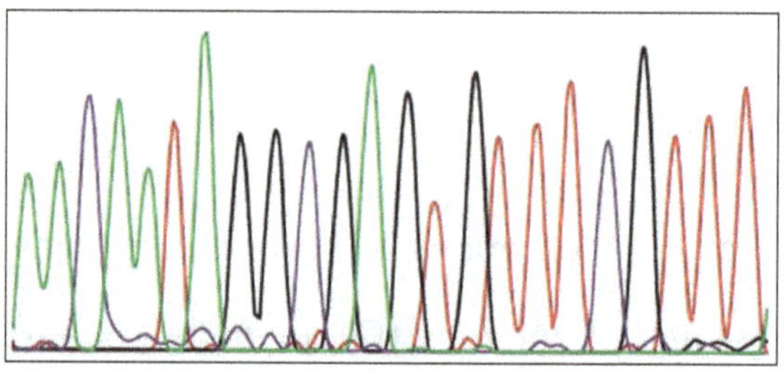

Electropherogram printout from automated sequencer for determining part of a DNA sequence.

DNA sequencing is the process of determining the nucleotide sequence of a given DNA fragment. The sequence of the DNA of a living thing encodes the necessary information for that living thing to survive and reproduce. Therefore, determining the sequence is useful in fundamental research into why and how organisms live, as well as in applied subjects. Because of the importance of DNA to living things, knowledge of a DNA sequence may be useful in practically any biological research. For example, in medicine it can be used to identify, diagnose and potentially develop treatments for genetic diseases. Similarly, research into pathogens may lead to treatments for contagious diseases. Biotechnology is a burgeoning discipline, with the potential for many useful products and services.

RNA is not sequenced directly. Instead, it is copied to a DNA by reverse transcriptase, and this DNA is then sequenced.

Current sequencing methods rely on the discriminatory ability of DNA polymerases, and therefore can only distinguish four bases. An inosine (created from adenosine during RNA editing) is read as a G, and 5-methyl-cytosine (created from cytosine by DNA methylation) is read as a C. With current technology, it is difficult to sequence small amounts of DNA, as the signal is too weak to measure. This is overcome by polymerase chain reaction (PCR) amplification.

Digital Representation

Once a nucleic acid sequence has been obtained from an organism, it is stored *in silico* in digital format. Digital genetic sequences may be stored in sequence databases, be analyzed, be digitally altered and be used as templates for creating new actual DNA using artificial gene synthesis.

Sequence Analysis

Digital genetic sequences may be analyzed using the tools of bioinformatics to attempt to determine its function.

Genetic Testing

The DNA in an organism's genome can be analyzed to diagnose vulnerabilities to inherited diseases, and can also be used to determine a child's paternity (genetic father) or a person's ancestry. Normally, every person carries two variations of every gene, one inherited from their mother, the other inherited from their father. The human genome is believed to contain around 20,000 - 25,000 genes. In addition to studying chromosomes to the level of individual genes, genetic testing in a broader sense includes biochemical tests for the possible presence of genetic diseases, or mutant forms of genes associated with increased risk of developing genetic disorders.

Genetic testing identifies changes in chromosomes, genes, or proteins. Usually, testing is used to find changes that are associated with inherited disorders. The results of a genetic test can confirm or rule out a suspected genetic condition or help determine a person's chance of developing or passing on a genetic disorder. Several hundred genetic tests are currently in use, and more are being developed.

Sequence Alignment

In bioinformatics, a sequence alignment is a way of arranging the sequences of DNA, RNA, or protein to identify regions of similarity that may be due to functional, structural, or evolutionary relationships between the sequences. If two sequences in an alignment share a common ancestor, mismatches can be interpreted as point mutations and gaps as insertion or deletion mutations (indels) introduced in one or both lineages in the time since they diverged from one another. In sequence alignments of proteins, the degree of similarity between amino acids occupying a particular position in the sequence can be interpreted as a rough measure of how conserved a particular region or sequence motif is among lineages. The absence of substitutions, or the presence of only very conservative substitutions (that is, the substitution of amino acids whose side chains have similar biochemical properties) in a particular region of the sequence, suggest that this region has structural or functional importance. Although DNA and RNA nucleotide bases are more similar to each other than are amino acids, the conservation of base pairs can indicate a similar functional or structural role.

Computational phylogenetics makes extensive use of sequence alignments in the construction and interpretation of phylogenetic trees, which are used to classify the evolutionary relationships between homologous genes represented in the genomes of divergent species. The degree to which sequences in a query set differ is qualitatively related to the sequences' evolutionary distance from one another. Roughly speaking, high sequence identity suggests that the sequences in question have a comparatively young most recent common ancestor, while low identity suggests that the divergence is more ancient. This approximation, which reflects the "molecular clock" hypothesis that a roughly constant rate of evolutionary change can be used to extrapolate the elapsed time since two genes first diverged (that is, the coalescence time), assumes that the effects of mutation and selection are constant across sequence lineages. Therefore, it does not account for possible difference among organisms or species in the rates of DNA repair or the possible functional conservation of specific regions in a sequence. (In the case of nucleotide sequences, the molecular clock hypothesis in its most basic form also discounts the difference in acceptance rates between silent mutations that do not alter the meaning of a given codon and other mutations that result in a different amino acid being incorporated into the protein.) More statistically accurate methods allow the evolutionary rate on each branch of the phylogenetic tree to vary, thus producing better estimates of coalescence times for genes.

Sequence Motifs

Frequently the primary structure encodes motifs that are of functional importance. Some examples of sequence motifs are: the C/D and H/ACA boxes of snoRNAs, Sm binding site found in spliceosomal RNAs such as U1, U2, U4, U5, U6, U12 and U3, the Shine-Dalgarno sequence, the Kozak consensus sequence and the RNA polymerase III terminator.

Long Range Correlations

Peng found the existence of long-range correlations in the non-coding base pair sequences of DNA. In contrast, such correlations seem not to appear in coding DNA sequences.

Sequence Entropy

In Bioinformatics, a sequence entropy, also known as sequence complexity or information profile, is a numerical sequence providing a quantitative measure of the local complexity of a DNA sequence, independently of the direction of processing. The manipulations of the information profiles enable the analysis of the sequences using alignment-free techniques, such as for example in motif and rearrangements detection.

Nucleic Acid Secondary Structure

Nucleic acid secondary structure is the basepairing interactions within a single nucleic acid polymer or between two polymers. It can be represented as a list of bases which are paired in a nucleic acid molecule. The secondary structures of biological DNA's and RNA's tend to be different biological DNA mostly exists as fully base paired double helices, while biological RNA is single stranded and often forms complex and intricate base-pairing interactions due to its increased ability to form hydrogen bonds stemming from the extra hydroxyl group in the ribose sugar.

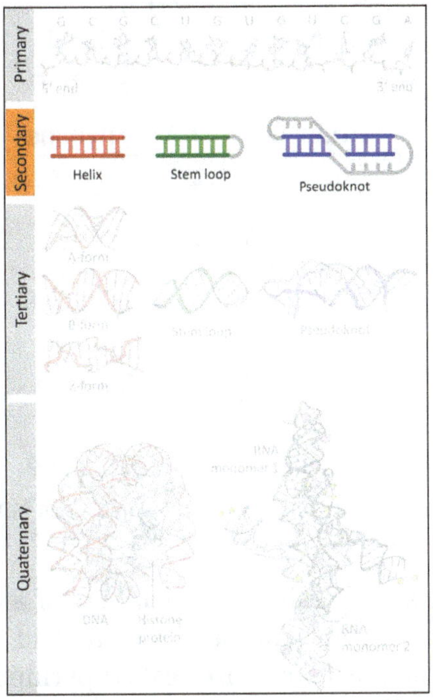

Interactive image of nucleic acid structure (primary, secondary, tertiary, and quaternary) using DNA helices and examples from the VS ribozyme and telomerase and nucleosome. (PDB: ADNA, 1BNA, 4OCB, 4R4V, 1YMO, 1EQZ)

In a non-biological context, secondary structure is a vital consideration in the nucleic acid design of nucleic acid structures for DNA nanotechnology and DNA computing, since the pattern of base-pairing ultimately determines the overall structure of the molecules.

Fundamental Concepts

Base Pairing

In molecular biology, two nucleotides on opposite complementary DNA or RNA strands that are connected via hydrogen bonds are called a base pair (often abbreviated bp). In the canonical Watson-Crick base pairing, adenine (A) forms a base pair with thymine (T) and guanine (G) forms one with cytosine (C) in DNA. In RNA, thymine is replaced by uracil (U). Alternate hydrogen bonding patterns, such as the wobble base pair and Hoogsteen base pair, also occur—particularly in RNA—giving rise to complex and functional tertiary structures. Importantly, pairing is the mechanism by which codons on messenger RNA molecules are recognized by anticodons on transfer RNA during protein translation. Some DNA- or RNA-binding enzymes can recognize specific base pairing patterns that identify particular regulatory regions of genes. Hydrogen bonding is the chemical mechanism that underlies the base-pairing rules described above. Appropriate geometrical correspondence of hydrogen bond donors and acceptors allows only the "right" pairs to form stably. DNA with high GC-content is more stable than DNA with low GC-content, but contrary to popular belief, the hydrogen bonds do not stabilize the DNA significantly and stabilization is mainly due to stacking interactions.

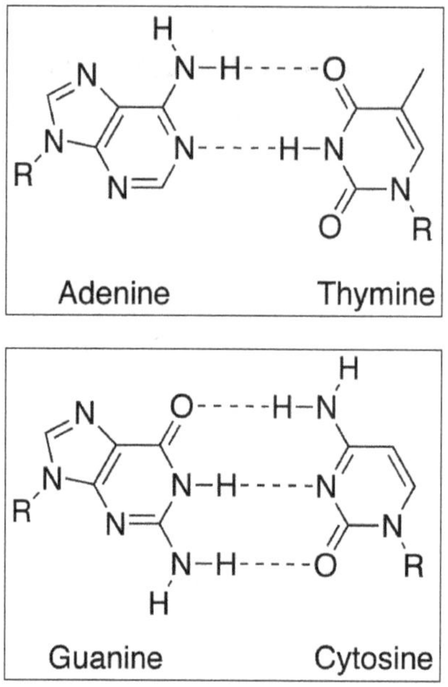

Top, an AT base pair demonstrating two intermolecular hydrogen bonds; *bottom*, a GC base pair demonstrating three intermolecular hydrogen bonds.

The larger nucleobases, adenine and guanine, are members of a class of doubly ringed chemical structures called purines; the smaller nucleobases, cytosine and thymine (and uracil), are members of a class of singly ringed chemical structures called pyrimidines. Purines are only complementary with pyrimidines: pyrimidine-pyrimidine pairings are energetically unfavorable because the molecules are too far apart for hydrogen bonding to be established; purine-purine pairings are energetically unfavorable because the molecules are too close, leading to overlap repulsion. The only other possible pairings are GT and AC; these pairings are mismatches because the pattern of hydrogen donors and acceptors do not correspond. The GU wobble base pair, with two hydrogen bonds, does occur fairly often in RNA.

Nucleic Acid Hybridization

Hybridization is the process of complementary base pairs binding to form a double helix. Melting is the process by which the interactions between the strands of the double helix are broken, separating the two nucleic acid strands. These bonds are weak, easily separated by gentle heating, enzymes, or physical force. Melting occurs preferentially at certain points in the nucleic acid. T and A rich sequences are more easily melted than C and G rich regions. Particular base steps are also susceptible to DNA melting, particularly T A and T G base steps. These mechanical features are reflected by the use of sequences such as TATAA at the start of many genes to assist RNA polymerase in melting the DNA for transcription.

Strand separation by gentle heating, as used in PCR, is simple providing the molecules have fewer than about 10,000 base pairs (10 kilobase pairs, or 10 kbp). The intertwining of the DNA strands makes long segments difficult to separate. The cell avoids this problem by allowing its DNA-melting enzymes (helicases) to work concurrently with topoisomerases, which can chemically cleave the phosphate backbone of one of the strands so that it can swivel around the other. Helicases unwind the strands to facilitate the advance of sequence-reading enzymes such as DNA polymerase.

Secondary Structure Motifs

Nucleic acid secondary structure is generally divided into helices (contiguous base pairs), and various kinds of loops (unpaired nucleotides surrounded by helices). Frequently these elements, or combinations of them, are further classified into additional categories including, for example, tetraloops, pseudoknots, and stem-loops.

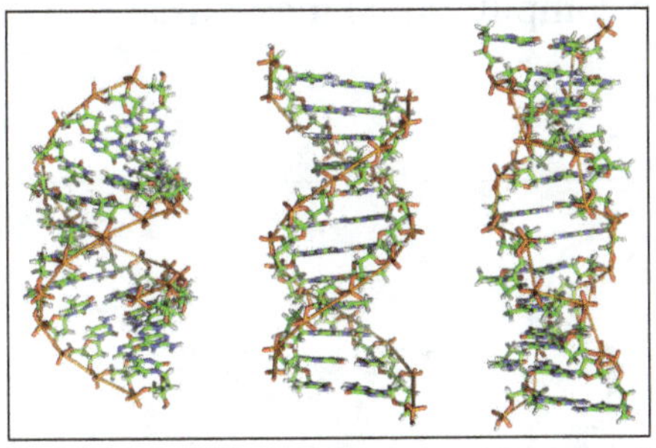

The main nucleic acid helix structures (A-, B- and Z-form).

Double Helix

The double helix is an important tertiary structure in nucleic acid molecules which is intimately connected with the molecule's secondary structure. A double helix is formed by regions of many consecutive base pairs.

The nucleic acid double helix is a spiral polymer, usually right-handed, containing two nucleotide strands which base pair together. A single turn of the helix constitutes about ten nucleotides, and contains a major groove and minor groove, the major groove being wider than the minor groove. Given the difference in widths of the major groove and minor groove, many proteins which bind to DNA do so through the wider major groove. Many double-helical forms are possible; for DNA the three biologically relevant forms are A-DNA, B-DNA, and Z-DNA, while RNA double helices have structures similar to the A form of DNA.

Stem-loop Structures

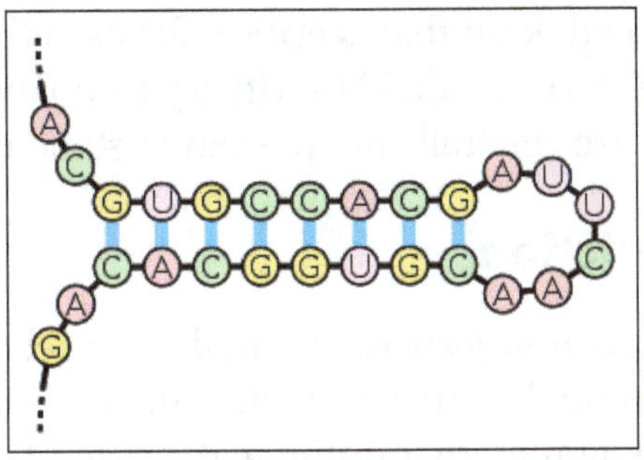

An RNA stem-loop secondary structure.

The secondary structure of nucleic acid molecules can often be uniquely decomposed into stems and loops. The stem-loop structure (also often referred to as an "hairpin"), in which a base-paired helix ends in a short unpaired loop, is extremely common and is a building block for larger structural motifs such as cloverleaf structures, which are four-helix junctions such as those found in transfer RNA. Internal loops (a short series of unpaired bases in a longer paired helix) and bulges (regions in which one strand of a helix has "extra" inserted bases with no counterparts in the opposite strand) are also frequent.

There are many secondary structure elements of functional importance to biological RNA's; some famous examples are the Rho-independent terminator stem-loops and the tRNA cloverleaf. Active research is on-going to determine the secondary structure of RNA molecules, with approaches including both experimental and computational methods.

Pseudoknots

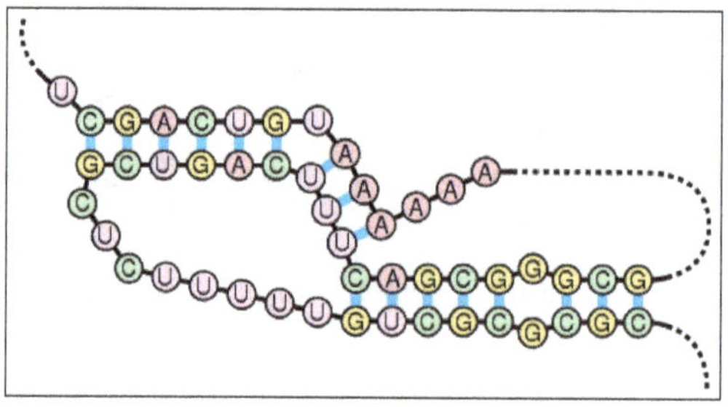

An RNA pseudoknot structure. For example, the RNA component of human telomerase.

A pseudoknot is a nucleic acid secondary structure containing at least two stem-loop structures in which half of one stem is intercalated between the two halves of another stem. Pseudoknots fold into knot-shaped three-dimensional conformations but are not true topological knots. The base pairing in pseudoknots is not well nested; that is, base pairs occur that "overlap" one another in sequence position. This makes the presence of general pseudoknots in nucleic acid sequences impossible to predict by the standard method of dynamic programming, which uses a recursive scoring system to identify paired stems and consequently cannot detect non-nested base pairs with common algorithms. However, limited subclasses of pseudoknots can be predicted using modified dynamic programs. Newer structure prediction techniques such as stochastic context-free grammars are also unable to consider pseudoknots.

Pseudoknots can form a variety of structures with catalytic activity and several important biological processes rely on RNA molecules that form pseudoknots. For example, the RNA component of the human telomerase contains a pseudoknot that is critical for its activity. The hepatitis delta virus ribozyme is a well known example of a catalytic RNA with a pseudoknot in its active site. Though DNA can also form pseudoknots, they are generally not present in standard physiological conditions.

Secondary Structure Prediction

Most methods for nucleic acid secondary structure prediction rely on a nearest neighbor thermodynamic model. A common method do determine the most probable structures given a sequence of nucleotides makes use of a dynamic programming algorithm that seeks to find structures with low free energy. Dynamic programming algorithms often forbid pseudoknots, or other cases in which base pairs are not fully nested, as considering these structures becomes computationally very expensive for even small nucleic acid molecules. Other methods, such as stochastic context-free grammars can also be used to predict nucleic acid secondary structure.

For many RNA molecules, the secondary structure is highly important to the correct function of the RNA – often more so than the actual sequence. This fact aids in the analysis of non-coding RNA sometimes termed "RNA genes". One application of bioinformatics uses predicted RNA secondary structures in searching a genome for noncoding but functional forms of RNA. For example, microRNAs have canonical long stem-loop structures interrupted by small internal loops.

RNA secondary structure applies in RNA splicing in certain species. In humans and other tetrapods, it has been shown that without the U2AF2 protein, the splicing process is inhibited. However, in zebrafish and other teleosts the RNA splicing process can still occur on certain genes in the absence of U2AF2. This may be because 10% of genes in zebrafish have alternating TG and AC base pairs at the 3' splice site (3'ss) and 5' splice site (5'ss) respectively on each intron, which alters the secondary structure of the RNA. This suggests that secondary structure of RNA can influence splicing, potentially without the use of proteins like U2AF2 that have been thought to be required for splicing to occur.

Nucleic Acid Tertiary Structure

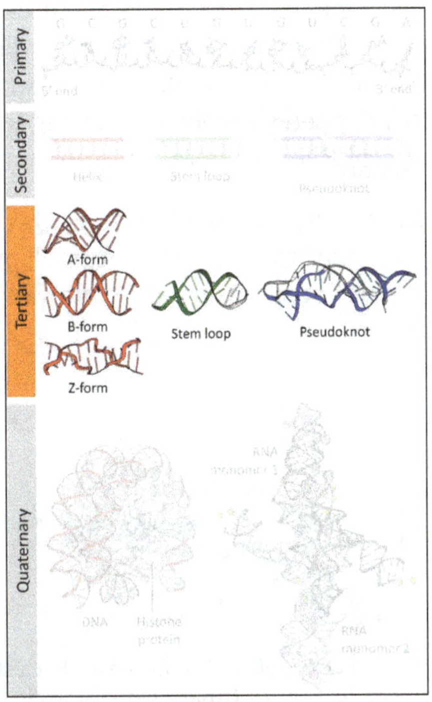

Interactive image of nucleic acid structure (primary, secondary, tertiary, and quaternary) using DNA helices and examples from the VS ribozyme and telomerase and nucleosome. (PDB: ADNA, 1BNA, 4OCB, 4R4V, 1YMO, 1EQZ)

Nucleic acid tertiary structure is the three-dimensional shape of a nucleic acid polymer. RNA and DNA molecules are capable of diverse functions ranging from molecular recognition to catalysis. Such functions require a precise three-dimensional tertiary structure. While such structures are diverse and seemingly complex, they are composed of recurring, easily recognizable tertiary structure motifs that serve as molecular building blocks. Some of the most common motifs for RNA and DNA tertiary structure are described below, but this information is based on a limited number of solved structures. Many more tertiary structural motifs will be revealed as new RNA and DNA molecules are structurally characterized.

Helical Structures

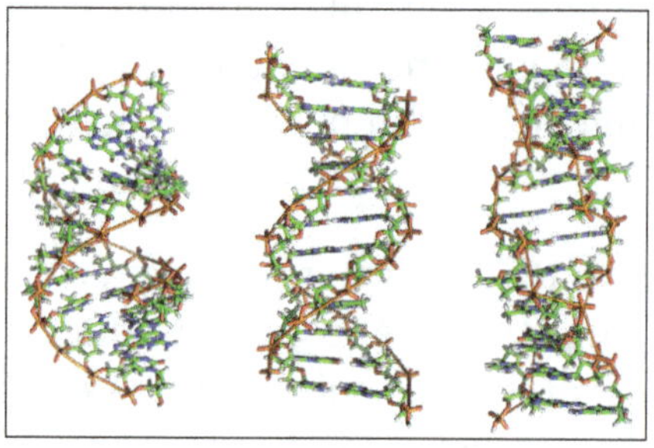

The structures of the A-, B-, and Z-DNA double helix structures.

Double Helix

The double helix is the dominant tertiary structure for biological DNA, and is also a possible structure for RNA. Three DNA conformations are believed to be found in nature, A-DNA, B-DNA, and Z-DNA. The "B" form described by James D. Watson and Francis Crick is believed to predominate in cells. James D. Watson and Francis Crick described this structure as a double helix with a radius of 10 Å and pitch of 34 Å, making one complete turn about its axis every 10 bp of sequence. The double helix makes one complete turn about its axis every 10.4–10.5 base pairs in solution. This frequency of twist (known as the helical *pitch*) depends largely on stacking forces that each base exerts on its neighbours in the chain. Double-helical RNA adopts a conformation similar to the A-form structure.

Other conformations are possible; in fact, only the letters F, Q, U, V, and Y are now available to describe any new DNA structure that may appear in the future. However, most of these forms have been created synthetically and have not been observed in naturally occurring biological systems.

RNA Triplexes

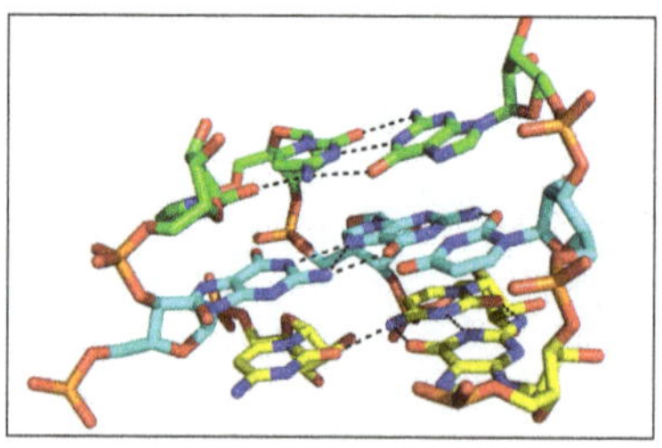

Major groove triples in the group II intron in *Oceanobacillus Iheyensis*. Each stacked layer is formed by one triplex with a different color scheme. Hydrogen bonds between triplexes are shown in black dashed lines. "N" atoms are colored in blue and "O" atoms in red. From top to bottom, the residues on the left side are G288, C289, and C377.

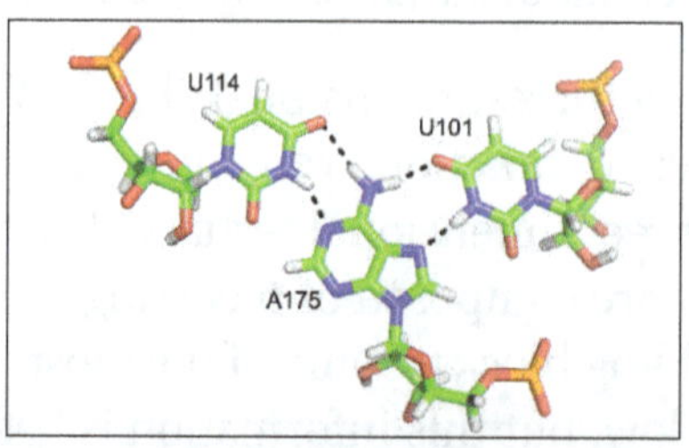

Close-up rendering of the U114:A175-U101 major groove (Hoogsteen base) triplex formed within the wild type pseudoknot of Human Telomerase RNA. Hydrogen bonds are shown in black dashed lines. "N" atoms are colored in blue and "o" atoms in red.

Major and Minor Groove Triplexes

The minor groove triple is a ubiquitous RNA structural motif. Because interactions with the minor groove are often mediated by the 2'-OH of the ribose sugar, this RNA motif looks very different from its DNA equivalent. The most common example of a minor loop triple is the A-minor motif, or the insertion of adenosine bases into the minor groove. However, this motif is not restricted to adenosines, as other nucleobases have also been observed to interact with the RNA minor groove.

The minor groove presents a near-perfect complement for an inserted base. This allows for optimal van der Waals contacts, extensive hydrogen bonding and hydrophobic surface burial, and creates a highly energetically favorable interaction. Because minor groove triples are capable of stably packing a free loop and helix, they are key elements in the structure of large ribonucleotides, including the group I intron, the group II intron, and the ribosome.

Although the major groove of standard A-form RNA is fairly narrow and therefore less available for triplex interaction than the minor groove, major groove triplex interactions can be observed in several RNA structures. These structures consist of several combinations of base pair and Hoogsteen interactions. For example, the GGC triplex (GGC amino(N-2)-N-7, imino-carbonyl, carbonyl-amino(N-4); Watson-Crick) observed in the 50S ribosome, composed of a Watson-Crick type G-C pair and an incoming G which forms a pseudo-Hoogsteen network of hydrogen bonding interactions between both bases involved in the canonical pairing. Other notable examples of major groove triplexes include (i) the catalytic core of the group II intron shown in the figure at left (ii) a catalytically essential triple helix observed in human telomerase RNA and (iii) the SAM-II riboswitch.

Triple-stranded DNA is also possible from Hoogsteen or reversed Hoogsteen hydrogen bonds in the major groove of B-form DNA.

Quadruplexes

Besides double helices and the above-mentioned triplexes, RNA and DNA can both also form quadruple helices. There are diverse structures of RNA base quadruplexes. Four consecutive guanine residues can form a quadruplex in RNA by Hoogsteen hydrogen bonds to form a "Hoogsteen ring". G-C and A-U pairs can also form base quadruplex with a combination of Watson-Crick pairing and noncanonical pairing in the minor groove.

The core of malachite green aptamer is also a kind of base quadruplex with a different hydrogen bonding pattern. The quadruplex can repeat several times consecutively, producing an immensely stable structure.

The unique structure of quadruplex regions in RNA may serve different functions in a biological system. Two important functions are the binding potential with ligands or proteins, and its ability to stabilize the whole tertiary structure of DNA or RNA. The strong structure can inhibit or modulate transcription and replication, such as in the telomeres of chromosomes and the UTR of mRNA. The base identity is important towards ligand binding. The G-quartet typically binds monovalent cations such as potassium, while other bases can bind numerous other ligands such as hypoxanthine in a U-U-C-U quadruplex.

Along with these functions, the G-quadruplex in the mRNA around the ribosome binding regions could serve as a regulator of gene expression in bacteria. There may be more interesting structures and functions yet to be discovered *in vivo*.

Quadruplexes

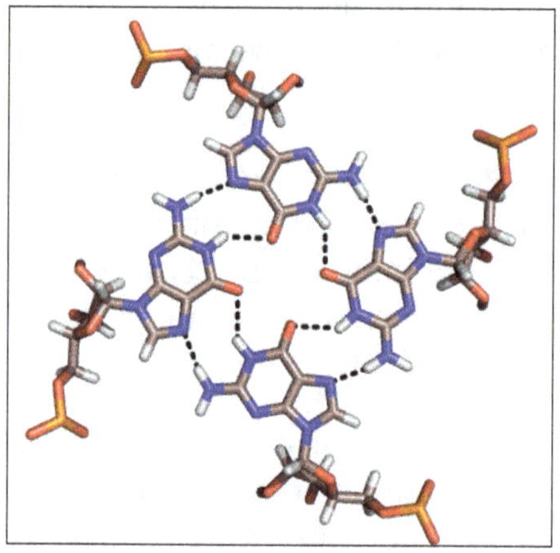

Above: Typical Ring Structure of a Hoogsteen paired G-quartet.

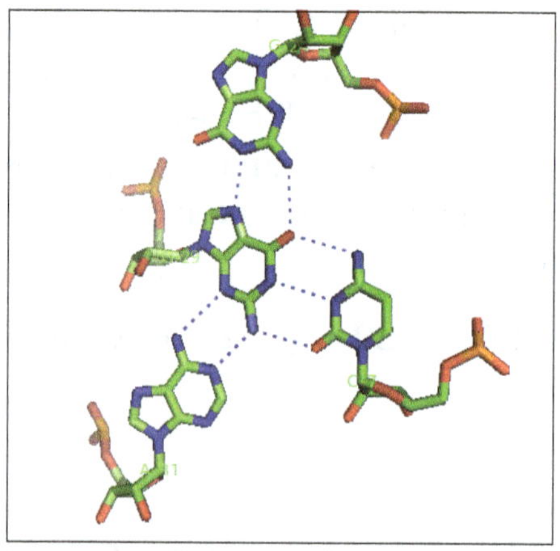

Above: Quadruplex seen in crystal structure of Malachite Green RNA aptamer. G29 involved in major groove, minor groove, and Watson-Crick hydrogen-bonding with three other bases.

Coaxial Stacking

Coaxial stacking, otherwise known as helical stacking, is a major determinant of higher order RNA tertiary structure. Coaxial stacking occurs when two RNA duplexes form a contiguous helix, which is stabilized by base stacking at the interface of the two helices. Coaxial stacking was noted in the crystal structure of tRNAPhe. More recently, coaxial stacking has been observed in higher order structures of many ribozymes, including many forms of the self-splicing group I and group II introns. Common coaxial stacking motifs include the kissing loop interaction and the pseudoknot. The stability of these interactions can be predicted by an adaptation of "Turner's rules".

In 1994, Walter and Turner determined the free energy contributions of nearest neighbor stacking interactions within a helix-helix interface by using a model system that created a helix-helix interface between a short oligomer and a four-nucleotide overhang at the end of a hairpin stem . Their experiments confirmed that the thermodynamic contribution of base-stacking between two helical secondary structures closely mimics the thermodynamics of standard duplex formation (nearest neighbor interactions predict the thermodynamic stability of the resulting helix). The relative stability of nearest neighbor interactions can be used to predict favorable coaxial stacking based on known secondary structure. Walter and Turner found that, on average, prediction of RNA structure improved from 67% to 74% accuracy when coaxial stacking contributions were included. Turner and colleagues later used an explicit quantification of stacking based on x-ray crystal structure databases to define such stacking and study kinetics in molecular dynamics simulations. Theories of coaxial stacking can be tested using the technique of helical fusion. This approach was used by Murphy and Cech to confirm a coaxial stacking interaction between the P4 and P6 helices within the catalytic center of the Tetrahymena group I intron.

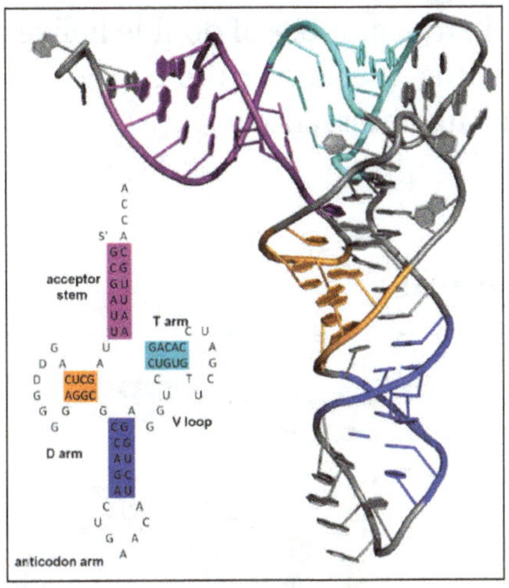

Secondary (inset) and tertiary structure of tRNA demonstrating coaxial stacking.

Most well-studied RNA tertiary structures contain examples of coaxial stacking. Some prominent examples are tRNA-Phe, group I introns, group II introns, and ribosomal RNAs. Crystal structures of tRNA revealed the presence of two extended helices that result from coaxial stacking of the amino-acid acceptor stem with the T-arm, and stacking of the D- and anticodon-arms. These interactions within tRNA orient the anticodon stem perpendicularly to the amino-acid stem, leading to the functional L-shaped tertiary structure. In group I introns, the P4 and P6 helices were shown to coaxially stack using a combination of biochemical and crystallographic methods. The P456 crystal structure provided a detailed view of how coaxial stacking stabilizes the packing of RNA helices into tertiary structures. In the self-splicing group II intron from Oceanobacillus iheyensis, the IA and IB stems coaxially stack and contribute to the relative orientation of the constituent helices of a five-way junction. This orientation facilitates proper folding of the active site of the functional ribozyme. The ribosome contains numerous examples of coaxial stacking, including stacked segments as long as 70 bp.

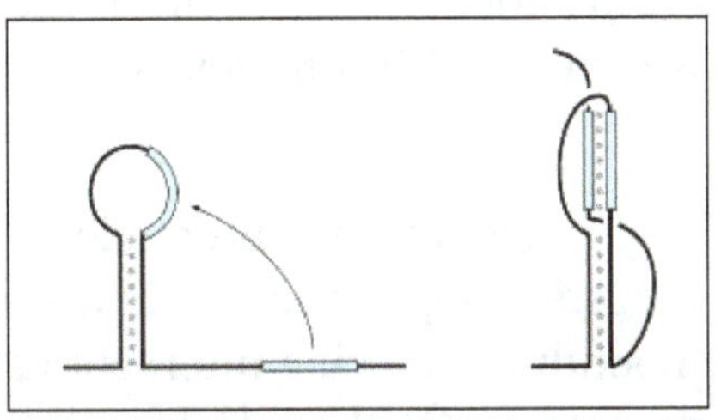

formation of a pseudoknot with coaxial stacking of the two helices.

Two common motifs involving coaxial stacking are kissing loops and pseudoknots. In kissing loop interactions, the single-stranded loop regions of two hairpins interact through base pairing, forming a composite, coaxially stacked helix. Notably, this structure allows all of the nucleotides in each loop to participate in base-pairing and stacking interactions. This motif was visualized and studied using NMR analysis by Lee and Crothers. The pseudoknot motif occurs when a single stranded region of a hairpin loop basepairs with an upstream or downstream sequence within the same RNA strand. The two resulting duplex regions often stack upon one another, forming a stable coaxially stacked composite helix. One example of a pseudoknot motif is the highly stable Hepatitis Delta virus ribozyme, in which the backbone shows an overall double pseudoknot topology.

An effect similar to coaxial stacking has been observed in rationally designed DNA structures. DNA origami structures contain a large number of double helixes with exposed blunt ends. These structures were observed to stick together along the edges that contained these exposed blunt ends, due to the hydrophobic stacking interactions.

Other Motifs

Tetraloop-receptor Interactions

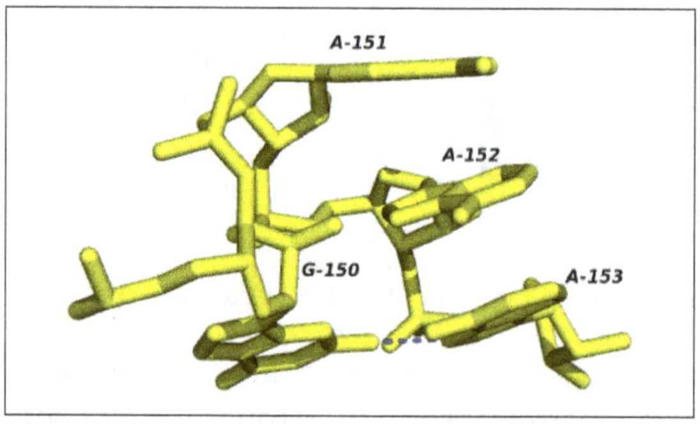

Stick representation of a GAAA tetraloop - an example from the GNRA tetraloop family.

Tetraloop-receptor interactions combine base-pairing and stacking interactions between the loop nucleotides of a tetraloop motif and a receptor motif located within an RNA duplex, creating a tertiary contact that stabilizes the global tertiary fold of an RNA molecule. Tetraloops are also possible structures in DNA duplexes.

Stem-loops can vary greatly in size and sequence, but tetraloops of four nucleotides are very common and they usually belong to one of three categories, based on sequence. These three families are the CUYG, UNCG and GNRA tetraloops. In each of these tetraloop families, the second and third nucleotides form a turn in the RNA strand and a base-pair between the first and fourth nucleotides stabilizes the stemloop structure. It has been determined, in general, that the stability of the tetraloop depends on the composition of bases within the loop and on the composition of this "closing base pair". The GNRA family of tetraloops is the most commonly observed within Tetraloop-receptor interactions.

"Tetraloop receptor motifs" are long-range tertiary interactions consisting of hydrogen bonding between the bases in the tetraloop to stemloop sequences in distal sections of the secondary RNA structure. In addition to hydrogen bonding, stacking interactions are an important component of these tertiary interactions. For example, in GNRA-tetraloop interactions, the second nucleotide of the tetraloop stacks directly on an A-platform motif within the receptor. The sequence of the tetraloop and its receptor often covary so that the same type of tertiary contact can be made with different isoforms of the tetraloop and its cognate receptor.

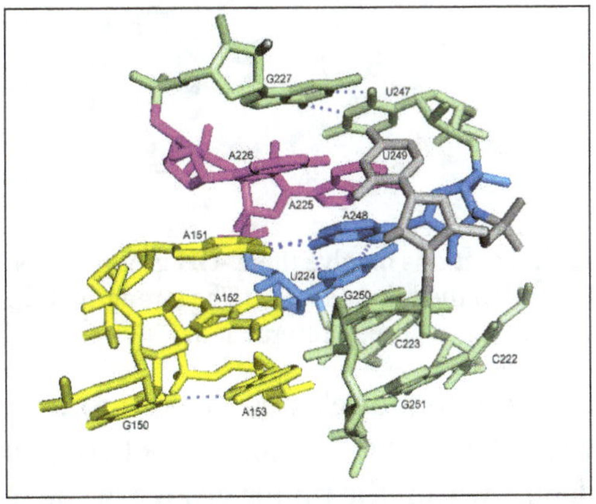

GAAA Tetraloop and Receptor: Stick representation of tetraloop (yellow) and its receptor, showing both Watson-Crick and Hoogsteen base-pairing.

For example, the self-splicing group I intron relies on tetraloop receptor motifs for its structure and function. Specifically, the three adenine residues of the canonical GAAA motif stack on top of the receptor helix and form multiple stabilizing hydrogen bonds with the receptor. The first adenine of the GAAA sequence forms a triple base-pair with the receptor AU bases. The second adenine is stabilized by hydrogen bonds with the same uridine, as well as via its 2'-OH with the receptor and via interactions with the guanine of the GAAA tetraloop. The third adenine forms a triple base pair.

A-minor Motif

A-minor Interactions

The A-minor motif is a ubiquitous RNA tertiary structural motif. It is formed by the insertion of an unpaired nucleoside into the minor groove of an RNA duplex. As such it is an example of a minor groove triple. Although guanosine, cytosine and uridine can also form minor groove triple interactions, minor groove interactions by adenine are very common. In the case of adenine, the N1-C2-N3 edge of the inserting base forms hydrogen bonds with one or both of the 2'-OH's of the duplex, as well as the bases of the duplex. The host duplex is often a G-C basepair.

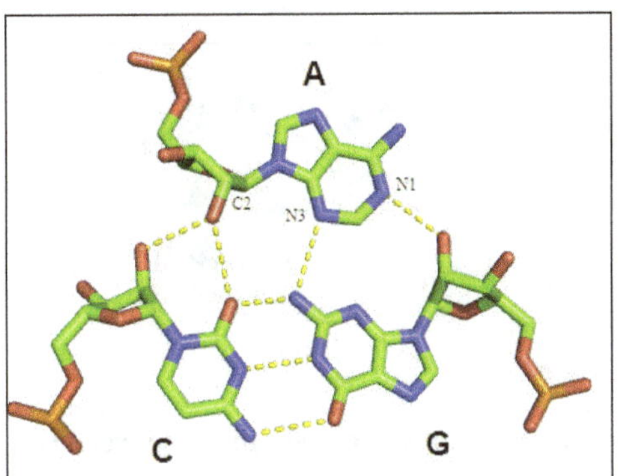

Type I A-minor interaction: Type I interactions are the most common, strongest A-minor interactions, as they involve numerous hydrogen bonds, and bury the incoming A base in the minor groove.

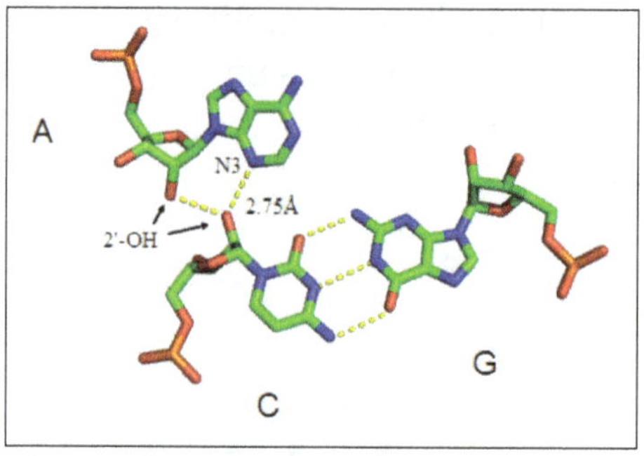

Type II A-minor interaction: Type II interactions involve the 2'-OH group and N3 of the adenosine. The adenosine interacts with the cytosine's 2'-OH group in the minor groove. The strength of this interaction is on the order of the Type I interaction.

A-minor motifs have been separated into four classes, types 0 to III, based upon the position of the inserting base relative to the two 2'-OH's of the Watson-Crick base pair. In type I and II A-minor motifs, N3 of adenine is inserted deeply within the minor groove of the duplex, and there is good shape complementarity with the base pair. Unlike types 0 and III, type I and II interactions are specific for adenine due to hydrogen bonding interactions. In the type III interaction, both the O2' and N3 of the inserting base are associated less closely with the minor groove of the duplex. Type 0 and III motifs are weaker and non-specific because they are mediated by interactions with a single 2'-OH.

The A-minor motif is among the most common RNA structural motifs in the ribosome, where it contributes to the binding of tRNA to the 23S subunit. They most often stabilize RNA duplex interactions in loops and helices, such as in the core of group II introns.

An interesting example of A-minor is its role in anticodon recognition. The ribosome must discriminate between correct and incorrect codon-anticodon pairs. It does so, in part, through the insertion of adenine bases into the minor groove. Incorrect codon-anticodon pairs will present distorted helical geometry, which will prevent the A-minor interaction from stabilizing the binding, and increase the dissociation rate of the incorrect tRNA.

An analysis of A-minor motifs in the 23S ribosomal RNA has revealed a hierarchical network of structural dependencies, suggested to be related to ribosomal evolution and to the order of events that led to the development of the modern bacterial large subunit.

Ribose Zipper

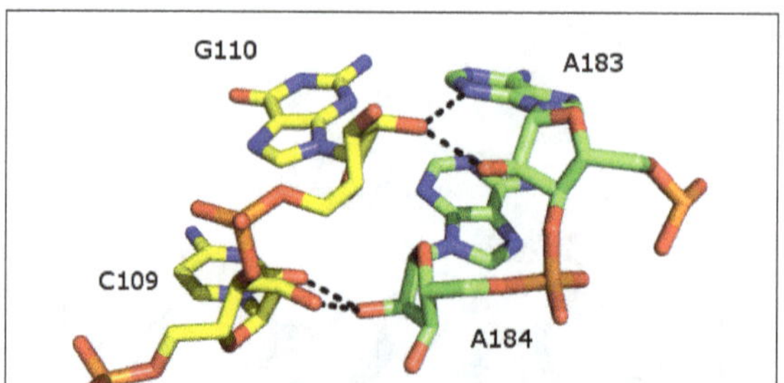

Ribose Zippers: View of a canonical ribose zipper between two RNA backbones.

The ribose zipper is an RNA tertiary structural element in which two RNA chains are held together by hydrogen bonding interactions involving the 2'OH of ribose sugars on different strands. The

2'OH can behave as both hydrogen bond donor and acceptor, which allows formation of bifurcated hydrogen bonds with another 2' OH.

Numerous forms of ribose zipper have been reported, but a common type involves four hydrogen bonds between 2'-OH groups of two adjacent sugars. Ribose zippers commonly occur in arrays that stabilize interactions between separate RNA strands. Ribose zippers are often observed as Stem-loop interactions with very low sequence specificity. However, in the small and large ribosomal subunits, there exists a propensity for ribose zippers of the CC/AA sequence- two cytosines on the first chain paired to two adenines on the second chain.

Role of Metal Ions

Functional RNAs are often folded, stable molecules with three-dimensional shapes rather than floppy, linear strands. Cations are essential for thermodynamic stabilization of RNA tertiary structures. Metal cations that bind RNA can be monovalent, divalent or trivalent. Potassium (K^+) is a common monovalent ion that binds RNA. A common divalent ion that binds RNA is magnesium (Mg^{2+}). Other ions including sodium (Na^+), calcium (Ca^{2+}) and manganese (Mn^{2+}) have been found to bind RNA *in vivo* and *in vitro*. Multivalent organic cations such as spermidine or spermine are also found in cells and these make important contributions to RNA folding. Trivalent ions such as cobalt hexamine or lanthanide ions such as terbium (Tb^{3+}) are useful experimental tools for studying metal binding to RNA.

A metal ion can interact with RNA in multiple ways. An ion can associate diffusely with the RNA backbone, shielding otherwise unfavorable electrostatic interactions. This charge screening is often fulfilled by monovalent ions. Site-bound ions stabilize specific elements of RNA tertiary structure. Site-bound interactions can be further subdivided into two categories depending on whether water mediates the metal binding. "Outer sphere" interactions are mediated by water molecules that surround the metal ion. For example, magnesium hexahydrate interacts with and stabilizes specific RNA tertiary structure motifs via interactions with guanosine in the major groove. Conversely, "inner sphere" interactions are directly mediated by the metal ion. RNA often folds in multiple stages and these steps can be stabilized by different types of cations. In the early stages, RNA forms secondary structures stabilized through the binding of monovalent cations, divalent cations and polyanionic amines in order to neutralize the polyanionic backbone. The later stages of this process involve the formation of RNA tertiary structure, which is stabilized almost largely through the binding of divalent ions such as magnesium with possible contributions from potassium binding.

Metal-binding sites are often localized in the deep and narrow major groove of the RNA duplex, coordinating to the Hoogsteen edges of purines. In particular, metal cations stabilize sites of backbone twisting where tight packing of phosphates results in a region of dense negative charge. There are several metal ion-binding motifs in RNA duplexes that have been identified in crystal structures. For instance, in the P4-P6 domain of the *Tetrahymena thermophila* group I intron, several ion-binding sites consist of tandem G-U wobble pairs and tandem G-A mismatches, in which divalent cations interact with the Hoogsteen edge of guanosine via O6 and N7. Another ion-binding motif in the *Tetrahymena* group I intron is the A-A platform motif, in which consecutive adenosines in the same strand of RNA form a non-canonical pseudobase pair. Unlike the tandem G-U motif, the A-A platform motif binds preferentially to monovalent cations. In many of these motifs, absence of the monovalent or divalent cations results in either greater flexibility or loss of tertiary structure.

Metal Ion Binding in the Group I Intron

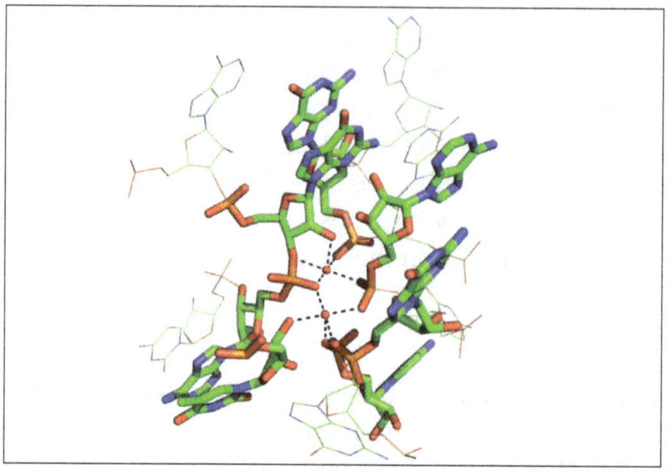

PDB rendering of Group I intron inner sphere magnesium coordination. The two red balls indicate magnesium ions and dashed lines coming from the ions indicate coordination with the respective groups on nucleotides. The color-coding scheme is as follows: green=carbon, orange=phosphate, pink=oxygen, blue=nitrogen.

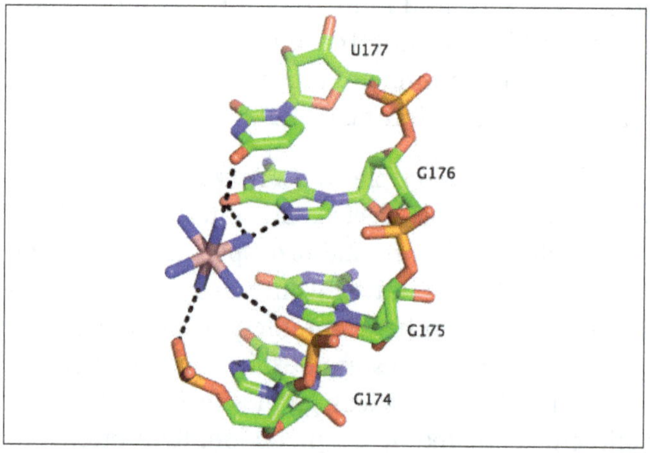

PDB rendering of Group 1 intron P5c binding pocket demonstrating outer sphere coordination. Here, the six amines of osmium hexamine(III) fulfill the role generally served by water molecules and mediate the ion's interaction with the major groove. Coordination via hydrogen bonds is indicated by dashed lines and osmium is rendered in pink, all other colors are as above.

Divalent metal ions, especially magnesium, have been found to be important for the structure of DNA junctions such as the Holliday junction intermediate in genetic recombination. The magnesium ion shields the negatively charged phosphate groups in the junction and allows them to be positioned closer together, allowing a stacked conformation rather than an unstacked conformation. Magnesium is vital in stabilizing these kinds of junctions in artificially designed structures used in DNA nanotechnology, such as the double crossover motif.

Nucleic Acid Quaternary Structure

Nucleic acid quaternary structure refers to the interactions between separate nucleic acid molecules, or between nucleic acid molecules and proteins. The concept is analogous to protein quaternary structure, but as the analogy is not perfect, the term is used to refer to a number of different concepts in nucleic acids and is less commonly encountered.

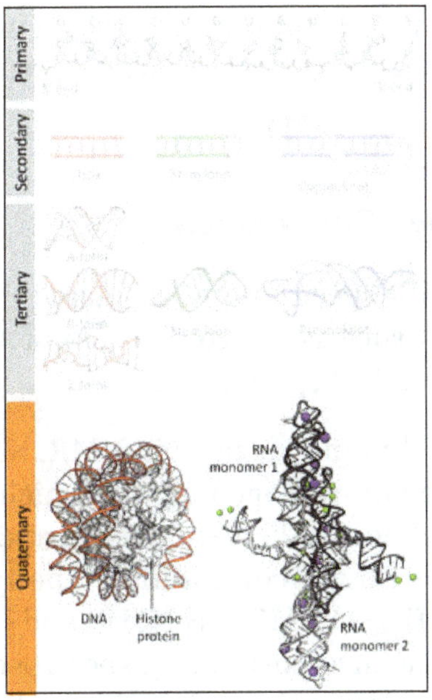

Interactive image of nucleic acid structure (primary, secondary, tertiary, and quaternary) using DNA helices and examples from the VS ribozyme and telomerase and nucleosome. (PDB: ADNA, 1BNA, 4OCB, 4R4V, 1YMO, 1EQZ)

DNA

DNA quaternary structure is used to refer to the binding of DNA to histones to form nucleosomes, and then their organisation into higher-order chromatin fibres. The quaternary structure of DNA strongly affects how accessible the DNA sequence is to the transcription machinery for expression of genes. DNA quaternary structure varies over time, as regions of DNA are condensed or exposed for transcription. The term has also been used to describe the hierarchical assembly of artificial nucleic acid building blocks used in DNA nanotechnology.

RNA

Symmetrical complexes of RNA molecules are extremely uncommon compared to protein oligomers. One example of an RNA homodimer is the VS ribozyme from Neurospora, with its two active sites consisting of nucleotides from both monomers.

The best known example of RNA forming quaternary structures with proteins is the ribosome, which consists of multiple rRNAs, supported by rProteins. Similar RNA-Protein complexes are also found in the spliceosome.

Nucleic Acid Structure Determination

Structure probing of nucleic acids is the process by which biochemical techniques are used to determine nucleic acid structure. This analysis can be used to define the patterns which can infer the molecular structure, experimental analysis of molecular structure and function, and further understanding on development of smaller molecules for further biological research. Structure probing analysis can be done through many different methods, which include chemical probing, hydroxyl radical probing, SHAPE, nucleotide analog interference mapping (NAIM), and in-line probing.

Physical Methods

Nuclear Magnetic Resonance Spectroscopy

Nucleic acid NMR is the use of NMR spectroscopy to obtain information about the structure and dynamics of nucleic acid molecules, such as DNA or RNA. As of 2003, nearly half of all known RNA structures had been determined by NMR spectroscopy.

Nucleic acid NMR uses similar techniques as protein NMR, but has several differences. Nucleic acids have a smaller percentage of hydrogen atoms, which are the atoms usually observed in NMR, and because nucleic acid double helices are stiff and roughly linear, they do not fold back on themselves to give "long-range" correlations. The types of NMR usually done with nucleic acids are ^{1}H or proton NMR, ^{13}C NMR, ^{15}N NMR, and ^{31}P NMR. Two-dimensional NMR methods are almost always used, such as correlation spectroscopy (COSY) and total coherence transfer spectroscopy (TOCSY) to detect through-bond nuclear couplings, and nuclear Overhauser effect spectroscopy (NOESY) to detect couplings between nuclei that are close to each other in space.

Parameters taken from the spectrum, mainly NOESY cross-peaks and coupling constants, can be used to determine local structural features such as glycosidic bond angles, dihedral angles (using the Karplus equation), and sugar pucker conformations. For large-scale structure, these local parameters must be supplemented with other structural assumptions or models, because errors add up as the double helix is traversed, and unlike with proteins, the double helix does not have a compact interior and does not fold back upon itself. NMR is also useful for investigating nonstandard geometries such as bent helices, non-Watson–Crick basepairing, and coaxial stacking. It has been especially useful in probing the structure of natural RNA oligonucleotides, which tend to adopt complex conformations such as stem-loops and pseudoknots. NMR is also useful for probing the binding of nucleic acid molecules to other molecules, such as proteins or drugs, by seeing which resonances are shifted upon binding of the other molecule.

RNA Sequencing

While methods for each type of probing differ in steps, the probing of secondary structure involves certain steps in order to determine the structure. The first step involved is submitting the structural RNA to the probe of interest and incubating over a certain amount of time to allow the reaction to occur. The RNA is then transcribed using reverse transcriptase PCR, where this results in different lengths of bands due to the modification of the RNA at specific sites, which causes the reverse transcriptase to fall off. These bands are then run on a gel with sequencing data determined from RNA sequencing.

Chemical Methods

RNA chemical probing can involve many different chemicals which serve to modify specific bases at certain sites to show certain locations available for specific modification type.

Hydroxyl Radical Probing

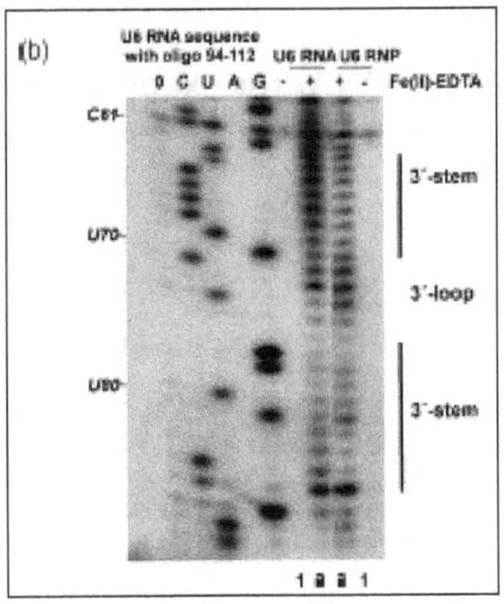

Hydroxyl radical probing gel showing bands at positions and dots indicating strength of protection.

Probing with hydroxyl radicals involves an additional step, as hydroxyl radicals are short lived in solution they need to be generated. This can be done using H_2O_2, ascorbic acid, and Fe(II)-EDTA complex which is attached to the backbone through EDTA. The Fe(II) along with ascorbic acid generates hydroxyl radicals which then can react with the nucleic acid molecules. Hydroxyl radical probing is often used in conjunction with chemical probing of nucleic acid molecules that are thought to associate with proteins, this is due to the modification done by hydroxyl radicals. Hydroxyl radicals attack the ribose/deoxyribose ring and this results in breaking of the phosphate backbone, which is independent of secondary structure, as all backbone is accessible, but is instead resultant of protein or tertiary structure protection. Probing with hydroxyl radicals shows the protection of structured nucleic acids by the proteins thought to be associated or folding on itself, where cleavage again results in a band formed through gel electrophoresis (after RT-PCR in the case of RNA) that is shorter than the full nucleic acid depending upon where it is cleaved. In this case since the last nucleotide is not modified the band length is indicative of the base that was cleaved. When examining the gel produced by running the gels on a band the areas of various strength of protection where areas of stronger protection for hydroxyl radicals can be said to have tighter association with a protein, or if no protein associates with the nucleic acid it can be caused by the tertiary fold.

DMS

Dimethyl sulfate, known as DMS, is a chemical that can be used to modify nucleic acids in order to determine secondary structure. DMS modifies certain bases through methylation. One set of methylation products that is used for RNA is the methylation of N1 adenosine and N3 of cytosine which prevents the natural hydrogen bonds to form between bases modified by DMS. This enables modification sites to be detected by RT-PCR, as modified sites cannot be basepaired, which results in the reverse transcriptase to fall off and produce different band sizes. DMS may modify adenines and cytosines that are single stranded, base paired at the end of a helix, or in a base pair next to a GU pair. Detection of these modifications is done by examination of a gel and looking at bands present. One thing of note is that since modification prevents the addition of nucleotides at the modification site each of the bands generated is shifted by one base down on the gel. Nucleotides may be protected from DMS modification by base pairing, tertiary contacts, or protein-RNA interactions. This can be used to begin to develop a model of secondary structure for the RNA molecule.

Structure probing by DMS allows for detection of secondary structure changes due to binding of RNA molecules along with detecting changes in tertiary structure.

DMS modification can also be used for DNA, for example in footprinting DNA-protein interactions.

CMCT

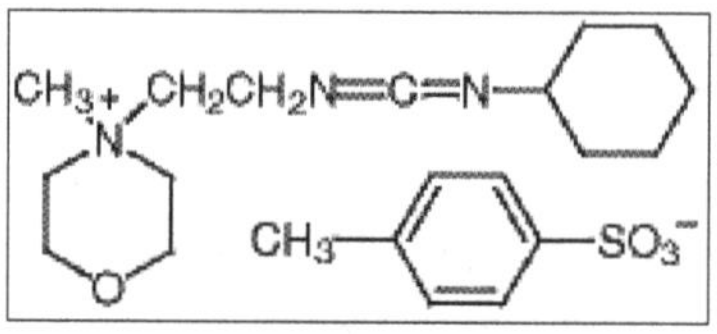

Structure of CMCT used in RNA structure probing.

1-cyclohexyl-(2-morpholinoethyl)carbodiimide metho-p-toluene sulfonate known as CMCT is another chemical that is used in structure probing. CMCT like DMS serves to modify the exposed bases of specific nucleotides, which are uridine, and to a smaller extent guanine. CMCT reacts primarily with N3 of uridine and N1 of guanine modifying two sites responsible for hydrogen bonding on the bases. Modification using CMCT is analogous in detection to DMS as modification prevents basepairing at specific nucleotides which are then detected using rt-PCR and running of the PCR products on an agarose gel as shown in Figure. Structure probing of RNA by CMCT indicates the presence of uridine and guanine in single stranded regions by accessibility to modification or the presence of uridine and guanine in double stranded regions by protection from CMCT, which is the absence of a band. CMCT, like DMS can detect secondary and tertiary structure changes, but still has the same weaknesses as the method of modification is the same as DMS.

Kethoxal

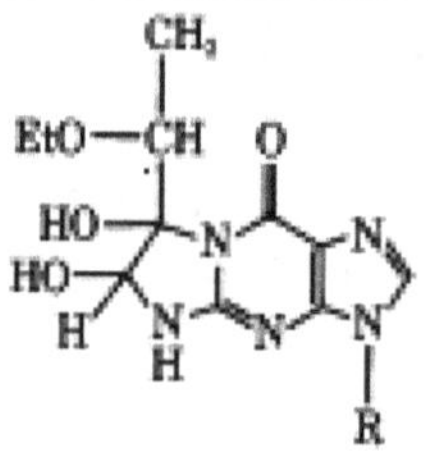

II

Binding of kethoxal to modified guanine preventing basepairing.

1,1-Dihydroxy-3-ethoxy-2-butanone, known as kethoxal, is used like DMS and CMCT, where treatment with kethoxal causes the modification of guanine, specifically altering the N1 and the exocyclic amino group simultaneously by covalent interaction. The modification by kethoxal prevents single stranded guanine nucleotides to become modified, so that when reverse transcriptase reaches the modified guanine it falls off resulting in a band. After rt-PCR the products will be of different lengths where a band present indicates modification of a guanine base, and the absence of a band indicates that the base was not available for modification. Using this gel in combination with DMS and CMCT a model for structural RNAs can be formed through comparison between the gels which indicate protected and accessible positions in all the bases, which can be used to form a preliminary model.

SHAPE

Selective 2′-hydroxyl acylation analyzed by primer extension, or SHAPE, takes advantage of reagents that preferentially modify the backbone of RNA in structurally flexible regions.

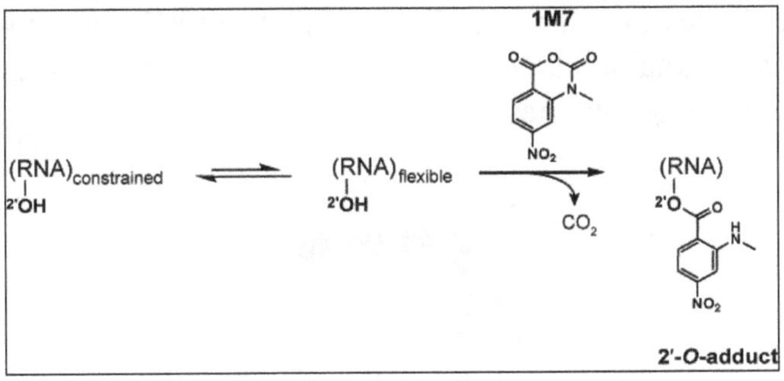

1-methyl-7-nitroisatoic anhydride (1M7) undergoes hydrolysis to form adducts on the backbone of unpaired RNA nucleotides.

Reagents such as N-methylisotoic anhydride (NMIA) and 1-methyl-7-nitroisatoic anhydride (1M7) react with the 2'-hydroxyl group to form adducts on the 2'-hydroxyl of the RNA backbone. Compared to the chemicals used in other RNA probing techniques, these reagents have the advantage of being largely unbiased to base identity, while remaining very sensitive to conformational dynamics. Nucleotides which are constrained (usually by base-pairing) show less adduct formation than nucleotides which are unpaired. Adduct formation is quantified for each nucleotide in a given RNA by extension of a complementary DNA primer with reverse transcriptase and comparison of the resulting fragments with those from an unmodified control. SHAPE therefore reports on RNA structure at the individual nucleotide level. This data can be used as input to generate highly accurate secondary structure models. SHAPE has been used to analyze diverse RNA structures, including that of an entire HIV-1 genome. The best approach is to use a combination of chemical probing reagents and experimental data. In SHAPE-Seq SHAPE is extended by bar-code based multiplexing combined with RNA-Seq and can be performed in a high-throughput fashion.

LASER

Light-Activated Structural Examination of RNA (LASER) probing utilizes UV light to active nicotinoyl azide (NAz), generating nitrenium cation, which reacts with guanosine and adenosine of RNA at C-8 position through a barrierless Friedel-Crafts reaction. This chemical probing method is light-controllable, and probes solvent accessbility of nucleobase, which has been shown to footprint RNA binding proteins inside cells.

In-line Probing

In-line probing does not involve treatment with any type of chemical or reagent to modify RNA structures. This type of probing assay uses the structure dependent cleavage of RNA; single stranded regions are more flexible and unstable and will degrade over time. The process of in-line probing is often used to determine changes in structure due to ligand binding. Binding of a ligand can result in different cleavage patterns. The process of in-line probing involves incubation of structural or functional RNAs over a long period of time. This period can be several days, but varies in each experiment. The incubated products are then run on a gel to visualize the bands. This experiment is often done using two different conditions: 1) with ligand and 2) in the absence of ligand. Cleavage results in shorter band lengths and is indicative of areas that are not basepaired, as basepaired regions tend to be less sensitive to spontaneous cleavage. In-line probing is a functional assay that can be used to determine structural changes in RNA in response to ligand binding. It can directly show the change in flexibility and binding of regions of RNA in response to a ligand, as well as compare that response to analogous ligands. This assay is commonly used in dynamic studies, specifically when examining riboswitches.

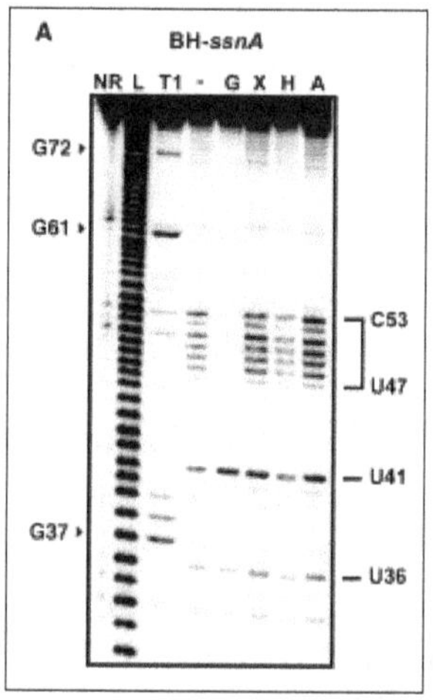

In-line probing assay of guanine riboswitches showing change in flexibility in response to various nucleotide ligands.

Nucleotide Analog Interference Mapping

Nucleotide analog interference mapping (NAIM) is the process of using nucleotide analogs, molecules that are similar in some ways to nucleotides but lack function, to determine the importance of a functional group at each location of an RNA molecule. The process of NAIM is to insert a single nucleotide analog into a unique site. This can be done by transcribing a short RNA using T7 RNA polymerase, then synthesizing a short oligonucleotide containing the analog in a specific position, then ligating them together on the DNA template using a ligase. The nucleotide analogs are tagged with a phosphorothioate, the active members of the RNA population are then distinguished from the inactive members, the inactive members then have the phosphorothioate tag removed and the analog sites are identified using gel electrophoresis and autoradiography. This indicates a functionally important nucleotide, as cleavage of the phosphorothioate by iodine results in an RNA that is cleaved at the site of the nucleotide analog insert. By running these truncated RNA molecules on a gel, the nucleotide of interest can be identified against a sequencing experiment Site directed incorporation results indicate positions of importance where when running on a gel, functional RNAs that have the analog incorporated at that position will have a band present, but if the analog results in non-functionality, when the functional RNA molecules are run on a gel there will be no band corresponding to that position on the gel. This process can be used to evaluate an entire area, where analogs are placed in site specific locations, differing by a single nucleotide, then when functional RNAs are isolated and run on a gel, all areas where bands are produced indicate non-essential nucleotides, but areas where bands are absent from the functional RNA indicate that inserting a nucleotide analog in that position caused the RNA molecule to become non-functional.

Nucleic Acid Structure Prediction

Nucleic acid structure prediction is a computational method to determine *secondary* and *tertiary* nucleic acid structure from its sequence. Secondary structure can be predicted from one or several nucleic acid sequences. Tertiary structure can be predicted from the sequence, or by comparative modeling (when the structure of a homologous sequence is known).

The problem of predicting nucleic acid secondary structure is dependent mainly on base pairing and base stacking interactions; many molecules have several possible three-dimensional structures, so predicting these structures remains out of reach unless obvious sequence and functional similarity to a known class of nucleic acid molecules, such as transfer RNA (tRNA) or microRNA (miRNA), is observed. Many secondary structure prediction methods rely on variations of dynamic programming and therefore are unable to efficiently identify pseudoknots.

While the methods are similar, there are slight differences in the approaches to RNA and DNA structure prediction. *In vivo*, DNA structures are more likely to be duplexes with full complementarity between two strands, while RNA structures are more likely to fold into complex secondary and tertiary structures such as in the ribosome, spliceosome, or transfer RNA. This is partly because the extra oxygen in RNA increases the propensity for hydrogen bonding in the nucleic acid backbone. The energy parameters are also different for the two nucleic acids.

Single Sequence Structure Prediction

A common problem for researchers working with RNA is to determine the three-dimensional structure of the molecule given only a nucleic acid sequence. However, in the case of RNA much of the final structure is determined by the secondary structure or intra-molecular base pairing interactions of the molecule. This is shown by the high conservation of base pairings across diverse species.

The Most Stable Structure

Secondary structure of small RNA molecules is largely determined by strong, local interactions such as hydrogen bonds and base stacking. Summing the free energy for such interactions should provide an approximation for the stability of a given structure. To predict the folding free energy of a given secondary structure, an empirical nearest-neighbor model is used. In the nearest neighbor model the free energy change for each motif depends on the sequence of the motif and of its closest base-pairs. The model and parameters of minimal energy for Watson–Crick pairs, GU pairs and loop regions were derived from empirical calorimetric experiments, the most up-to-date parameters were published in 2004, although most software packages use the prior set assembled in 1999.

The simplest way to find the lowest free energy structure would be to generate all possible structures and calculate the free energy for it, but the number of possible structures for a sequence increases exponentially with the length of RNA: number of secondary structures = $(1,8)^N$, N- number of nucleotides. For longer molecules, the number of possible secondary structures is huge: a sequence of 100 nucleotides has more than 10^{25} possible secondary structures.

Dynamic Programming Algorithms

The first and the most popular method for finding the most stable structure is a dynamic programming algorithm. One of the first attempts to predict RNA secondary structure was made by Ruth Nussinov and co-workers who used the dynamic programming method for maximizing the number of base-pairs. However, there are several issues with this approach: most importantly, the solution is not unique. Nussinov et al. published an adaptation of their approach using a simple nearest-neighbour energy model in 1980. In 1981, Michael Zuker and Patrick Stiegler proposed using a slightly refined dynamic programming approach to modelling nearest neighbor energy interactions that directly incorporates stacking into the prediction.

Dynamic programming algorithms provide a means to implicitly check all variants of possible RNA secondary structures without explicitly generating the structures. First, the lowest conformational free energy is determined for each possible sequence fragment starting with the shortest fragments and then for longer fragments. For longer fragments, recursion on the optimal free energy changes determined for shorter sequences speeds the determination of the lowest folding free energy. Once the lowest free energy of the complete sequence is calculated, the exact structure of RNA molecule is determined.

Dynamic programming algorithms are commonly used to detect base pairing patterns that are "well-nested", that is, form hydrogen bonds only to bases that do not overlap one another in sequence position. Secondary structures that fall into this category include double helices, stem-loops, and variants of the "cloverleaf" pattern found in transfer RNA molecules. These methods rely on pre-calculated parameters which estimate the free energy associated with certain types of base-pairing interactions, including Watson-Crick and Hoogsteen base pairs. Depending on the complexity of the method, single base pairs may be considered, and short two- or three-base segments, to incorporate the effects of base stacking. This method cannot identify pseudoknots, which are not well nested, without substantial algorithmic modifications that are computationally very costly.

Suboptimal Structures

The accuracy of RNA secondary structure prediction from one sequence by free energy minimization is limited by several factors:

1. The free energy value's list in nearest neighbor model is incomplete.
2. Not all known RNA folds in such a way as to conform with the thermodynamic minimum.
3. Some RNA sequences have more than one biologically active conformation (i.e., riboswitches).

For this reason, the ability to predict structures which have similar low free energy can provide significant information. Such structures are termed suboptimal structures. MFOLD is one program that generates suboptimal structures.

Predicting Pseudoknots

One of the issues when predicting RNA secondary structure is that the standard free energy minimization and statistical sampling methods can not find pseudoknots. The major problem is that the usual dynamic programing algorithms, when predicting secondary structure, consider only the interactions between the closest nucleotides, while pseudoknotted structures are formed due to interactions between distant nucleotides. Rivas and Eddy published a dynamic programming algorithm for predicting pseudoknots. However, this dynamic programming algorithm is very slow. The standard dynamic programming algorithm for free energy minimization scales $O(N^3)$ in time (N is the number of nucleotides in the sequence), while the Rivas and Eddy algorithm scales $O(N^6)$ in time. This has prompted several researchers to implement versions of the algorithm that restrict classes of pseudoknots, resulting in performance gains. For example, pknotsRG tool includes only the class of simple recursive pseudoknots and scales O(N4) in time.

Other Approaches for RNA Secondary Structure Prediction

Another approach for RNA secondary structure determination is to sample structures from the Boltzmann ensemble, as exemplified by the program SFOLD. The program generates a statistical sample of all possible RNA secondary structures. The algorithm samples secondary structures according to the Boltzmann distribution. The sampling method offers an appealing solution to the problem of uncertainties in folding.

Comparative Secondary Structure Prediction

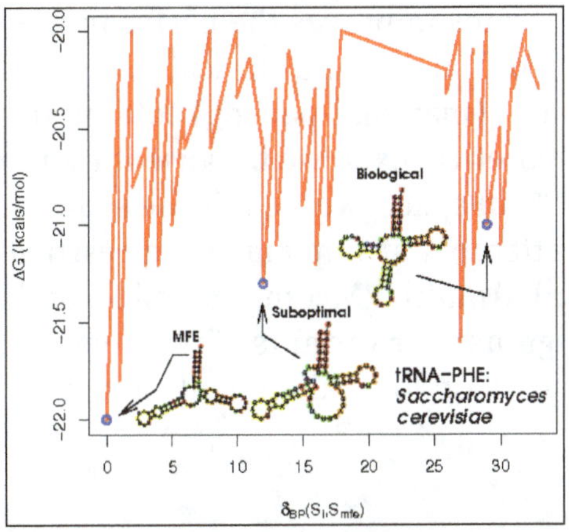

S. cerevisiae tRNA-PHE structure space: the energies and structures were calculated using RNAsubopt and the structure distances computed using RNAdistance.

Sequence covariation methods rely on the existence of a data set composed of multiple homologous RNA sequences with related but dissimilar sequences. These methods analyze the covariation of individual base sites in evolution; maintenance at two widely separated sites of a pair of base-pairing nucleotides indicates the presence of a structurally required hydrogen bond between those positions. The general problem of pseudoknot prediction has been shown to be NP-complete.

In general, the problem of alignment and consensus structure prediction are closely related. Three different approaches to the prediction of consensus structures can be distinguished:

1. Folding of alignment.
2. Simultaneous sequence alignment and folding.
3. Alignment of predicted structures.

Align then Fold

A practical heuristic approach is to use multiple sequence alignment tools to produce an alignment of several RNA sequences, to find consensus sequence and then fold it. The quality of the alignment determines the accuracy of the consensus structure model. Consensus sequences are folded using various approaches similarly as in individual structure prediction problem. The thermodynamic folding approach is exemplified by RNAalifold program. The different approaches are exemplified by Pfold and ILM programs. Pfold program implements a SCFGs. ILM (iterated loop matching) unlike the other algorithms for folding of alignments, can return pseudoknoted structures. It uses combination of thermodynamics and mutual information content scores.

Align and Fold

Evolution frequently preserves functional RNA structure better than RNA sequence. Hence, a common biological problem is to infer a common structure for two or more highly diverged but homologous RNA sequences. In practice, sequence alignments become unsuitable and do not help to improve the accuracy of structure prediction, when sequence similarity of two sequences is less than 50%.

Structure-based alignment programs improves the performance of these alignments and most of them are variants of the Sankoff algorithm. Basically, Sankoff algorithm is a merger of sequence alignment and Nussinov (maximal-pairing) folding dynamic programming method. Sankoff algorithm itself is a theoretical exercise because it requires extreme computational resources ($O^{(n3m)}$ in time, and $O^{(n2m)}$ in space, where n is the sequence length and m is the number of sequences). Some notable attempts at implementing restricted versions of Sankoff's algorithm are Foldalign, Dynalign, PMmulti/PMcomp, Stemloc, and Murlet. In these implementations the maximal length of alignment or variants of possible consensus structures are restricted. For example, Foldalign focuses on local alignments and restricts the possible length of the sequences alignment.

Fold then Align

A less widely used approach is to fold the sequences using single sequence structure prediction methods and align the resulting structures using tree-based metrics. The fundamental weakness with this approach is that single sequence predictions are often inaccurate, thus all further analyses are affected.

Tertiary Structure Prediction

Once secondary structure of RNA is known, the next challenge is to predict tertiary structure. The biggest problem is to determine the structure of regions between double stranded helical regions. Also RNA molecules often contain posttranscriptionally modified nucleosides, which because of new possible non-canonical interactions, cause a lot of troubles for tertiary structure prediction.

Nucleic Acid Quantitation

In molecular biology, quantitation of nucleic acids is commonly performed to determine the average concentrations of DNA or RNA present in a mixture, as well as their purity. Reactions that use nucleic acids often require particular amounts and purity for optimum performance. To date, there are two main approaches used by scientists to quantitate, or establish the concentration, of nucleic acids (such as DNA or RNA) in a solution. These are spectrophotometric quantification and UV fluorescence tagging in presence of a DNA dye.

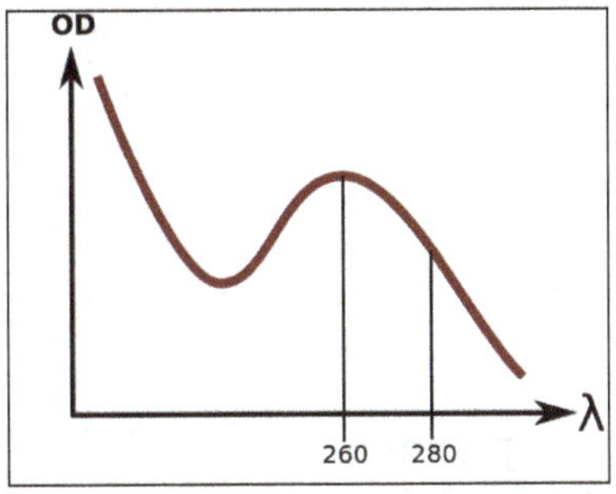

Optical density of ribosome sample. The important wavelengths of 260nm and 280nm are labeled.

Spectrophotometric Analysis

One of the more commonly used practices to quantitate DNA or RNA is the use of spectrophotometric analysis using a spectrophotometer. A spectrophotometer is able to determine the average concentrations of the nucleic acids DNA or RNA present in a mixture, as well as their purity.

Spectrophotometric analysis is based on the principles that nucleic acids absorb ultraviolet light in a specific pattern. In the case of DNA and RNA, a sample is exposed to ultraviolet light at a wavelength of 260 nanometres (nm) and a photo-detector measures the light that passes through the sample. Some of the ultraviolet light will pass through and some will be absorbed by the DNA / RNA. The more light absorbed by the sample, the higher the nucleic acid concentration in the sample. The resulting effect is that less light will strike the photodetector and this will produce a higher optical density (OD).

Using the Beer-Lambert Law it is possible to relate the amount of light absorbed to the concentration of the absorbing molecule. At a wavelength of 260 nm, the average extinction coefficient for double-stranded DNA is 0.020 $(\mu g/ml)^{-1}\ cm^{-1}$, for single-stranded DNA it is 0.027 $(\mu g/ml)^{-1}\ cm^{-1}$, for single-stranded RNA it is 0.025 $(\mu g/ml)^{-1}\ cm^{-1}$ and for short single-stranded oligonucleotides it is dependent on the length and base composition. Thus, an Absorbance (A) of 1 corresponds to a concentration of 50 μg/ml for double-stranded DNA. This method of calculation is valid for up to an A of at least 2. A more accurate extinction coefficient may be needed for oligonucleotides; these can be predicted using the nearest-neighbor model.

Calculations

The optical density is generated from equation:

Optical Density= Log (Intensity of Incident Light / Intensity of Transmitted Light)

In practical terms, a sample that contains no DNA or RNA should not absorb any of the ultraviolet light and therefore produce an OD of 0.

Optical Density=Log (100/100)=0

When using spectrophotometric analysis to determine the concentration of DNA or RNA, the Beer-Lambert law is used to determine unknown concentrations without the need for standard curves. In essence, the Beer Lambert Law makes it possible to relate the amount of light absorbed to the concentration of the absorbing molecule. The following absorbance units to nucleic acid concentration conversion factors are used to convert OD to concentration of unknown nucleic acid samples:

A260 dsDNA = 50 µg/ml

A260 ssDNA = 33 µg/ml

A260 ssRNA = 40 µg/ml

Conversion Factors

When using a 10 mm path length, simply multiply the OD by the conversion factor to determine the concentration. Example, a 2.0 OD dsDNA sample corresponds to a sample with a 100 ug/ml concentration.

When using a path length that is shorter than 10mm, the resultant OD will be reduced by a factor of 10/path length. Using the example above with a 3 mm path length, the OD for the 100 ug/ml sample would be reduced to 0.6. To normalize the concentration to a 10mm equivalent, the following is done:

0.6 OD X (10/3) * 50 ug/ml=100 ug/ml

Most spectrophotometers allow selection of the nucleic acid type and path length such that resultant concentration is normalized to the 10 mm path length which is based on the principles of Beer's law.

A260 as Quantity Measurement

The "A260 unit" is used as a quantity measure for nucleic acids. One A260 unit is the amount of nucleic acid contained in 1 mL and producing an OD of 1. The same conversion factors apply, and therefore, in such contexts:

1 A260 unit dsDNA = 50 µg

1 A260 unit ssDNA = 33 µg

1 A260 unit ssRNA = 40 µg

Sample Purity (260:280 / 260:230 Ratios)

It is common for nucleic acid samples to be contaminated with other molecules (i.e. proteins, organic compounds, other). The secondary benefit of using spectrophotometric analysis for nucleic acid quantitation is the ability to determine sample purity using the 260 nm:280 nm calculation. The ratio of the absorbance at 260 and 280 nm ($A_{260/280}$) is used to assess the purity of nucleic acids. For pure DNA, $A_{260/280}$ is widely considered ~1.8 but has been argued to translate - due to numeric errors in the original Warburg paper - into a mix of 60% protein and 40% DNA. The ratio for pure RNA $A_{260/280}$ is ~2.0. These ratios are commonly used to assess the amount of protein contamination that is left from the nucleic acid isolation process since proteins absorb at 280 nm.

The ratio of absorbance at 260 nm vs 280 nm is commonly used to assess DNA contamination of protein solutions, since proteins (in particular, the aromatic amino acids) absorb light at 280 nm. The reverse, however, is not true — it takes a relatively large amount of protein contamination to significantly affect the 260:280 ratio in a nucleic acid solution.

260:280 ratio has high sensitivity for nucleic acid contamination in protein:

% protein	% nucleic acid	260:280 ratio
100	0	0.57
95	5	1.06
90	10	1.32
70	30	1.73

260:230 ratio lacks sensitivity for protein contamination in nucleic acids (table shown for RNA, 100% DNA is approximately 1.8):

% nucleic acid	% protein	260:230 ratio
100	0	2.00
95	5	1.99
90	10	1.98
70	30	1.94

This difference is due to the much higher mass attenuation coefficient nucleic acids have at 260 nm and 280 nm, compared to that of proteins. Because of this, even for relatively high concentrations of protein, the protein contributes relatively little to the 260 and 280 absorbance. While the protein contamination cannot be reliably assessed with a 260:280 ratio, this also means that it contributes little error to DNA quantity estimation.

Contamination Identification

Examination of sample spectra may be useful in identifying that a problem with sample purity exists.

Table of Potential Contamination Factors		
	Low Reading	High Reading
A260/A230 Ratio	• Carbohydrate carryover (often a problem with plants). • Residual phenol from nucleic acid extraction. • Residual guanidine (often used in column based kits). • Glycogen used for precipitation.	• Making a Blank measurement on a dirty pedestal. • Using an inappropriate solution for the Blank measurement. The blank solution should be the same pH and of a similar ionic strength as the sample solution. Example: Using water for the Blank measurement for samples dissolved in TE may result in low 260/230 ratios.

A260/A280 Ratio	• Residual phenol or other reagent associated with the extraction protocol. • A very low concentration(< 10 ng/ul).of nucleic acid.	• Residual RNA from nucleic acid extraction. * High 260/280 purity ratios are not normally indicative of any issues.

Other Common Contaminants

- Contamination by phenol, which is commonly used in nucleic acid purification, can significantly throw off quantification estimates. Phenol absorbs with a peak at 270 nm and a $A_{260/280}$ of 1.2. Nucleic acid preparations uncontaminated by phenol should have a $A_{260/280}$ of around 2. Contamination by phenol can significantly contribute to overestimation of DNA concentration.
- Absorption at 230 nm can be caused by contamination by phenolate ion, thiocyanates, and other organic compounds. For a pure RNA sample, the $A_{230:260:280}$ should be around 1:2:1, and for a pure DNA sample, the $A_{230:260:280}$ should be around 1:1.8:1.
- Absorption at 330 nm and higher indicates particulates contaminating the solution, causing scattering of light in the visible range. The value in a pure nucleic acid sample should be zero.
- Negative values could result if an incorrect solution was used as blank. Alternatively, these values could arise due to fluorescence of a dye in the solution.

Analysis with Fluorescent Dye Tagging

An alternative method to assess DNA and RNA concentration is to tag the sample with a Fluorescent tag, which is a fluorescent dye used to measure the intensity of the dyes that bind to nucleic acids and selectively fluoresce when bound (e.g. Ethidium bromide). This method is useful for cases where concentration is too low to accurately assess with spectrophotometry and in cases where contaminants absorbing at 260 nm make accurate quantitation by that method impossible. The benefit of fluorescence quantitation of DNA and RNA is the improved sensitivity over spectrophotometric analysis. Although, that increase in sensitivity comes at the cost of a higher price per sample and a lengthier sample preparation process.

There are two main ways to approach this. "Spotting" involves placing a sample directly onto an agarose gel or plastic wrap. The fluorescent dye is either present in the agarose gel, or is added in appropriate concentrations to the samples on the plastic film. A set of samples with known concentrations are spotted alongside the sample. The concentration of the unknown sample is then estimated by comparison with the fluorescence of these known concentrations. Alternatively, one may run the sample through an agarose or polyacrylamide gel, alongside some samples of known concentration. As with the spot test, concentration is estimated through comparison of fluorescent intensity with the known samples.

If the sample volumes are large enough to use microplates or cuvettes, the dye-loaded samples can also be quantified with a fluorescence photometer. Minimum sample volume starts at 0.3 μl.

To date there is no fluorescence method to determine protein contamination of a DNA sample that is similar to the 260 nm/280 nm spectrophotometric version.

Nucleic Acid Double Helix

In molecular biology, the term double helix refers to the structure formed by double-stranded molecules of nucleic acids such as DNA. The double helical structure of a nucleic acid complex arises as a consequence of its secondary structure, and is a fundamental component in determining its tertiary structure.

The DNA double helix polymer of nucleic acid, held together by nucleotides which base pair together. In B-DNA, the most common double helical structure found in nature, the double helix is right-handed with about 10–10.5 base pairs per turn. The double helix structure of DNA contains a *major groove* and *minor groove*. In B-DNA the major groove is wider than the minor groove. Given the difference in widths of the major groove and minor groove, many proteins which bind to B-DNA do so through the wider major groove.

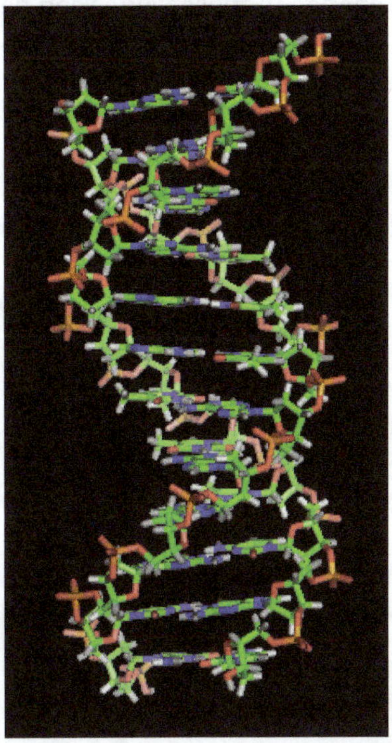

Two complementary regions of nucleic acid molecules will bind and form a double helical structure held together by base pairs.

Nucleic Acid Hybridization

Hybridization is the process of complementary base pairs binding to form a double helix. Melting is the process by which the interactions between the strands of the double helix are broken, separating the two nucleic acid strands. These bonds are weak, easily separated by gentle heating, enzymes, or physical force. Melting occurs preferentially at certain points in the nucleic acid. *T* and *A* rich regions are more easily melted than *C* and *G* rich regions. Some base steps (pairs) are also susceptible to DNA melting, such as *T A* and *T G*. These mechanical features are reflected by the use of sequences such as *TATA* at the start of many genes to assist RNA polymerase in melting the DNA for transcription.

Strand separation by gentle heating, as used in polymerase chain reaction (PCR), is simple, providing the molecules have fewer than about 10,000 base pairs (10 kilobase pairs, or 10 kbp). The intertwining of the DNA strands makes long segments difficult to separate. The cell avoids this problem by allowing its DNA-melting enzymes (helicases) to work concurrently with topoisomerases, which can chemically cleave the phosphate backbone of one of the strands so that it can swivel around the other. Helicases unwind the strands to facilitate the advance of sequence-reading enzymes such as DNA polymerase.

Base Pair Geometry

The geometry of a base, or base pair step can be characterized by 6 coordinates: shift, slide, rise, tilt, roll, and twist. These values precisely define the location and orientation in space of every base or base pair in a nucleic acid molecule relative to its predecessor along the axis of the helix. Together, they characterize the helical structure of the molecule. In regions of DNA or RNA where the *normal* structure is disrupted, the change in these values can be used to describe such disruption.

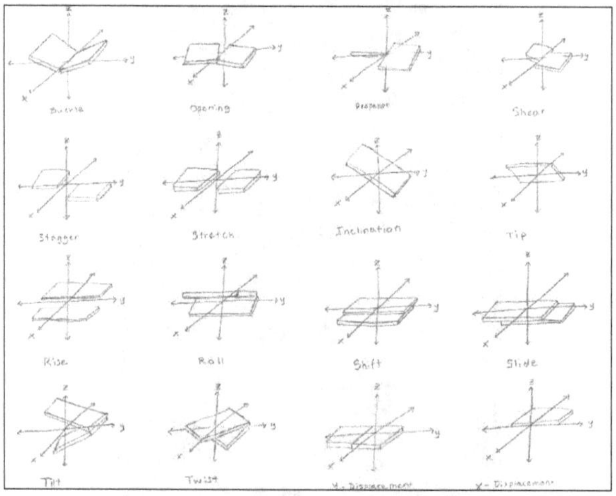

Base pair geometries.

For each base pair, considered relative to its predecessor, there are the following base pair geometries to consider:

- Shear.
- Stretch.
- Stagger.
- Buckle.
- Propeller: rotation of one base with respect to the other in the same base pair.
- Opening.
- Shift: displacement along an axis in the base-pair plane perpendicular to the first, directed from the minor to the major groove.
- Slide: displacement along an axis in the plane of the base pair directed from one strand to the other.

- Rise: displacement along the helix axis.
- Tilt: rotation around the shift axis.
- Roll: rotation around the slide axis.
- Twist: rotation around the rise axis.
- x-displacement.
- y-displacement.
- inclination.
- tip.
- pitch: the number of base pairs per complete turn of the helix.

Rise and twist determine the handedness and pitch of the helix. The other coordinates, by contrast, can be zero. Slide and shift are typically small in B-DNA, but are substantial in A- and Z-DNA. Roll and tilt make successive base pairs less parallel, and are typically small.

Note that "tilt" has often been used differently in the scientific literature, referring to the deviation of the first, inter-strand base-pair axis from perpendicularity to the helix axis. This corresponds to slide between a succession of base pairs, and in helix-based coordinates is properly termed "inclination".

Helix Geometries

At least three DNA conformations are believed to be found in nature, A-DNA, B-DNA, and Z-DNA. The *B* form described by James Watson and Francis Crick is believed to predominate in cells. It is 23.7 Å wide and extends 34 Å per 10 bp of sequence. The double helix makes one complete turn about its axis every 10.4-10.5 base pairs in solution. This frequency of twist (termed the helical *pitch*) depends largely on stacking forces that each base exerts on its neighbours in the chain. The absolute configuration of the bases determines the direction of the helical curve for a given conformation.

A-DNA and Z-DNA differ significantly in their geometry and dimensions to B-DNA, although still form helical structures. It was long thought that the A form only occurs in dehydrated samples of DNA in the laboratory, such as those used in crystallographic experiments, and in hybrid pairings of DNA and RNA strands, but DNA dehydration does occur in vivo, and A-DNA is now known to have biological functions. Segments of DNA that cells have been methylated for regulatory purposes may adopt the Z geometry, in which the strands turn about the helical axis the opposite way to A-DNA and B-DNA. There is also evidence of protein-DNA complexes forming Z-DNA structures.

Other conformations are possible; A-DNA, B-DNA, C-DNA, E-DNA, L-DNA (the enantiomeric form of D-DNA), P-DNA, S-DNA, Z-DNA, etc. have been described so far. In fact, only the letters F, Q, U, V, and Y are now available to describe any new DNA structure that may appear in the future. However, most of these forms have been created synthetically and have not been observed in naturally occurring biological systems. There are also triple-stranded DNA forms and quadruplex forms such as the G-quadruplex.

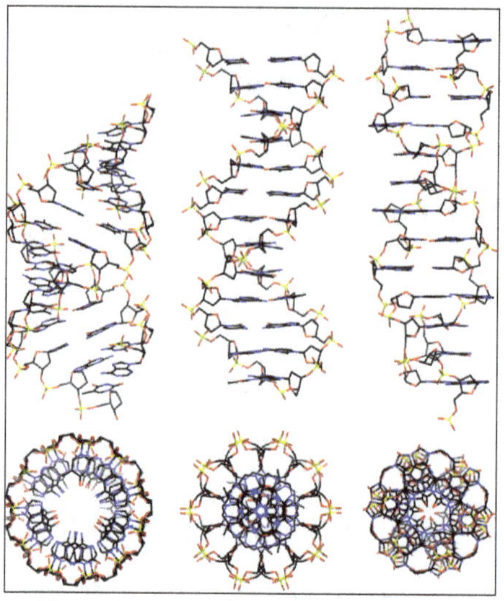

The structures of A-, B-, and Z-DNA.

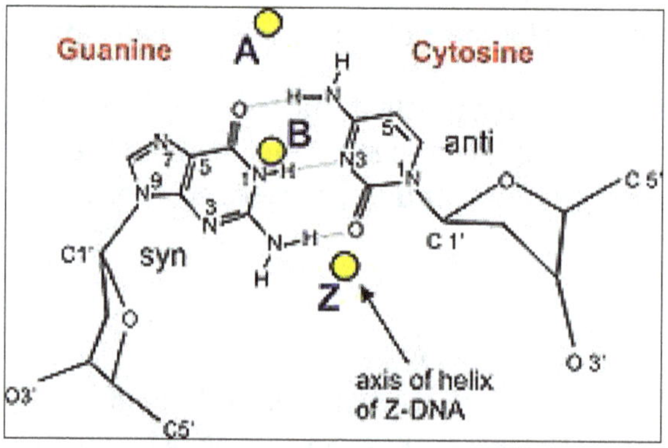

The helix axis of A-, B-, and Z-DNA.

Structural features of the three major forms of DNA			
Geometry attribute	A-DNA	B-DNA	Z-DNA
Helix sense	right-handed	right-handed	left-handed
Repeating unit	1 bp	1 bp	2 bp
Rotation/bp	32.7°	34.3°	60°/2
bp/turn	11	10.5	12
Inclination of bp to axis	+19°	−1.2°	−9°
Rise/bp along axis	2.3 Å (0.23 nm)	3.32 Å (0.332 nm)	3.8 Å (0.38 nm)
Pitch/turn of helix	28.2 Å (2.82 nm)	33.2 Å (3.32 nm)	45.6 Å (4.56 nm)
Mean propeller twist	+18°	+16°	0°
Glycosyl angle	anti	anti	C: anti, G: syn
Sugar pucker	C3'-endo	C2'-endo	C: C2'-endo, G: C2'-exo
Diameter	23 Å (2.3 nm)	20 Å (2.0 nm)	18 Å (1.8 nm)

Grooves

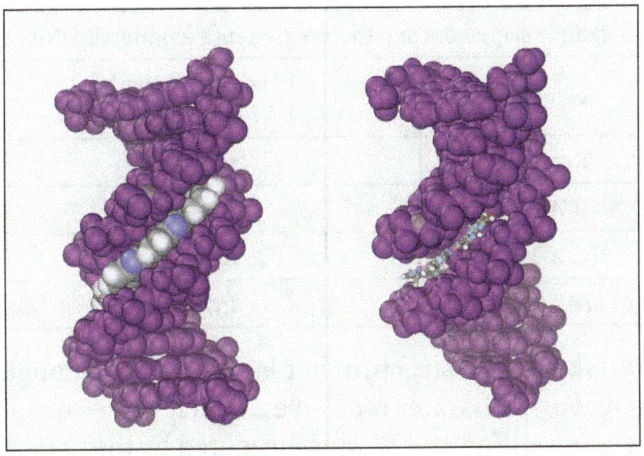

Major and minor grooves of DNA. Minor groove is a binding site for the dye Hoechst 33258.

Twin helical strands form the DNA backbone. Another double helix may be found by tracing the spaces, or grooves, between the strands. These voids are adjacent to the base pairs and may provide a binding site. As the strands are not directly opposite each other, the grooves are unequally sized. One groove, the major groove, is 22 Å wide and the other, the minor groove, is 12 Å wide. The narrowness of the minor groove means that the edges of the bases are more accessible in the major groove. As a result, proteins like transcription factors that can bind to specific sequences in double-stranded DNA usually make contacts to the sides of the bases exposed in the major groove. This situation varies in unusual conformations of DNA within the cell, but the major and minor grooves are always named to reflect the differences in size that would be seen if the DNA is twisted back into the ordinary B form.

Non-double Helical Forms

Alternative non-helical models were briefly considered in the late 1970s as a potential solution to problems in DNA replication in plasmids and chromatin. However, the models were set aside in favor of the double-helical model due to subsequent experimental advances such as X-ray crystallography of DNA duplexes and later the nucleosome core particle, and the discovery of topoisomerases. Also, the non-double-helical models are not currently accepted by the mainstream scientific community.

Single-stranded nucleic acids (ssDNA) do not adopt a helical formation, and are described by models such as the random coil or worm-like chain.

Bending

DNA is a relatively rigid polymer, typically modelled as a worm-like chain. It has three significant degrees of freedom; bending, twisting, and compression, each of which cause certain limits on what is possible with DNA within a cell. Twisting-torsional stiffness is important for the circularisation of DNA and the orientation of DNA bound proteins relative to each other and bending-axial stiffness is important for DNA wrapping and circularisation and protein interactions. Compression-extension is relatively unimportant in the absence of high tension.

Persistence Length and Axial Stiffness

Example sequences and their persistence lengths (B DNA)	
Sequence	Persistence length / base pairs
Random	154±10
$(CA)_{repeat}$	133±10
$(CAG)_{repeat}$	124±10
$(TATA)_{repeat}$	137±10

DNA in solution does not take a rigid structure but is continually changing conformation due to thermal vibration and collisions with water molecules, which makes classical measures of rigidity impossible. Hence, the bending stiffness of DNA is measured by the persistence length, defined as:

> "The length of DNA over which the time-averaged orientation of the polymer becomes uncorrelated by a factor of *e*".

This value may be directly measured using an atomic force microscope to directly image DNA molecules of various lengths. In an aqueous solution, the average persistence length is 46-50 nm or 140-150 base pairs (the diameter of DNA is 2 nm), although can vary significantly. This makes DNA a moderately stiff molecule.

The persistence length of a section of DNA is somewhat dependent on its sequence, and this can cause significant variation. The variation is largely due to base stacking energies and the residues which extend into the minor and major grooves.

Models for DNA Bending

Stacking stability of base steps (B DNA)	
Step	Stacking ΔG /kcal mol^{-1}
T A	-0.19
T G or C A	-0.55
C G	-0.91
A G or C T	-1.06
A A or T T	-1.11
A T	-1.34
G A or T C	-1.43
C C or G G	-1.44
A C or G T	-1.81
G C	-2.17

The entropic flexibility of DNA is remarkably consistent with standard polymer physics models, such as the *Kratky-Porod* worm-like chain model. Consistent with the worm-like chain model is the observation that bending DNA is also described by Hooke's law at very small (sub-piconewton) forces. However, for DNA segments less than the persistence length, the bending force is approximately constant and behaviour deviates from the worm-like chain predictions.

This effect results in unusual ease in circularising small DNA molecules and a higher probability of finding highly bent sections of DNA.

Bending Preference

DNA molecules often have a preferred direction to bend, i.e., anisotropic bending. This is, again, due to the properties of the bases which make up the DNA sequence - a random sequence will have no preferred bend direction, i.e., isotropic bending.

Preferred DNA bend direction is determined by the stability of stacking each base on top of the next. If unstable base stacking steps are always found on one side of the DNA helix then the DNA will preferentially bend away from that direction. As bend angle increases then steric hindrances and ability to roll the residues relative to each other also play a role, especially in the minor groove. *A* and *T* residues will be preferentially be found in the minor grooves on the inside of bends. This effect is particularly seen in DNA-protein binding where tight DNA bending is induced, such as in nucleosome particles.

DNA molecules with exceptional bending preference can become intrinsically bent. This was first observed in trypanosomatid kinetoplast DNA. Typical sequences which cause this contain stretches of 4-6 *T* and *A* residues separated by *G* and *C* rich sections which keep the A and T residues in phase with the minor groove on one side of the molecule. For example:

```
                                                   AATGCCAA
G ATTCCCAAAAATGTCAAAA AATAGGCAAAA AAAATCCC
                                                     AAAC
```

The intrinsically bent structure is induced by the 'propeller twist' of base pairs relative to each other allowing unusual bifurcated Hydrogen-bonds between base steps. At higher temperatures this structure is denatured, and so the intrinsic bend, is lost.

All DNA which bends anisotropically has, on average, a longer persistence length and greater axial stiffness. This increased rigidity is required to prevent random bending which would make the molecule act isotropically.

Circularization

DNA circularization depends on both the axial (bending) stiffness and torsional (rotational) stiffness of the molecule. For a DNA molecule to successfully circularize it must be long enough to easily bend into the full circle and must have the correct number of bases so the ends are in the correct rotation to allow bonding to occur. The optimum length for circularization of DNA is around 400 base pairs (136 nm), with an integral number of turns of the DNA helix, i.e., multiples of 10.4 base pairs. Having a non integral number of turns presents a significant energy barrier for circularization, for example a 10.4 x 30 = 312 base pair molecule will circularize hundreds of times faster than 10.4 x 30.5 ≈ 317 base pair molecule.

Stretching

Elastic Stretching Regime

Longer stretches of DNA are entropically elastic under tension. When DNA is in solution, it undergoes continuous structural variations due to the energy available in the thermal bath of the solvent. This is due to the thermal vibration of the molecule combined with continual collisions with water molecules. For entropic reasons, more compact relaxed states are thermally accessible than stretched out states, and so DNA molecules are almost universally found in a tangled relaxed layouts. For this reason, one molecule of DNA will stretch under a force, straightening it out. Using optical tweezers, the entropic stretching behavior of DNA has been studied and analyzed from a polymer physics perspective, and it has been found that DNA behaves largely like the *Kratky-Porod* worm-like chain model under physiologically accessible energy scales.

Phase Transitions Under Stretching

Under sufficient tension and positive torque, DNA is thought to undergo a phase transition with the bases splaying outwards and the phosphates moving to the middle. This proposed structure for overstretched DNA has been called *P-form DNA*, in honor of Linus Pauling who originally presented it as a possible structure of DNA.

Evidence from mechanical stretching of DNA in the absence of imposed torque points to a transition or transitions leading to further structures which are generally referred to as *S-form DNA*. These structures have not yet been definitively characterised due to the difficulty of carrying out atomic-resolution imaging in solution while under applied force although many computer simulation studies have been made.

Proposed S-DNA structures include those which preserve base-pair stacking and hydrogen bonding (GC-rich), while releasing extension by tilting, as well as structures in which partial melting of the base-stack takes place, while base-base association is nonetheless overall preserved (AT-rich).

Periodic fracture of the base-pair stack with a break occurring once per three bp (therefore one out of every three bp-bp steps) has been proposed as a regular structure which preserves planarity of the base-stacking and releases the appropriate amount of extension , with the term "Σ-DNA" introduced as a mnemonic, with the three right-facing points of the Sigma character serving as a reminder of the three grouped base pairs. The Σ form has been shown to have a sequence preference for GNC motifs which are believed under the GNC_hypothesis to be of evolutionary importance .

Supercoiling and Topology

The B form of the DNA helix twists 360° per 10.4-10.5 bp in the absence of torsional strain. But many molecular biological processes can induce torsional strain. A DNA segment with excess or insufficient helical twisting is referred to, respectively, as positively or negatively *supercoiled*. DNA *in vivo* is typically negatively supercoiled, which facilitates the unwinding (melting) of the double-helix required for RNA transcription.

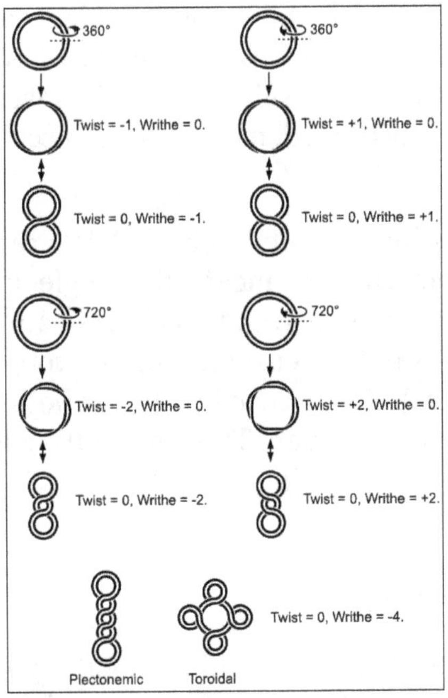

Supercoiled structure of circular DNA molecules with low writhe. The helical aspect of the DNA duplex is omitted for clarity.

Within the cell most DNA is topologically restricted. DNA is typically found in closed loops (such as plasmids in prokaryotes) which are topologically closed, or as very long molecules whose diffusion coefficients produce effectively topologically closed domains. Linear sections of DNA are also commonly bound to proteins or physical structures (such as membranes) to form closed topological loops.

Francis Crick was one of the first to propose the importance of linking numbers when considering DNA supercoils. In a paper published in 1976, Crick outlined the problem as follows:

In considering supercoils formed by closed double-stranded molecules of DNA certain mathematical concepts, such as the linking number and the twist, are needed. The meaning of these for a closed ribbon is explained and also that of the writhing number of a closed curve. Some simple examples are given, some of which may be relevant to the structure of chromatin.

Analysis of DNA topology uses three values:

- L = linking number - the number of times one DNA strand wraps around the other. It is an integer for a closed loop and constant for a closed topological domain.
- T = twist - total number of turns in the double stranded DNA helix. This will normally tend to approach the number of turns that a topologically open double stranded DNA helix makes free in solution: number of bases/10.5, assuming there are no intercalating agents (e.g., ethidium bromide) or other elements modifying the stiffness of the DNA.
- W = writhe - number of turns of the double stranded DNA helix around the superhelical axis.
- $L = T + W$ and $\Delta L = \Delta T + \Delta W$

Any change of T in a closed topological domain must be balanced by a change in W, and vice versa. This results in higher order structure of DNA. A circular DNA molecule with a writhe of 0 will be circular. If the twist of this molecule is subsequently increased or decreased by supercoiling then the writhe will be appropriately altered, making the molecule undergo plectonemic or toroidal superhelical coiling.

When the ends of a piece of double stranded helical DNA are joined so that it forms a circle the strands are topologically knotted. This means the single strands cannot be separated any process that does not involve breaking a strand (such as heating). The task of un-knotting topologically linked strands of DNA falls to enzymes termed topoisomerases. These enzymes are dedicated to un-knotting circular DNA by cleaving one or both strands so that another double or single stranded segment can pass through. This un-knotting is required for the replication of circular DNA and various types of recombination in linear DNA which have similar topological constraints.

The Linking Number Paradox

For many years, the origin of residual supercoiling in eukaryotic genomes remained unclear. This topological puzzle was referred to by some as the "linking number paradox". However, when experimentally determined structures of the nucleosome displayed an over-twisted left-handed wrap of DNA around the histone octamer, this *paradox* was considered to be solved by the scientific community.

Nucleic Acid Metabolism

Nucleic acid metabolism is the process by which nucleic acids (DNA and RNA) are synthesized and degraded. Nucleic acids are polymers of nucleotides. Nucleotide synthesis is an anabolic mechanism generally involving the chemical reaction of phosphate, pentose sugar, and a nitrogenous base. Destruction of nucleic acid is a catabolic reaction. Additionally, parts of the nucleotides or nucleobases can be salvaged to recreate new nucleotides. Both synthesis and degradation reactions require enzymes to facilitate the event. Defects or deficiencies in these enzymes can lead to a variety of diseases.

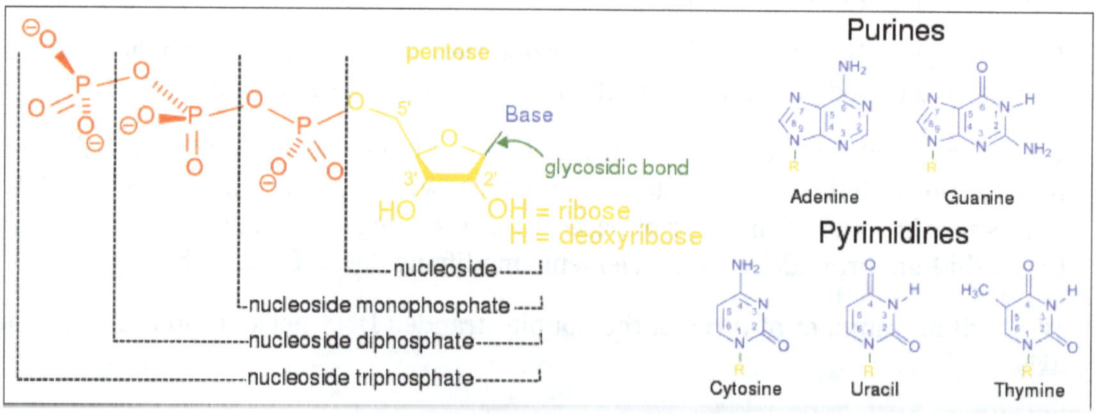

Composition of nucleotides, which make up nucleic acids.

Synthesis of Nucleic Acids

Nucleotides can be separated into purines and pyrimidines. They are both primarily produced in the liver. They both contain a sugar and a phosphate, but have nitrogenous bases that are different sizes. Because of this, the two different groups are synthesized in different ways. However, all nucleotide synthesis requires the use of phosphoribosyl pyrophosphate (PRPP) which donates the ribose and phosphate necessary to create a nucleotide.

Purine Synthesis

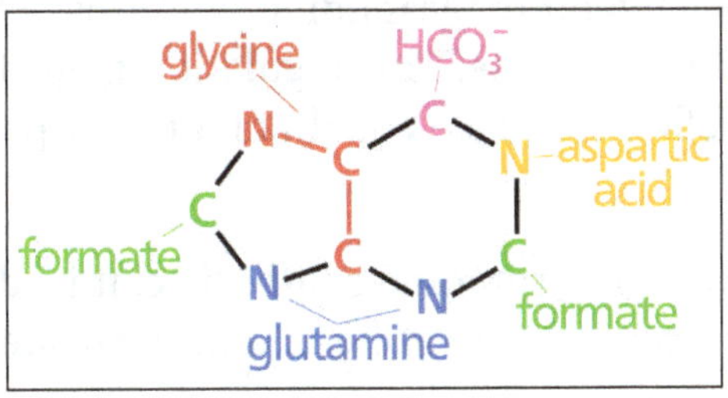

The origin of atoms that make up purine bases.

Adenine and guanine are the two nucleotides classified as purines. In purine synthesis, PRPP is turned into inosine monophosphate, or IMP. Production of IMP from PRPP requires glutamine, glycine, aspartate, and 6 ATP, among other things. IMP is then converted to AMP (adenosine monophosphate) using GTP and aspartate, which is converted into fumarate. While IMP can be directly converted to AMP, synthesis of GMP (guanosine monophosphate) requires an intermediate step, in which NAD+ is used to form the intermediate xanthosine monophosphate, or XMP. XMP is then converted into GMP by using the hydrolysis of 1 ATP and the conversion of glutamine to glutamate. AMP and GMP can then be converted into ATP and GTP, respectively, by kinases that add additional phosphates.

ATP stimulates production of GTP, while GTP stimulates production of ATP. This cross regulation keeps the relative amounts of ATP and GTP the same. Excess of either nucleotide could increase the likelihood of DNA mutations, where the wrong purine nucleotide is inserted.

Lesch-Nyhan syndrome is caused by a deficiency in hypoxanthine-guanine phosphoribosyltransferase or HGPRT, the enzyme that catalyzes the reversible reaction of producing guanine from GMP. This is a sex-linked congenital defect that causes overproduction of uric acid along with mental retardation, spasticity, and an urge to self-mutilate.

Pyrimidine Synthesis

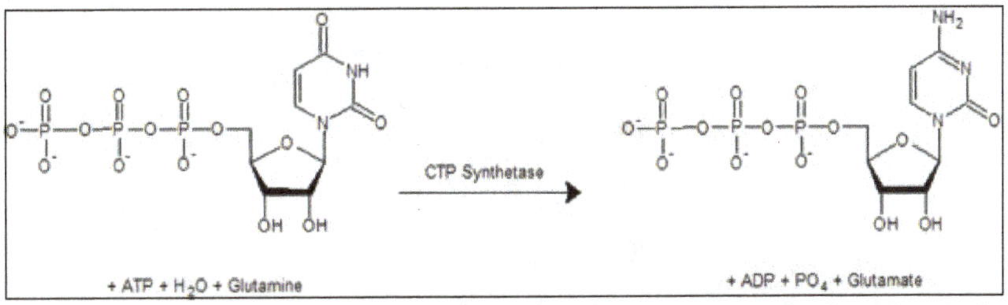

Uridine-triphosphate (UTP), at left, reacts with glutamine and other chemicals to form cytidine-triphosphate (CTP), on the right.

Pyrimidine nucleotides include cytidine, uridine, and thymidine. The synthesis of any pyrimidine nucleotide begins with the formation of uridine. This reaction requires aspartate, glutamine, bicarbonate, and 2 ATP molecules (to provide energy), as well as PRPP which provides the ribose-monophosphate. Unlike in purine synthesis, the sugar/phosphate group from PRPP is not added to the nitrogenous base until towards the end of the process. After uridine-monophosphate is synthesized, it can react with 2 ATP to form uridine-triphosphate or UTP. UTP can be converted to CTP (cytidine-triphosphate) in a reaction catalyzed by CTP synthetase. Thymidine synthesis first requires reduction of the uridine to deoxyuridine, before the base can be methylated to produce thymidine.

ATP, a purine nucleotide, is an activator of pyrimidine synthesis, while CTP, a pyrimidine nucleotide, is an inhibitor of pyrimidine synthesis. This regulation helps to keep the purine/pyrimidine amounts similar, which is beneficial because equal amounts of purines and pyrimidines are required for DNA synthesis.

Deficiencies of enzymes involved in pyrimidine synthesis can lead to the genetic disease Orotic aciduria which causes excessive excretion of orotic acid in the urine.

Converting Nucleotides to Deoxynucleotides

Nucleotides are initially made with ribose as the sugar component, which is a feature of RNA. DNA, however, requires *deoxy*ribose, which is missing the 2'-hydroxyl (-OH group) on the ribose. The reaction to remove this -OH is catalyzed by ribonucleotide reductase. This enzyme converts NDPs (nucleoside-diphosphate) to dNDPs (deoxynucleoside-diphosphate). The nucleotides must be in the diphosphate form for the reaction to occur.

In order to synthesize thymidine, a component of DNA which only exists in the deoxy form, uridine is converted to deoxyuridine (by ribonucleotide reductase), and then is methylated by thymidylate synthase to create thymidine.

Degradation of Nucleic Acids

The breakdown of DNA and RNA is occurring continuously in the cell. Purine and pyrimidine nucleosides can either be degraded to waste products and excreted or can be salvaged as nucleotide components.

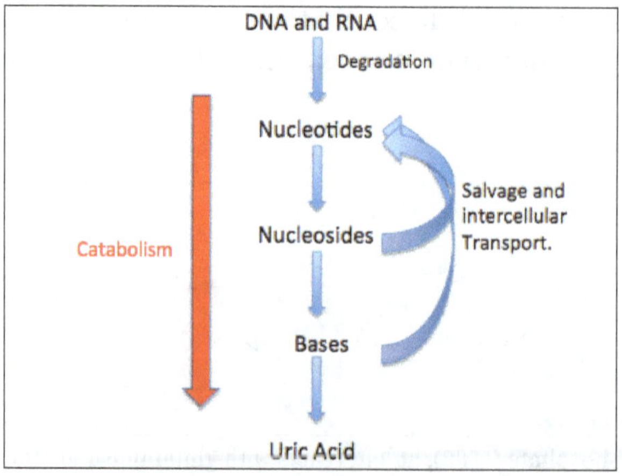

General outline of nucleic acid degradation for purines.

Pyrimidine Catabolism

Cytosine and uracil are converted into beta-alanine and later to malonyl-CoA which is needed for fatty acid synthesis, among other things. Thymine, on the other hand, is converted into β-aminoisobutyric acid which is then used to form methylmalonyl-CoA. The leftover carbon skeletons such as acetyl-CoA and Succinyl-CoA can then be oxidized by the citric acid cycle. Pyrimidine degradation ultimately ends in the formation of ammonium, water, and carbon dioxide. The ammonium can then enter the urea cycle which occurs in the cytosol and the mitochondria of cells.

Pyrimidine bases can also be salvaged. For example, the uracil base can be combined with ribose-1-phosphate to create uridine monophosphate or UMP. A similar reaction can also be done with thymine and deoxyribose-1-phosphate.

Deficiencies in enzymes involved in pyrimidine catabolism can lead to diseases such as Dihydropyrimidine dehydrogenase deficiency which has negative neurological effects.

Purine Catabolism

Purine degradation takes place mainly in the liver of humans and requires an assortment of enzymes to degrade purines to uric acid. First, the nucleotide will lose its phosphate through 5'-nucleotidase. The nucleoside, adenosine, is then deaminated and hydrolyzed to form hypoxanthine via adenosine deaminase and nucleosidase respectively. Hypoxanthine is then oxidized to form xanthine and then uric acid through the action of xanthine oxidase. The other purine nucleoside, guanosine, is cleaved to form guanine. Guanine is then deaminated via guanine deaminase to form xanthine which is then converted to uric acid. Oxygen is the final electron acceptor in the degradation of both purines. Uric acid is then excreted from the body in different forms depending on the animal.

Free purine and pyrimidine bases that are released into the cell are typically transported intercellularly across membranes and salvaged to create more nucleotides via nucleotide salvage. For example, adenine + PRPP --> AMP + PPi. This reaction requires the enzyme adenine phosphoribosyltransferase. Free guanine is salvaged in the same way except it requires hypoxanthine-guanine phosphoribosyltransferase.

Defects in purine catabolism can result in a variety of diseases including gout, which stems from an accumulation of uric acid crystals in various joints, and adenosine deaminase deficiency, which causes immunodeficiency.

Interconversion of Nucleotides

Once the nucleotides are synthesized they can exchange phosphates among one another in order to create mono-, di-, and tri-phosphate molecules. The conversion of a nucleoside-diphosphate (NDP) to a nucleoside-triphosphate (NTP) is catalyzed by nucleoside diphosphate kinase, which uses ATP as the phosphate donor. Similarly, nucleoside-monophosphate kinase carries out the phosphorylation of nucleside-monophosphates. Adenylate kinase is a specific nucleoside-monophosphate kinase that functions only on adenosine-monophosphate.

References

- Anthony-Cahill SJ, Mathews CK, van Holde KE, Appling DR (2012). Biochemistry (4th Edition). Englewood Cliffs, N.J: Prentice Hall. ISBN 0-13-800464-1
- National Institute of Health (September 28, 2007). "Discovering DNA: Friedrich Miescher and the early years of nucleic acid research". Hum. Genet. nih.gov. 122(6): 565–81. doi:10.1007/s00439-007-0433-0. PMID 17901982
- Alberts B, Johnson A, Lewis J, Raff M, Roberts K, Wlater P (2002). Molecular Biology of the Cell (4th ed.). New York NY: Garland Science. ISBN 0-8153-3218-1
- Brock, Thomas D.; Madigan, Michael T. (2009). Brock biology of microorganisms. Pearson / Benjamin Cummings. ISBN 0-321-53615-0
- Venter, JC; et al. (2001). "The sequence of the human genome" (PDF). Science. 291 (5507): 13041351. Bibcode:2001Sci...291.1304V. doi:10.1126/science.1058040. PMID 11181995
- Richard Owczarzy (2008-08-28). "DNA melting temperature - How to calculate it?". High-throughput DNA biophysics. owczarzy.net. Retrieved 2008-10-02
- Bailey, CJ (2009). "Orotic aciduria and uridine monophosphate synthase: a reappraisal". Journal of Inherited Metabolic Disease. 32: S227–33. doi:10.1007/s10545-009-1176-y. PMID 19562503
- Budowle B, van Daal A (April 2009). "Extracting evidence from forensic DNA analyses: future molecular biology directions". BioTechniques. 46 (5): 339–40, 342–50. doi:10.2144/000113136. PMID 19480629
- Owczarzy, Richard (2008-08-28). "DNA melting temperature - How to calculate it?". High-throughput DNA biophysics. owczarzy.net. Retrieved 2008-10-02
- Roe DR, Chaka AM (2009). "Structural basis of pathway-dependent force profiles in stretched DNA". Journal of Physical Chemistry B. 113: 15364–15371. doi:10.1021/jp906749j. PMID 19845321
- Krieger M, Scott MP, Matsudaira PT, Lodish HF, Darnell JE, Lawrence Z, Kaiser C, Berk A (2004). "Section 4.1: Structure of Nucleic Acids". Molecular cell biology. New York: W.H. Freeman and CO. ISBN 0-7167-4366-3
- International Human Genome Sequencing Consortium (2001). "Initial sequencing and analysis of the human genome" (PDF). Nature. 409 (6822): 860–921. Bibcode:2001Natur.409..860L. doi:10.1038/35057062. PMI D 11237011
- "Definitions of Genetic Testing". Definitions of Genetic Testing (Jorge Sequeiros and Bárbara Guimarães). EuroGentest Network of Excellence Project. 2008-09-11. Archived from the original on February 4, 2009. Retrieved 2008-08-10
- "The Analysis of DNA or RNA using Its Wavelengths: 230 nm, 260 nm, 280 nm". Bioteachnology.com. 2010-01-13. Archived from the original on 2012-09-06. Retrieved 2010-03-12

6

Role of Enzymes as a Catalyst

Enzymes are the biological catalysts which accelerate the rate of reactions without getting used in the reactions and are proteinaceous in nature. Some enzymes catalyze the proteolysis of proteins and are known as proteolytic enzymes. The topics elaborated in this chapter will help in gaining a better perspective about enzymes and proteolysis.

Enzymes

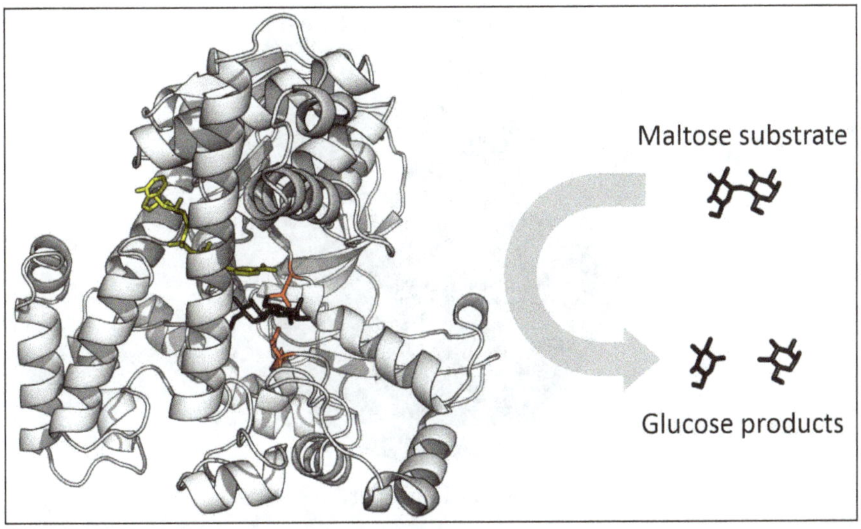

The enzyme glucosidase converts sugar maltose to two glucosesugars. Active site residues in red, maltose substrate in black, and NAD cofactor in yellow. (PDB: 1OBB)

Enzymes accelerate, or catalyze, chemical reactions. The molecules at the beginning of the process upon which enzymes may act are called substrates and the enzyme converts these into different molecules, called products. Almost all metabolic processes in the cell need enzymes in order to occur at rates fast enough to sustain life. The set of enzymes made in a cell determines which metabolic pathways occur in that cell. The study of enzymes is called enzymology.

Enzymes are known to catalyze more than 5,000 biochemical reaction types. Most enzymes are proteins, although a few are catalytic RNA molecules. Enzymes' specificity comes from their unique three-dimensional structures.

Like all catalysts, enzymes increase the rate of a reaction by lowering its activation energy. Some enzymes can make their conversion of substrate to product occur many millions of times faster. An extreme example is orotidine 5'-phosphate decarboxylase, which allows a reaction that would otherwise take millions of years to occur in milliseconds. Chemically, enzymes are like any catalyst and are not consumed in chemical reactions, nor do they alter the equilibrium of a reaction. Enzymes differ from most other catalysts by being much more specific. Enzyme activity can be affected by other molecules: inhibitors are molecules that decrease enzyme activity, and activators are molecules that increase activity. Many drugs and poisons are enzyme inhibitors. An enzyme's activity decreases markedly outside its optimal temperature and pH.

Some enzymes are used commercially, for example, in the synthesis of antibiotics. Some household products use enzymes to speed up chemical reactions: enzymes in biological washing powders break down protein, starch or fat stains on clothes, and enzymes in meat tenderizer break down proteins into smaller molecules, making the meat easier to chew.

Etymology and History

Eduard Buchner.

By the late 17th and early 18th centuries, the digestion of meat by stomach secretions and the conversion of starch to sugars by plant extracts and saliva were known but the mechanisms by which these occurred had not been identified.

French chemist Anselme Payen was the first to discover an enzyme, diastase, in 1833. A few decades later, when studying the fermentation of sugar to alcohol by yeast, Louis Pasteur concluded that this fermentation was caused by a vital force contained within the yeast cells called "ferments", which were thought to function only within living organisms. He wrote that "alcoholic fermentation is an act correlated with the life and organization of the yeast cells, not with the death or putrefaction of the cells."

In 1877, German physiologist Wilhelm Kühne (1837–1900) first used the term enzyme, to describe this process. The word enzyme was used later to refer to nonliving substances such as pepsin, and the word ferment was used to refer to chemical activity produced by living organisms.

Eduard Buchner submitted his first paper on the study of yeast extracts in 1897. In a series of experiments at the University of Berlin, he found that sugar was fermented by yeast extracts even when there were no living yeast cells in the mixture. He named the enzyme that brought about the fermentation of sucrose "zymase". In 1907, he received the Nobel Prize in Chemistry for "his discovery of cell-free fermentation". Following Buchner's example, enzymes are usually named according to the reaction they carry out: the suffix -ase is combined with the name of the substrate (e.g., lactase is the enzyme that cleaves lactose) or to the type of reaction (e.g., DNA polymerase forms DNA polymers).

The biochemical identity of enzymes was still unknown in the early 1900s. Many scientists observed that enzymatic activity was associated with proteins, but others (such as Nobel laureate Richard Willstätter) argued that proteins were merely carriers for the true enzymes and that proteins per se were incapable of catalysis. In 1926, James B. Sumner showed that the enzyme urease was a pure protein and crystallized it; he did likewise for the enzyme catalase in 1937. The conclusion that pure proteins can be enzymes was definitively demonstrated by John Howard Northrop and Wendell Meredith Stanley, who worked on the digestive enzymes pepsin (1930), trypsin and chymotrypsin. These three scientists were awarded the 1946 Nobel Prize in Chemistry.

The discovery that enzymes could be crystallized eventually allowed their structures to be solved by x-ray crystallography. This was first done for lysozyme, an enzyme found in tears, saliva and egg whites that digests the coating of some bacteria; the structure was solved by a group led by David Chilton Phillips and published in 1965. This high-resolution structure of lysozyme marked the beginning of the field of structural biology and the effort to understand how enzymes work at an atomic level of detail.

Structure

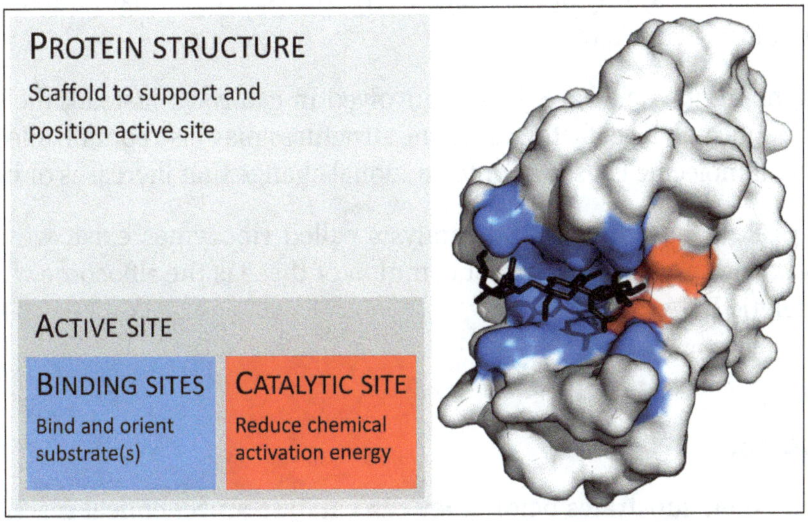

Organisation of enzyme structure and lysozyme example. Binding sites in blue, catalytic site in red and peptidoglycan substrate in black. (PDB: 9LYZ)

Enzymes are generally globular proteins, acting alone or in larger complexes. Like all proteins, enzymes are linear chains of amino acids that fold to produce a three-dimensional structure. The sequence of the amino acids specifies the structure which in turn determines the catalytic activity of the enzyme. Although structure determines function, a novel enzyme's activity cannot yet be predicted from its structure alone. Enzyme structures unfold (denature) when heated or exposed to chemical denaturants and this disruption to the structure typically causes a loss of activity. Enzyme denaturation is normally linked to temperatures above a species' normal level; as a result, enzymes from bacteria living in volcanic environments such as hot springs are prized by industrial users for their ability to function at high temperatures, allowing enzyme-catalysed reactions to be operated at a very high rate.

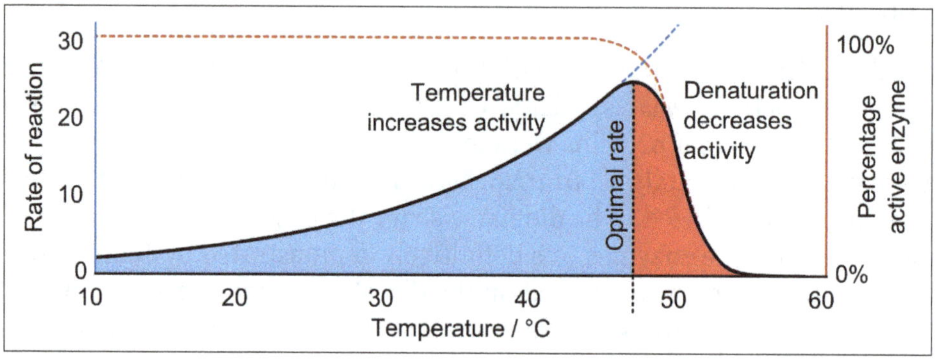

Enzyme activity initially increases with temperature (Q10 coefficient) until the enzyme's structure unfolds (denaturation), leading to an optimal rate of reaction at an intermediate temperature.

Enzymes are usually much larger than their substrates. Sizes range from just 62 amino acid residues, for the monomer of 4-oxalocrotonate tautomerase, to over 2,500 residues in the animal fatty acid synthase. Only a small portion of their structure (around 2–4 amino acids) is directly involved in catalysis: the catalytic site. This catalytic site is located next to one or more binding sites where residues orient the substrates. The catalytic site and binding site together comprise the enzyme's active site. The remaining majority of the enzyme structure serves to maintain the precise orientation and dynamics of the active site.

In some enzymes, no amino acids are directly involved in catalysis; instead, the enzyme contains sites to bind and orient catalytic cofactors. Enzyme structures may also contain allosteric sites where the binding of a small molecule causes a conformational change that increases or decreases activity.

A small number of RNA-based biological catalysts called ribozymes exist, which again can act alone or in complex with proteins. The most common of these is the ribosome which is a complex of protein and catalytic RNA components.

Mechanism

Substrate Binding

Enzymes must bind their substrates before they can catalyse any chemical reaction. Enzymes are usually very specific as to what substrates they bind and then the chemical reaction catalysed. Specificity is achieved by binding pockets with complementary shape, charge and hydrophilic/hydrophobic characteristics to the substrates. Enzymes can therefore distinguish between very similar substrate molecules to be chemoselective, regioselective and stereospecific.

Some of the enzymes showing the highest specificity and accuracy are involved in the copying and expression of the genome. Some of these enzymes have "proof-reading" mechanisms. Here, an enzyme such as DNA polymerase catalyzes a reaction in a first step and then checks that the product is correct in a second step. This two-step process results in average error rates of less than 1 error in 100 million reactions in high-fidelity mammalian polymerases.·· Similar proofreading mechanisms are also found in RNA polymerase, aminoacyl tRNA synthetases and ribosomes.

Conversely, some enzymes display enzyme promiscuity, having broad specificity and acting on a range of different physiologically relevant substrates. Many enzymes possess small side activities which arose fortuitously (i.e. neutrally), which may be the starting point for the evolutionary selection of a new function.

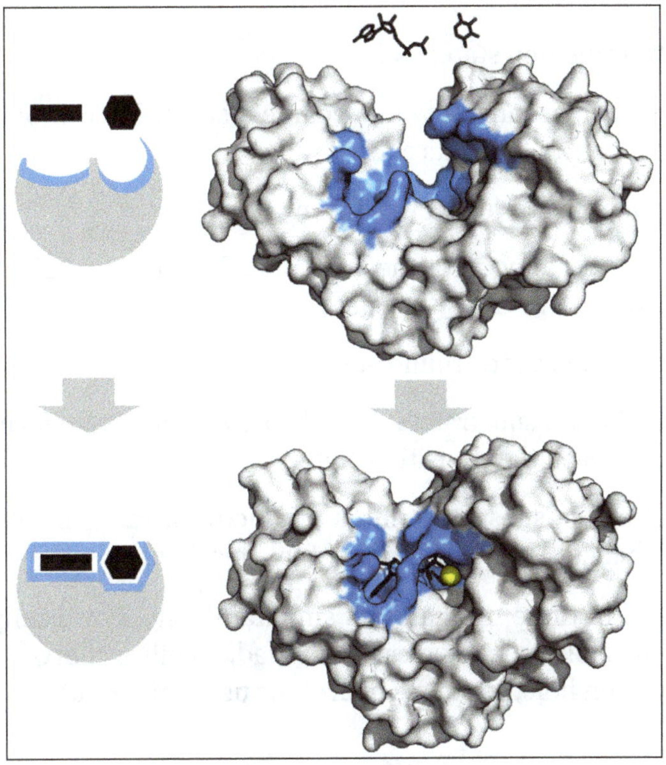

Enzyme changes shape by induced fit upon substrate binding to form enzyme-substrate complex. Hexokinase has a large induced fit motion that closes over the substrates adenosine triphosphate and xylose. Binding sites in blue, substrates in black and Mg^{+} cofactor in yellow. (PDB: 2E2N, 2E2Q)

"Lock and Key" Model

To explain the observed specificity of enzymes, in 1894 Emil Fischer proposed that both the enzyme and the substrate possess specific complementary geometric shapes that fit exactly into one another. This is often referred to as "the lock and key" model.·· This early model explains enzyme specificity, but fails to explain the stabilization of the transition state that enzymes achieve.

Induced Fit Model

In 1958, Daniel Koshland suggested a modification to the lock and key model: since enzymes are rather flexible structures, the active site is continuously reshaped by interactions with the substrate as the substrate interacts with the enzyme. As a result, the substrate does not simply bind to a rigid active site; the amino acid side-chains that make up the active site are molded into the precise positions that enable the enzyme to perform its catalytic function. In some cases, such as glycosidases, the substrate molecule also changes shape slightly as it enters the active site. The active site continues to change until the substrate is completely bound, at which point the final shape and charge distribution is determined. Induced fit may enhance the fidelity of molecular recognition in the presence of competition and noise via the conformational proofreading mechanism.

Catalysis

Enzymes can accelerate reactions in several ways, all of which lower the activation energy ($\Delta G^{\ddagger}$, Gibbs free energy).

1. By stabilizing the transition state:
   - Creating an environment with a charge distribution complementary to that of the transition state to lower its energy.
2. By providing an alternative reaction pathway:
   - Temporarily reacting with the substrate, forming a covalent intermediate to provide a lower energy transition state.
3. By destabilising the substrate ground state:
   - Distorting bound substrate(s) into their transition state form to reduce the energy required to reach the transition state.
   - By orienting the substrates into a productive arrangement to reduce the reaction entropy change. The contribution of this mechanism to catalysis is relatively small.

Enzymes may use several of these mechanisms simultaneously. For example, proteases such as trypsin perform covalent catalysis using a catalytic triad, stabilise charge build-up on the transition states using an oxyanion hole, complete hydrolysis using an oriented water substrate.

Dynamics

Enzymes are not rigid, static structures; instead they have complex internal dynamic motions – that is, movements of parts of the enzyme's structure such as individual amino acid residues, groups of residues forming a protein loop or unit of secondary structure, or even an entire protein domain. These motions give rise to a conformational ensemble of slightly different structures that interconvert with one another at equilibrium. Different states within this ensemble may be associated with different aspects of an enzyme's function. For example, different conformations of the enzyme dihydrofolate reductase are associated with the substrate binding, catalysis, cofactor release, and product release steps of the catalytic cycle.

Allosteric Modulation

Allosteric sites are pockets on the enzyme, distinct from the active site, that bind to molecules in the cellular environment. These molecules then cause a change in the conformation or dynamics of the enzyme that is transduced to the active site and thus affects the reaction rate of the enzyme. In this way, allosteric interactions can either inhibit or activate enzymes. Allosteric interactions with metabolites upstream or downstream in an enzyme's metabolic pathway cause feedback regulation, altering the activity of the enzyme according to the flux through the rest of the pathway.

Cofactors

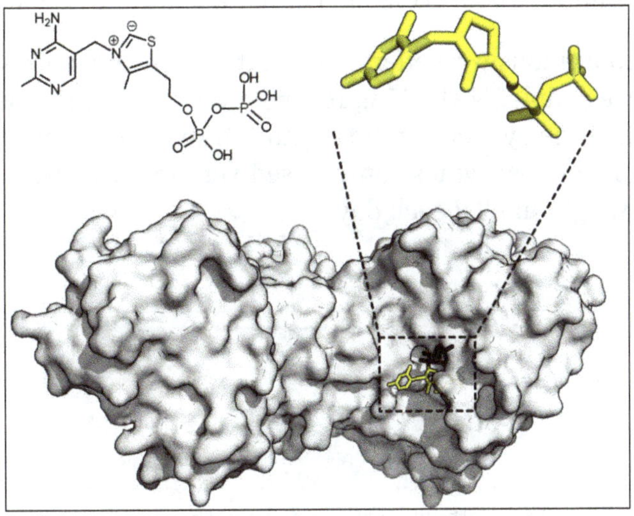

Chemical structure for thiamine pyrophosphate and protein structure of transketolase. Thiamine pyrophosphate cofactor in yellow and xylulose 5-phosphate substrate in black. (PDB: 4KXV)

Some enzymes do not need additional components to show full activity. Others require non-protein molecules called cofactors to be bound for activity. Cofactors can be either inorganic (e.g., metal ions and iron-sulfur clusters) or organic compounds (e.g., flavin and heme). Organic cofactors can be either coenzymes, which are released from the enzyme's active site during the reaction, or prosthetic groups, which are tightly bound to an enzyme. Organic prosthetic groups can be covalently bound (e.g., biotin in enzymes such as pyruvate carboxylase).

An example of an enzyme that contains a cofactor is carbonic anhydrase, which is shown in the ribbon diagram above with a zinc cofactor bound as part of its active site. These tightly bound ions or molecules are usually found in the active site and are involved in catalysis.For example, flavin and heme cofactors are often involved in redox reactions.

Enzymes that require a cofactor but do not have one bound are called apoenzymes or apoproteins. An enzyme together with the cofactor(s) required for activity is called a holoenzyme (or haloenzyme). The term holoenzyme can also be applied to enzymes that contain multiple protein subunits, such as the DNA polymerases; here the holoenzyme is the complete complex containing all the subunits needed for activity.

Coenzymes

Coenzymes are small organic molecules that can be loosely or tightly bound to an enzyme. Coenzymes transport chemical groups from one enzyme to another. Examples include NADH, NADPH and adenosine triphosphate (ATP). Some coenzymes, such as riboflavin, thiamine and folic acid, are vitamins, or compounds that cannot be synthesized by the body and must be acquired from the diet. The chemical groups carried include the hydride ion (H^-) carried by NAD or $NADP^+$, the phosphate group carried by adenosine triphosphate, the acetyl group carried by coenzyme A, formyl, methenyl or methyl groups carried by folic acid and the methyl group carried by S-adenosylmethionine.

Since coenzymes are chemically changed as a consequence of enzyme action, it is useful to consider coenzymes to be a special class of substrates, or second substrates, which are common to many different enzymes. For example, about 1000 enzymes are known to use the coenzyme NADH.

Coenzymes are usually continuously regenerated and their concentrations maintained at a steady level inside the cell. For example, NADPH is regenerated through the pentose phosphate pathway and S-adenosylmethionine by methionine adenosyltransferase. This continuous regeneration means that small amounts of coenzymes can be used very intensively. For example, the human body turns over its own weight in ATP each day.

Kinetics

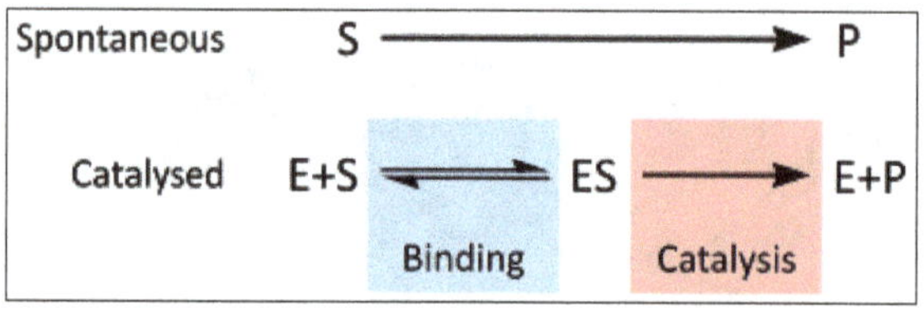

A chemical reaction mechanism with or without enzyme catalysis. The enzyme (E) binds substrate (S) to produce product (P).

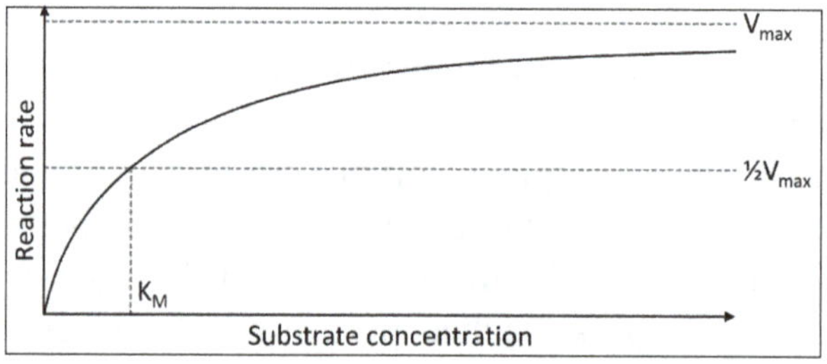

Saturation curve for an enzyme reaction showing the relation between the substrate concentration and reaction rate.

Enzyme kinetics is the investigation of how enzymes bind substrates and turn them into products. The rate data used in kinetic analyses are commonly obtained from enzyme assays. In 1913 Leonor Michaelis and Maud Leonora Menten proposed a quantitative theory of enzyme kinetics, which is referred to as Michaelis–Menten kinetics. The major contribution of Michaelis and Menten was to think of enzyme reactions in two stages. In the first, the substrate binds reversibly to the enzyme, forming the enzyme-substrate complex. This is sometimes called the Michaelis-Menten complex in their honor. The enzyme then catalyzes the chemical step in the reaction and releases the product. This work was further developed by G. E. Briggs and J. B. S. Haldane, who derived kinetic equations that are still widely used today.

Enzyme rates depend on solution conditions and substrate concentration. To find the maximum speed of an enzymatic reaction, the substrate concentration is increased until a constant rate of product formation is seen. This is shown in the saturation curve on the right. Saturation happens because, as substrate concentration increases, more and more of the free enzyme is converted into the substrate-bound ES complex. At the maximum reaction rate (V_{max}) of the enzyme, all the enzyme active sites are bound to substrate, and the amount of ES complex is the same as the total amount of enzyme.

V_{max} is only one of several important kinetic parameters. The amount of substrate needed to achieve a given rate of reaction is also important. This is given by the Michaelis-Menten constant

(K_m), which is the substrate concentration required for an enzyme to reach one-half its maximum reaction rate; generally, each enzyme has a characteristic K_m for a given substrate. Another useful constant is k_{cat}, also called the turnover number, which is the number of substrate molecules handled by one active site per second.

The efficiency of an enzyme can be expressed in terms of k_{cat}/K_m. This is also called the specificity constant and incorporates the rate constants for all steps in the reaction up to and including the first irreversible step. Because the specificity constant reflects both affinity and catalytic ability, it is useful for comparing different enzymes against each other, or the same enzyme with different substrates. The theoretical maximum for the specificity constant is called the diffusion limit and is about 10 to 10 (M^{-} s^{-}). At this point every collision of the enzyme with its substrate will result in catalysis, and the rate of product formation is not limited by the reaction rate but by the diffusion rate. Enzymes with this property are called catalytically perfect or kinetically perfect. Example of such enzymes are triose-phosphate isomerase, carbonic anhydrase, acetylcholinesterase, catalase, fumarase, β-lactamase, and superoxide dismutase.·· The turnover of such enzymes can reach several million reactions per second.

Michaelis–Menten kinetics relies on the law of mass action, which is derived from the assumptions of free diffusion and thermodynamically driven random collision. Many biochemical or cellular processes deviate significantly from these conditions, because of macromolecular crowding and constrained molecular movement. More recent, complex extensions of the model attempt to correct for these effects.

Inhibition

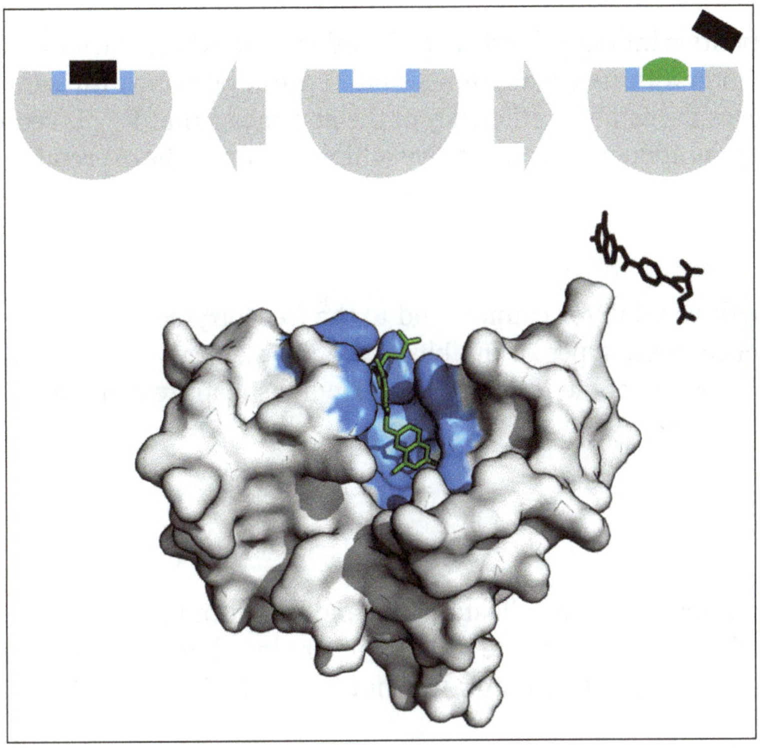

An enzyme binding site that would normally bind substrate can alternatively bind a competitive inhibitor, preventing substrate access. Dihydrofolate reductase is inhibited by methotrexate which prevents binding of its substrate, folic acid. Binding site in blue, inhibitor in green, and substrate in black. (PDB: 4QI9)

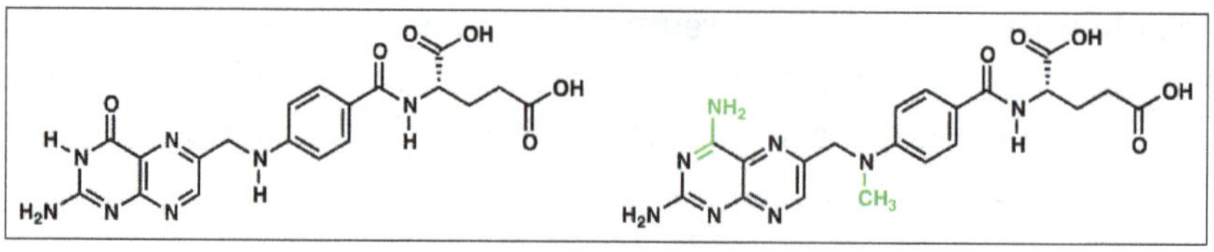

The coenzyme folic acid (left) and the anti-cancer drug methotrexate (right) are very similar in structure (differences show in green). As a result, methotrexate is a competitive inhibitor of many enzymes that use folates.

Enzyme reaction rates can be decreased by various types of enzyme inhibitors.

Types of Inhibition

Competitive

A competitive inhibitor and substrate cannot bind to the enzyme at the same time. Often competitive inhibitors strongly resemble the real substrate of the enzyme. For example, the drug methotrexate is a competitive inhibitor of the enzyme dihydrofolate reductase, which catalyzes the reduction of dihydrofolate to tetrahydrofolate. The similarity between the structures of dihydrofolate and this drug are shown in the accompanying figure. This type of inhibition can be overcome with high substrate concentration. In some cases, the inhibitor can bind to a site other than the binding-site of the usual substrate and exert an allosteric effect to change the shape of the usual binding-site.

Non-Competitive

A non-competitive inhibitor binds to a site other than where the substrate binds. The substrate still binds with its usual affinity and hence K_m remains the same. However the inhibitor reduces the catalytic efficiency of the enzyme so that V_{max} is reduced. In contrast to competitive inhibition, non-competitive inhibition cannot be overcome with high substrate concentration.

Uncompetitive

An uncompetitive inhibitor cannot bind to the free enzyme, only to the enzyme-substrate complex; hence, these types of inhibitors are most effective at high substrate concentration. In the presence of the inhibitor, the enzyme-substrate complex is inactive. This type of inhibition is rare.

Mixed

A mixed inhibitor binds to an allosteric site and the binding of the substrate and the inhibitor affect each other. The enzyme's function is reduced but not eliminated when bound to the inhibitor. This type of inhibitor does not follow the Michaelis-Menten equation.

Irreversible

An irreversible inhibitor permanently inactivates the enzyme, usually by forming a covalent bond to the protein. Penicillin and aspirin are common drugs that act in this manner.

Functions of Inhibitors

In many organisms, inhibitors may act as part of a feedback mechanism. If an enzyme produces too much of one substance in the organism, that substance may act as an inhibitor for the enzyme at the beginning of the pathway that produces it, causing production of the substance to slow down or stop when there is sufficient amount. This is a form of negative feedback. Major metabolic pathways such as the citric acid cycle make use of this mechanism.

Since inhibitors modulate the function of enzymes they are often used as drugs. Many such drugs are reversible competitive inhibitors that resemble the enzyme's native substrate, similar to methotrexate above; other well-known examples include statins used to treat high cholesterol, and protease inhibitors used to treat retroviral infections such as HIV. A common example of an irreversible inhibitor that is used as a drug is aspirin, which inhibits the COX-1 and COX-2 enzymes that produce the inflammation messenger prostaglandin. Other enzyme inhibitors are poisons. For example, the poison cyanide is an irreversible enzyme inhibitor that combines with the copper and iron in the active site of the enzyme cytochrome c oxidase and blocks cellular respiration.

Biological Function

Enzymes serve a wide variety of functions inside living organisms. They are indispensable for signal transduction and cell regulation, often via kinases and phosphatases. They also generate movement, with myosin hydrolyzing ATP to generate muscle contraction, and also transport cargo around the cell as part of the cytoskeleton. Other ATPases in the cell membrane are ion pumps involved in active transport. Enzymes are also involved in more exotic functions, such as luciferase generating light in fireflies. Viruses can also contain enzymes for infecting cells, such as the HIV integrase and reverse transcriptase, or for viral release from cells, like the influenza virus neuraminidase.

An important function of enzymes is in the digestive systems of animals. Enzymes such as amylases and proteases break down large molecules (starch or proteins, respectively) into smaller ones, so they can be absorbed by the intestines. Starch molecules, for example, are too large to be absorbed from the intestine, but enzymes hydrolyze the starch chains into smaller molecules such as maltose and eventually glucose, which can then be absorbed. Different enzymes digest different food substances. In ruminants, which have herbivorous diets, microorganisms in the gut produce another enzyme, cellulase, to break down the cellulose cell walls of plant fiber.

Metabolism

Several enzymes can work together in a specific order, creating metabolic pathways. In a metabolic pathway, one enzyme takes the product of another enzyme as a substrate. After the catalytic reaction, the product is then passed on to another enzyme. Sometimes more than one enzyme can catalyze the same reaction in parallel; this can allow more complex regulation: with, for example, a low constant activity provided by one enzyme but an inducible high activity from a second enzyme.

Enzymes determine what steps occur in these pathways. Without enzymes, metabolism would neither progress through the same steps and could not be regulated to serve the needs of the cell. Most central metabolic pathways are regulated at a few key steps, typically through enzymes whose activity involves the hydrolysis of ATP. Because this reaction releases so much energy, other reactions that are thermodynamically unfavorable can be coupled to ATP hydrolysis, driving the overall series of linked metabolic reactions.

Control of Activity

There are five main ways that enzyme activity is controlled in the cell.

Regulation

Enzymes can be either activated or inhibited by other molecules. For example, the end product(s) of a metabolic pathway are often inhibitors for one of the first enzymes of the pathway (usually the first irreversible step, called committed step), thus regulating the amount of end product made by the pathways. Such a regulatory mechanism is called a negative feedback mechanism, because the amount of the end product produced is regulated by its own concentration.⁻ Negative feedback mechanism can effectively adjust the rate of synthesis of intermediate metabolites according to the demands of the cells. This helps with effective allocations of materials and energy economy, and it prevents the excess manufacture of end products. Like other homeostatic devices, the control of enzymatic action helps to maintain a stable internal environment in living organisms.

Post-Translational Modification

Examples of post-translational modification include phosphorylation, myristoylation and glycosylation.⁻ For example, in the response to insulin, the phosphorylation of multiple enzymes, including glycogen synthase, helps control the synthesis or degradation of glycogen and allows the cell to respond to changes in blood sugar. Another example of post-translational modification is the cleavage of the polypeptide chain. Chymotrypsin, a digestive protease, is produced in inactive form as chymotrypsinogen in the pancreas and transported in this form to the stomach where it is activated. This stops the enzyme from digesting the pancreas or other tissues before it enters the gut. This type of inactive precursor to an enzyme is known as a zymogen⁻ or proenzyme.

Quantity

Enzyme production (transcription and translation of enzyme genes) can be enhanced or diminished by a cell in response to changes in the cell's environment. This form of gene regulation is called enzyme induction. For example, bacteria may become resistant to antibiotics such as penicillin because enzymes called beta-lactamases are induced that hydrolyse the crucial beta-lactam ring within the penicillin molecule. Another example comes from enzymes in the liver called cytochrome P450 oxidases, which are important in drug metabolism. Induction or inhibition of these enzymes can cause drug interactions. Enzyme levels can also be regulated by changing the rate of enzyme degradation.

Subcellular Distribution

Enzymes can be compartmentalized, with different metabolic pathways occurring in different cellular compartments. For example, fatty acids are synthesized by one set of enzymes in the cytosol, endoplasmic reticulum and Golgi and used by a different set of enzymes as a source of energy in the mitochondrion, through β-oxidation. In addition, trafficking of the enzyme to different compartments may change the degree of protonation (cytoplasm neutral and lysosome acidic) or oxidative state [e.g., oxidized (periplasm) or reduced (cytoplasm)] which in turn affects enzyme activity.

Organ Specialization

In multicellular eukaryotes, cells in different organs and tissues have different patterns of gene expression and therefore have different sets of enzymes (known as isozymes) available for metabolic reactions. This provides a mechanism for regulating the overall metabolism of the organism. For example, hexokinase, the first enzyme in the glycolysis pathway, has a specialized form called glucokinase expressed in the liver and pancreas that has a lower affinity for glucose yet is more sensitive to glucose concentration. This enzyme is involved in sensing blood sugar and regulating insulin production.

Involvement in Disease

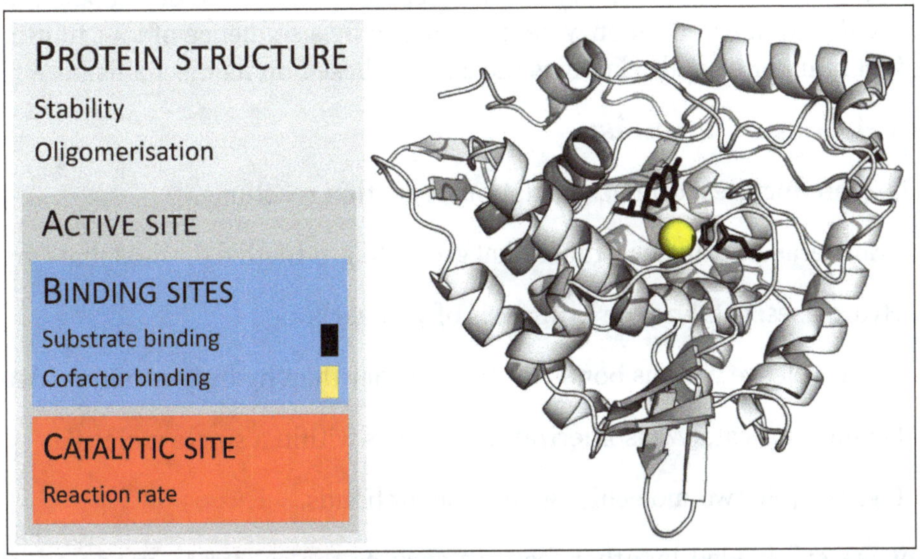

In phenylalanine hydroxylase over 300 different mutations throughout the structure cause phenylketonuria. Phenylalanine substrate and tetrahydrobiopterin coenzyme in black, and Fe^{+} cofactor in yellow. (PDB: 1KWO)

Since the tight control of enzyme activity is essential for homeostasis, any malfunction (mutation, overproduction, underproduction or deletion) of a single critical enzyme can lead to a genetic disease. The malfunction of just one type of enzyme out of the thousands of types present in the human body can be fatal. An example of a fatal genetic disease due to enzyme insufficiency is Tay-Sachs disease, in which patients lack the enzyme hexosaminidase.

One example of enzyme deficiency is the most common type of phenylketonuria. Many different single amino acid mutations in the enzyme phenylalanine hydroxylase, which catalyzes the first step in the degradation of phenylalanine, result in build-up of phenylalanine and related products. Some mutations are in the active site, directly disrupting binding and catalysis, but many are far from the active site and reduce activity by destabilising the protein structure, or affecting correct oligomerisation. This can lead to intellectual disability if the disease is untreated. Another example is pseudocholinesterase deficiency, in which the body's ability to break down choline ester drugs is impaired. Oral administration of enzymes can be used to treat some functional enzyme deficiencies, such as pancreatic insufficiency and lactose intolerance.

Another way enzyme malfunctions can cause disease comes from germline mutations in genes coding for DNA repair enzymes. Defects in these enzymes cause cancer because cells are less able to repair mutations in their genomes. This causes a slow accumulation of mutations and results in the development of cancers. An example of such a hereditary cancer syndrome is xeroderma pigmentosum, which causes the development of skin cancers in response to even minimal exposure to ultraviolet light.

Naming Conventions

An enzyme's name is often derived from its substrate or the chemical reaction it catalyzes, with the word ending in -ase. Examples are lactase, alcohol dehydrogenase and DNA polymerase. Different enzymes that catalyze the same chemical reaction are called isozymes.

The International Union of Biochemistry and Molecular Biology have developed a nomenclature for enzymes, the EC numbers; each enzyme is described by a sequence of four numbers preceded by "EC". The first number broadly classifies the enzyme based on its mechanism.

The top-Level Classification is:

- EC 1, Oxidoreductases: catalyze oxidation/reduction reactions.
- EC 2, Transferases: transfer a functional group (e.g. a methyl or phosphate group).
- EC 3, Hydrolases: catalyze the hydrolysis of various bonds.
- EC 4, Lyases: cleave various bonds by means other than hydrolysis and oxidation.
- EC 5, Isomerases: catalyze isomerization changes within a single molecule.
- EC 6, Ligases: join two molecules with covalent bonds.

These sections are subdivided by other features such as the substrate, products, and chemical mechanism. An enzyme is fully specified by four numerical designations. For example, hexokinase (EC 2.7.1.1) is a transferase (EC 2) that adds a phosphate group (EC 2.7) to a hexose sugar, a molecule containing an alcohol group (EC 2.7.1).

Industrial Applications

Enzymes are used in the chemical industry and other industrial applications when extremely specific catalysts are required. Enzymes in general are limited in the number of reactions they have evolved to catalyze and also by their lack of stability in organic solvents and at high temperatures. As a consequence, protein engineering is an active area of research and involves attempts to create new enzymes with novel properties, either through rational design or in vitro evolution. These efforts have begun to be successful, and a few enzymes have now been designed "from scratch" to catalyze reactions that do not occur in nature.

Application	Enzymes used	Uses
Biofuel industry	Cellulases	Break down cellulose into sugars that can be fermented to produce cellulosic ethanol.
	Ligninases	Pretreatment of biomass for biofuel production.
Biological detergent	Proteases, amylases, lipases	Remove protein, starch, and fat or oil stains from laundry and dishware.
	Mannanases	Remove food stains from the common food additive guar gum.
Brewing industry	Amylase, glucanases, proteases	Split polysaccharides and proteins in the malt.
	Betaglucanases	Improve the wort and beer filtration characteristics.
	Amyloglucosidase and pullulanases	Make low-calorie beer and adjust fermentability.
	Acetolactate decarboxylase (ALDC)	Increase fermentation efficiency by reducing diacetyl formation.
Culinary uses	Papain	Tenderize meat for cooking.
Dairy industry	Rennin	Hydrolyze protein in the manufacture of cheese.
	Lipases	Produce Camembert cheese and blue cheeses such as Roquefort.
Food processing	Amylases	Produce sugars from starch, such as in making high-fructose corn syrup.
	Proteases	Lower the protein level of flour, as in biscuit-making.
	Trypsin	Manufacture hypoallergenic baby foods.
	Cellulases, pectinases	Clarify fruit juices.
Molecular biology	Nucleases, DNA ligase and polymerases	Use restriction digestion and the polymerase chain reaction to create recombinant DNA.
Paper industry	Xylanases, hemicellulases and lignin peroxidases	Remove lignin from kraft pulp.
Personal care	Proteases	Remove proteins on contact lenses to prevent infections.
Starch industry	Amylases	Convert starch into glucose and various syrups.

Proteolysis

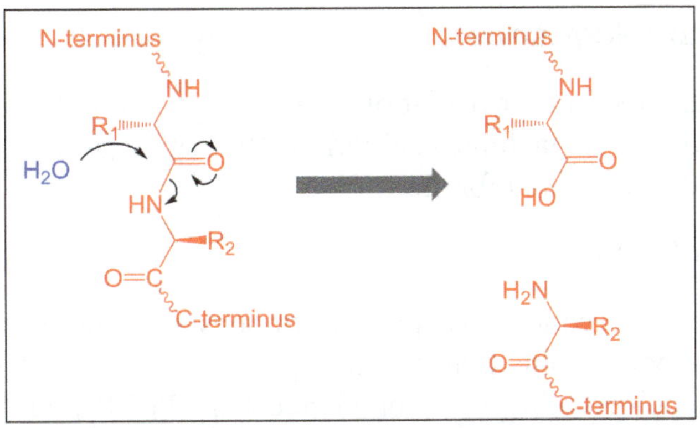

The hydrolysis of a protein (red) by thenucleophilic attack of water (blue). The uncatalysed half-life is several hundred years.

Proteolysis is the breakdown of proteins into smaller polypeptides or amino acids. Uncatalysed, the hydrolysis of peptide bonds is extremely slow, taking hundreds of years. Proteolysis is typically catalysed by cellular enzymes called proteases, but may also occur by intra-molecular digestion. Low pH or high temperatures can also cause proteolysis non-enzymatically.

Proteolysis in organisms serves many purposes; for example, digestive enzymes break down proteins in food to provide amino acids for the organism, while proteolytic processing of a polypeptide chain after its synthesis may be necessary for the production of an active protein. It is also important in the regulation of some physiological and cellular processes, as well as preventing the accumulation of unwanted or abnormal proteins in cells. Consequently, dis-regulation of proteolysis can cause diseases and is used in some venoms to damage their prey.

Proteolysis is important as an analytical tool for studying proteins in the laboratory, as well as industrially, for example in food processing and stain removal.

Biological Functions

Post-Translational Proteolytic Processing

Limited proteolysis of a polypeptide during or after translation in protein synthesis often occurs for many proteins. This may involve removal of the N-terminal methionine, signal peptide, and/or the conversion of an inactive or non-functional protein to an active one. The precursor to the final functional form of protein is termed proprotein, and these proproteins may be first synthesized as preproprotein. For example, albumin is first synthesized as preproalbumin and contains an uncleaved signal peptide. This forms the proalbumin after the signal peptide is cleaved, and a further processing to remove the N-terminal 6-residue propeptide yields the mature form of the protein.

Removal of N-Terminal Methionine

The initiating methionine (and, in prokaryotes, fMet) may be removed during translation of the nascent protein. For E. coli, fMet is efficiently removed if the second residue is small and uncharged, but not if the second residue is bulky and charged. In both prokaryotes and eukaryotes, the exposed N-terminal residue may determine the half-life of the protein according to the N-end rule.

Removal of the Signal Sequence

Proteins that are to be targeted to a particular organelle or for secretion have an N-terminal signal peptide that directs the protein to its final destination. This signal peptide is removed by proteolysis after their transport through a membrane.

Cleavage of Polyproteins

Some proteins and most eukaryotic polypeptide hormones are synthesized as a large precursor polypeptide known as a polyprotein that requires proteolytic cleavage into individual smaller polypeptide chains. The polyprotein pro-opiomelanocortin (POMC) contains many polypeptide hormones. The cleavage pattern of POMC, however, may vary between different tissues, yielding different sets of polypeptide hormones from the same polyprotein.

Many viruses also produce their proteins initially as a single polypeptide chain that were translated from a polycistronic mRNA. This polypeptide is subsequently cleaved into individual polypeptide chains.

Cleavage of Precursor Proteins

Many proteins and hormones are synthesized in the form of their precursors - zymogens, proenzymes, and prehormones. These proteins are cleaved to form their final active structures. Insulin, for example, is synthesized as preproinsulin, which yields proinsulin after the signal peptide has been cleaved. To form the mature insulin, the proinsulin is then cleaved at two positions to yield two polypeptide chains linked by 2 disulphide bonds. Proinsulin is necessary for the folding of the polypeptide chain, as the 2 polypeptide chains of insulin may not correctly assemble into the correct form, whereas its precursor proinsulin does.

Proteases in particular are synthesized in the inactive form so that they may be safely stored in cells, and ready for release in sufficient quantity when required. This is to ensure that the protease is activated only in the correct location or context, as inappropriate activation of these proteases can be very destructive for an organism. Proteolysis of the zymogen yields an active protein; for example, when trypsinogen is cleaved to form trypsin, a slight rearrangement of the protein structure that completes the active site of the protease occurs, thereby activating the protein.

Proteolysis can, therefore, be a method of regulating biological processes by turning inactive proteins into active ones. A good example is the blood clotting cascade whereby an initial event triggers a cascade of sequential proteolytic activation of many specific proteases, resulting in blood coagulation. The complement system of the immune response also involves a complex sequential proteolytic activation and interaction that result in an attack on invading pathogens.

Protein Degradation

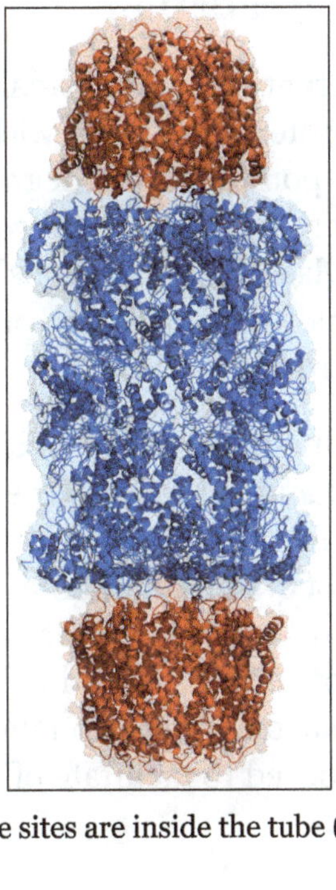

Structure of a proteasome. Its active sites are inside the tube (blue) where proteins are degraded.

Protein degradation may take place intracellularly or extracellularly. In digestion of food, digestive enzymes may be released into the environment for extracellular digestion whereby proteolytic cleavage breaks down proteins into smaller peptides and amino acids so that they may be absorbed and used by an organism. In animals the food may be processed extracellularly in specialized digestive organs or guts, but in many bacteria the food may be internalized into the cell via phagocytosis. Microbial degradation of protein in the environment can be regulated by nutrient availability. For example, limitation for major elements in proteins (carbon, nitrogen, and sulfur) has been shown to induce proteolytic activity in the fungus Neurospora crassa as well as in whole communities of soil organisms.

Proteins in cells are also constantly being broken down into amino acids. This intracellular degradation of protein serves a number of functions: It removes damaged and abnormal protein and prevent their accumulation, and it also serves to regulate cellular processes by removing enzymes and regulatory proteins that are no longer needed. The amino acids may then be reused for protein synthesis.

Lysosome and Proteasome

The intracellular degradation of protein may be achieved in two ways - proteolysis in lysosome, or a ubiquitin-dependent process that targets unwanted proteins to proteasome. The autophagy-lysosomal pathway is normally a non-selective process, but it may become selective upon starvation whereby proteins with peptide sequence KFERQ or similar are selectively broken down. The lysosome contains a large number of proteases such as cathepsins.

The ubiquitin-mediated process is selective. Proteins marked for degradation are covalently linked to ubiquitin. Many molecules of ubiquitin may be linked in tandem to a protein destined for degradation. The polyubiquinated protein is targeted to an ATP-dependent protease complex, the proteasome. The ubiquitin is released and reused, while the targeted protein is degraded.

Rate of Intracellular Protein Degradation

Different proteins are degraded at different rates. Abnormal proteins are quickly degraded, whereas the rate of degradation of normal proteins may vary widely depending on their functions. Enzymes at important metabolic control points may be degraded much faster than those enzymes whose activity is largely constant under all physiological conditions. One of the most rapidly degraded proteins is ornithine decarboxylase, which has a half-life of 11 minutes. In contrast, other proteins like actin and myosin have a half-life of a month or more, while, in essence, haemoglobin lasts for the entire life-time of an erythrocyte.

The N-end rule may partially determine the half-life of a protein, and proteins with segments rich in proline, glutamic acid, serine, and threonine (the so-called PEST proteins) have short half-life. Other factors suspected to affect degradation rate include the rate deamination of glutamine and asparagine and oxidation of cystein, histidine, and methionine, the absence of stabilizing ligands, the presence of attached carbohydrate or phosphate groups, the presence of free α-amino group, the negative charge of protein, and the flexibility and stability of the protein. Proteins with larger degrees of intrinsic disorder also tend to have short cellular half-life, with disordered segments having been proposed to facilitate efficient initiation of degradation by the proteasome.

The rate of proteolysis may also depend on the physiological state of the organism, such as its hormonal state as well as nutritional status. In time of starvation, the rate of protein degradation increases.

Digestion

In human digestion, proteins in food are broken down into smaller peptide chains by digestive enzymes such as pepsin, trypsin, chymotrypsin, and elastase, and into amino acids by various enzymes such as carboxypeptidase, aminopeptidase, and dipeptidase. It is necessary to break down proteins into small peptides (tripeptides and dipeptides) and amino acids so they can be absorbed by the intestines, and the absorbed tripeptides and dipeptides are also further broken into amino acids intracellularly before they enter the bloodstream. Different enzymes have different specificity for their substrate; trypsin, for example, cleaves the peptide bond after a positively charged residue (arginine and lysine); chymotrypsin cleaves the bond after an aromatic residue (phenylalanine, tyrosine, and tryptophan); elastase cleaves the bond after a small non-polar residue such as alanine or glycine.

In order to prevent inappropriate or premature activation of the digestive enzymes (they may, for example, trigger pancreatic self-digestion causing pancreatitis), these enzymes are secreted as inactive zymogen. The precursor of pepsin, pepsinogen, is secreted by the stomach, and is activated only in the acidic environment found in stomach. The pancreas secretes the precursors of a number of proteases such as trypsin and chymotrypsin. The zymogen of trypsin is trypsinogen, which is activated by a very specific protease, enterokinase, secreted by the mucosa of the duodenum. The trypsin, once activated, can also cleave other trypsinogens as well as the precursors of other proteases such as chymotrypsin and carboxypeptidase to activate them.

In bacteria, a similar strategy of employing an inactive zymogen or prezymogen is used. Subtilisin, which is produced by Bacillus subtilis, is produced as preprosubtilisin, and is released only if the signal peptide is cleaved and autocatalytic proteolytic activation has occurred.

Cellular Regulation

Proteolysis is also involved in the regulation of many cellular processes by activating or deactivating enzymes, transcription factors, and receptors, for example in the biosynthesis of cholesterol, or the mediation of thrombin signalling through protease-activated receptors.

Some enzymes at important metabolic control points such as ornithine decarboxylase is regulated entirely by its rate of synthesis and its rate of degradation. Other rapidly degraded proteins include the protein products of proto-oncogenes, which play central roles in the regulation of cell growth.

Cell cycle Regulation

Cyclins are a group of proteins that activate kinases involved in cell division. The degradation of cyclins is the key step that governs the exit from mitosis and progress into the next cell cycle. Cyclins accumulate in the course the cell cycle, then abruptly disappear just before the anaphase of mitosis. The cyclins are removed via a ubiquitin-mediated proteolytic pathway.

Apoptosis

Caspases are an important group of proteases involved in apoptosis. The precursors of caspase, procaspase, may be activated by proteolysis through its association with a protein complex that forms apoptosome, or by granzyme B, or via the death receptor pathways.

Proteolysis and Diseases

Abnormal proteolytic activity is associated with many diseases. In pancreatitis, leakage of proteases and their premature activation in the pancreas results in the self-digestion of the pancreas. People with diabetes mellitus may have increased lysosomal activity and the degradation of some proteins can increase significantly. Chronic inflammatory diseases such as rheumatoid arthritis may involve the release of lysosomal enzymes into extracellular space that break down surrounding tissues. Abnormal proteolysis and generation of peptides that aggregate in cells and their ineffective removal may result in many age-related neurological diseases such as Alzheimer's.

Proteases may be regulated by antiproteases or protease inhibitors, and imbalance between proteases and antiproteases can result in diseases, for example, in the destruction of lung tissues in emphysema brought on by smoking tobacco. Smoking is thought to increase the neutrophils and macrophages in the lung which release excessive amount of proteolytic enzymes such as elastase, such that they can no longer be inhibited by serpins such as α_1-antitrypsin, thereby resulting in the breaking down of connective tissues in the lung. Other proteases and their inhibitors may also be involved in this disease, for example matrix metalloproteinases (MMPs) and tissue inhibitors of metalloproteinases (TIMPs).

Other diseases linked to aberrant proteolysis include muscular dystrophy, degenerative skin disorders, respiratory and gastrointestinal diseases, and malignancy.

Non-Enzymatic Proteolysis

Protein backbones are very stable in water at neutral pH and room temperature, although the rate of hydrolysis of different peptide bonds can vary. The half life of a peptide bond under normal conditions can range from 7 years to 350 years, even higher for peptides protected by modified terminus or within the protein interior. The rate of proteolysis however can be significantly increased by extremes of pH and heat.

Strong mineral acids can readily hydrolyse the peptide bonds in a protein (acid hydrolysis). The standard way to hydrolyze a protein or peptide into its constituent amino acids for analysis is to heat it to 105 °C for around 24 hours in 6M hydrochloric acid. However, some proteins are resistant to acid hydrolysis. One well-known example is ribonuclease A, which can be purified by treating crude extracts with hot sulphuric acid so that other proteins become degraded while ribonuclease A is left intact.

Certain chemicals cause proteolysis only after specific residues, and these can be used to selectively break down a protein into smaller polypeptides for laboratory analysis. For example, cyanogen bromide cleaves the peptide bond after a methionine. Similar methods may be used to specifically cleave tryptophanyl, aspartyl, cysteinyl, and asparaginyl peptide bonds. Acids such as trifluoroacetic acid and formic acid may also be used.

Like other biomolecules, proteins can also be broken down by high heat alone. At 250 °C, the peptide bond may be easily hydrolyzed, with its half-life dropping to about a minute. Protein may also be broken down without hydrolysis through pyrolysis; small heterocyclic compounds may start to form upon degradation, above 500 °C, polycyclic aromatic hydrocarbon may also form, which is of interest in the study of generation of carcinogens in tobacco smoke and cooking at high heat.

Laboratory Applications

Proteolysis is also used in research and diagnostic applications:

- Cleavage of fusion protein so that the fusion partner and protein tag used in protein expression and purification may be removed. The proteases used have high degree of specificity, such as thrombin, enterokinase, and TEV protease, so that only the targeted sequence may be cleaved.
- Complete inactivation of undesirable enzymatic activity or removal of unwanted proteins. For example, proteinase K, a broad-spectrum proteinase stable in urea and SDS, is often used in the preparation of nucleic acids to remove unwanted nuclease contaminants that may otherwise degrade the DNA or RNA.
- Partial inactivation, or changing the functionality, of specific protein. For example, treatment of DNA polymerase I with subtilisin yields the Klenow fragment, which retains its polymerase function but lacks 5'-exonuclease activity.
- Digestion of proteins in solution for proteome analysis by liquid chromatography-mass spectrometry (LC-MS). This may also be done by in-gel digestion of proteins after separation by gel electrophoresis for the identification by mass spectrometry.
- Analysis of the stability of folded domain under a wide range of conditions.
- Increasing success rate of crystallisation projects.
- Production of digested protein used in growth media to culture bacteria and other organisms, e.g. tryptone in Lysogeny Broth.

Protease Enzymes

Proteases may be classified according to the catalytic group involved in its active site.

- Cysteine protease
- Serine protease
- Threonine protease
- Aspartic protease
- Glutamic protease
- Metalloprotease
- Asparagine peptide lyase

Venoms

Certain types of venom, such as those produced by venomous snakes, can also cause proteolysis. These venoms are, in fact, complex digestive fluids that begin their work outside of the body. Proteolytic venoms cause a wide range of toxic effects, including effects that are:

- cytotoxic (cell-destroying)
- hemotoxic (blood-destroying)
- myotoxic (muscle-destroying)
- hemorrhagic (bleeding)

References

- De Strooper B. (2010). "Proteases and proteolysis in Alzheimer disease: a multifactorial view on the disease process". Physiological Reviews. 90 (2): 465–94. doi:10.1152/physrev.00023.2009. PMID 20393191
- Fabbri D, Adamiano A, Torri C (2010). "GC-MS determination of polycyclic aromatic hydrocarbons evolved from pyrolysis of biomass". Anal Bioanal Chem. 397 (1): 309–17. doi:10.1007/s00216-010-3563-5. PMID 20213167.
- Bernard Testa, Joachim M. Mayer (1 July 2003). Hydrolysis in Drug and Prodrug Metabolism. Wiley VCH. pp. 270–288. ISBN 978-3906390253.
- Bryan John Smith (2002). "Chapter 71-75". In John M. Walker. The Protein Protocols Handbook (2 ed.). Humana Press. pp. 485–510. doi:10.1385/1592591698. ISBN 978-0-89603-940-7.
- Kohei Oda (2012). "New families of carboxyl peptidases: serine-carboxyl peptidases and glutamic peptidases". Journal of Biochemistry. 151 (1): 13–25. doi:10.1093/jb/mvr129. PMID 22016395.
- Inobe, Tomonao; Matouschek, Andreas (2014-02-01). "Paradigms of protein degradation by the proteasome". Current Opinion in Structural Biology. 24: 156–164. doi:10.1016/j.sbi.2014.02.002. ISSN 1879-033X. PMC 4010099. PMID 24632559.

Permissions

All chapters in this book are published with permission under the Creative Commons Attribution Share Alike License or equivalent. Every chapter published in this book has been scrutinized by our experts. Their significance has been extensively debated. The topics covered herein carry significant information for a comprehensive understanding. They may even be implemented as practical applications or may be referred to as a beginning point for further studies.

We would like to thank the editorial team for lending their expertise to make the book truly unique. They have played a crucial role in the development of this book. Without their invaluable contributions this book wouldn't have been possible. They have made vital efforts to compile up to date information on the varied aspects of this subject to make this book a valuable addition to the collection of many professionals and students.

This book was conceptualized with the vision of imparting up-to-date and integrated information in this field. To ensure the same, a matchless editorial board was set up. Every individual on the board went through rigorous rounds of assessment to prove their worth. After which they invested a large part of their time researching and compiling the most relevant data for our readers.

The editorial board has been involved in producing this book since its inception. They have spent rigorous hours researching and exploring the diverse topics which have resulted in the successful publishing of this book. They have passed on their knowledge of decades through this book. To expedite this challenging task, the publisher supported the team at every step. A small team of assistant editors was also appointed to further simplify the editing procedure and attain best results for the readers.

Apart from the editorial board, the designing team has also invested a significant amount of their time in understanding the subject and creating the most relevant covers. They scrutinized every image to scout for the most suitable representation of the subject and create an appropriate cover for the book.

The publishing team has been an ardent support to the editorial, designing and production team. Their endless efforts to recruit the best for this project, has resulted in the accomplishment of this book. They are a veteran in the field of academics and their pool of knowledge is as vast as their experience in printing. Their expertise and guidance has proved useful at every step. Their uncompromising quality standards have made this book an exceptional effort. Their encouragement from time to time has been an inspiration for everyone.

The publisher and the editorial board hope that this book will prove to be a valuable piece of knowledge for students, practitioners and scholars across the globe.

Index

www.ingramcontent.com/pod-product-compliance
Lightning Source LLC
LaVergne TN
LVHW080722210926
848485LV00003B/508